Filter Design With Time Domain Mask Constraints: Theory and Applications

Applied Optimization

Volume 56

The titles published in this series are listed at the end of this volume.

Filter Design With Time Domain Mask Constraints: Theory and Applications

by

Ba-Ngu Vo

The University of Melbourne, Australia

Antonio Cantoni

The University of Western Australia, Australia

and

Kok Lay Teo

The Hong Kong Polytechnic University, Hong Kong

KLUWER ACADEMIC PUBLISHERS

DORDRECHT / BOSTON / LONDON

A C.I.P. Catalogue record for this book is available from the Library of Congress.

ISBN 0-7923-7138-0

Published by Kluwer Academic Publishers,
P.O. Box 17, 3300 AA Dordrecht, The Netherlands.

Sold and distributed in North, Central and South America
by Kluwer Academic Publishers,
101 Philip Drive, Norwell, MA 02061, U.S.A.

In all other countries, sold and distributed
by Kluwer Academic Publishers,
P.O. Box 322, 3300 AH Dordrecht, The Netherlands.

Printed on acid-free paper

Printed in the Netherlands.

CONTENTS

List of Figures

List of Tables

Preface

In the signal processing literature, filter design is a well-established area and there has been a considerable amount of work devoted to it. Whether it's classical or at the leading edge of research, the literature primarily concentrates on frequency domain constraints (such as passband ripples, stopband attenuation). However, very little attention has been directed towards filter design with time domain constraints in the form of envelopes or masks, which have become increasingly important in the performance specification of modern telecommunication systems. Time domain filter design problems are commonly known in the literature as time-domain synthesis problems and deal with finding a network to give a prescribed response for a given excitation, both of which are specified as functions of time. This class of problems arises in pulse-shaping circuits, pulse transmission systems, delay network, transmission channel equalization, video distribution systems, and the like.

In time domain synthesis problems, often the performance criterion is the mean square error between the filter output and some desired signal. In many practical signal processing problems, this soft least-square approach is artificial. The approach may also yield unsatisfactory results because large narrow excursions from the desired shape occur, and the norm of the filter can be large. In addition, the choice of an appropriate weighting function is not obvious. Moreover, the solution can be sensitive to the detailed structure of the desired pulse, and it is usually not obvious how the desired pulse should be chosen so as to obtain the best possible solution. The distinctive feature of the hard envelope-constrained filter formulation is that the output waveform is specified to lie within an envelope defined by a set of inequality constraints, rather than attempting to match it with a specific desired pulse. Therefore, in the hard envelope-constrained filter formulation, we deal with a whole set of allowable outputs, and the objective is to seek the one for which the corresponding filter results in minimum enhancement of filter input noise. The hard envelope-constrained filter formulation is more relevant than the soft least-square approach in a number of signal processing fields. For example, in TV channel equalization, it is required that the shaped signal simply fits into a prescribed envelope called a K-mask. In digital transmission, the performance of a digital link is often specified in terms of a mask applied to the received test signal. The envelope-constrained approach is directly applicable for shaping test signals into given masks. For pulse compression applications in sonar and radar detection, envelope-constrained filters can also be used to suppress sidelobes.

This monograph has evolved from work on envelope constrained filter design problems that has origins dating back to the late 60's. Over the years, there have been numerous fundamental contributions reported in the literature. In this monograph, our modest intention is to present a unified formulation that covers digital, analog and hybrid (consisting of both analog and digital parts) filters. This formulation considers not only envelope-constrained filters but also problems with more general response constraints and wider range of cost functions. Motivated by practical realization issues, finite dimensional approximations to this general problem are investigated and a number of realizable filter structures are suggested. Furthermore, the important issue of designing envelope constrained filters that are robust to implementation errors are addressed. A class of iterative algorithms is developed for solving this envelope-constrained filtering problem with finite filter structures. Convergence properties of these algorithms are addressed. In continuous-time, this technique yields optimum EC filters with analog or hybrid realizations. This is achieved without discretizing the constraints of the filter. Practical real examples are included to illustrate the effectiveness of each of the algorithms developed in this monograph. Most of the material contained in this monograph is based on the research carried out by the authors and their collaborators during the last several years.

This monograph can also be viewed as an application of constrained optimization theory to a class of engineering problems in signal processing. It takes the reader through the three phases of problem solving: problem formulation, characterization of solution, and numerical techniques for solving the problem.

In the formulation phase (Chapter 2) we see how a raw engineering problem is formulated by taking into account real world constraints. We also see that the first iteration of the formulation yields a solution, which is not directly usable. The next iteration in the formulation process is to improve and refine the formulation and the mathematical model. The solution characterization phase (Chapter 3) marches through the mathematical analysis of optimization theory to derive the forms that the solutions take on. Interpretations of what these results mean in signal processing terms are then made.

Numerical algorithms for solving these problems are perhaps the most important phase in engineering, because if we can't find the solution to the problem, the other two phases are of little use to the designer. This phase is covered in Chapters 4, 5 and 6. Chapters 4 and 5 present algorithms for discrete-time and continuous-time problems respectively. We look at the discrete-time problem first because it is simpler. This also makes it easier to grasp the continuous-time problem. Chapter 6 considers the robustness problem.

This monograph is intended for both engineers and applied mathematicians. The authors believe that both engineers and applied mathematicians can make further contributions to the subject. With the background acquired from this monograph, engineers will learn the mathematical rigor in problem formulation and the development of solution methods. On the other hand, applied mathematicians will find this monograph as a stepping stone towards a research area in signal processing. Moreover, the authors also believe that the monograph will be useful as a reference to practicing engineers and scientists as well as mathematicians.

The background required for understanding the mathematical formulation, the algorithms and the application of these algorithms to solve practical problems is advanced calculus. However, to analyze the convergence properties of these algorithms, some results in real and functional analysis and optimization theory are required. For the convenience of the reader, essential mathematical concepts and facts in real and functional analysis are stated without proof in Appendix A, while optimization theory, in particular, convex mathematical programming, are reviewed in Appendix B.

To ensure readability of the monograph amongst both engineers and applied mathematicians, the proofs are given in the appendix of each chapter so as not to disrupt the continuity of flow. Those who are interested in the mathematical details can take excursions to the appendices. The authors believe that in this way both classes of reader can gain a better appreciation of the subject.

It is a pleasure to express our gratitude to R. Evans, W. X. Zheng, G. Lin, Z. Zang, C. H. Tseng, and H. H. Dam. They have made a significant contribution to the material presented in this monograph through collaborative research projects. We wish to thank Australian Telecommunications Research Institute of Curtin University of Technology and the Co-operative Research Centre for Broadband Telecommunications and Networking for the stimulating research environment and we also appreciate the financial supports provided by the Australian Research Council, the Department of Electrical and Electronic Engineering of the University of Western Australia, and the Department of Electrical and Electronic Engineering of the University of Melbourne. Also the financial supports of the Research Committee, the Centre for Multimedia Signal Processing of the Department of Electronics and Information Engineering, the Department of Applied Mathematics, all with The Hong Kong Polytechnic University, and the Research Grant Committee of Hong Kong are gratefully acknowledged. Furthermore, we wish to express our appreciation to John R. Martindale, Senior Publishing Editor of Kluwer Academic Publishers, for his encouragement, enthusiasm and collaboration. Our sincere thanks also go to our families for their support, patience and understanding.

B. Vo
A. Cantoni
K. L. Teo

February 2001

CHAPTER 1 INTRODUCTION

In signal processing, the design of many filters can often be cast as a constrained optimization problem where the constraints are defined by the specifications of the filter. These specifications can arise either from practical considerations or from the standards set by certain regulatory bodies. Traditional filter design techniques are developed in the frequency domain where the response of a filter at a particular frequency is dependent solely on the excitation of that frequency. This is not true with time domain problems. Clearly, an excitation applied at any time will have effects on the response at all times after the application of the excitation. Conversely, the response at any particular instant depends on all the excitations that have been applied to the filter *prior to* that instant. This difference manifests mathematically in the form of multiplication and convolution.

Time domain filter design problems are commonly known in the literature as *time-domain synthesis problems,* which deal with the finding of a network to give a prescribed response d for a given excitation s, both of which are specified as functions of time. This class of problems arises in pulse-shaping circuits, pulse transmission systems, delay networks, transmission line equalization, video coupling systems, and the like.

Time-domain synthesis generally involves two separate steps: the approximation and the realization. The former deals with finding a mathematical filter that, under certain criteria, best meets the requirement for which the synthesis is being performed. The latter deals with finding a network or networks that realize the mathematical filter.

Envelope constrained filter design can be interpreted as the approximation step of the time domain synthesis problem. However, instead of finding a filter whose response approximates (in some sense) the prescribed response d for a given excitation s, we seek a filter such that its response stays within a given tolerance band about d, while the noise enhancement is minimized.

1.1 Applications

Often in a time domain synthesis problem, the performance criterion is the mean square error between the filter output and some desired signal. In many practical applications, this "soft" approach is not the most suitable and it becomes necessary to use "hard" constraints such as envelope constraints. Moreover, problems often arise in practice where it is very crucial to demand that the shaped signal simply fits into a prescribed envelope. To illustrate these points, a number of applications are considered.

Pulse compression

In many radar and communication systems, narrow pulses are required for range resolution and clutter reduction, but it is also important that the transmitted pulses contain sufficient energy for long-range detection. These two conflicting requirements would seem to dictate narrow pulses with high peak power. However, because of transmitter design considerations, peak power is limited, pulses of relatively long duration are transmitted and an operation known as *pulse compression* is performed at the receiver as illustrated in Figure 1.1.1. This compression is most commonly achieved with a *matched filter* [15], [32], [44], [56], [94], [105], [112], [130], that is, by correlating the incoming signal with the time-shifted and/or frequency shifted copies of the transmitted waveform. Matched filtering is well known to be optimal with respect to various performance criteria.

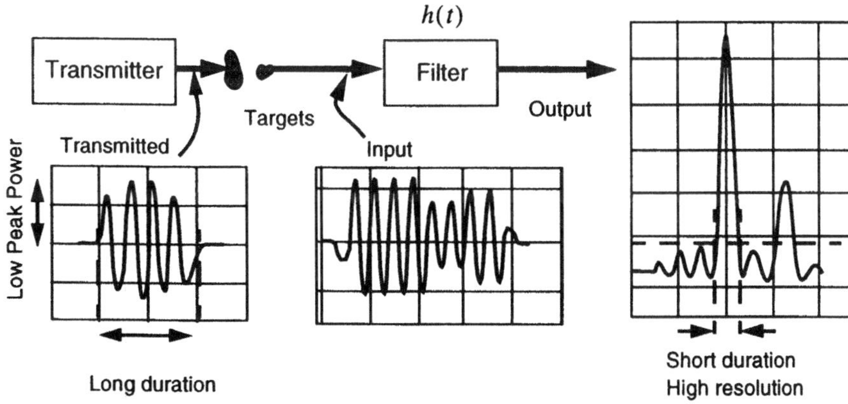

Figure 1.1.1. Pulse Compression

The presence of a signal is detected when the matched filter output exceeds a threshold value, and parameters such as time delay and (Doppler) frequency shift are *estimated* by locating the peak output in time and frequency. One example of such a system is a linear frequency modulated (FM) or *chirp* radar [15], [94], [105], where the frequencies of the transmitted pulses are swept linearly with time and the receiver has a linear time delay versus frequency characteristic of the opposite slope.

If a signal s, non-zero on the interval $(t_0, t_0 + T)$, appears at the input to a pulse-compression filter, the output typically consists of a main peak surrounded by sidelobes, as shown in Figure 1.1.1. When a radar must distinguish among multiple targets, these sidelobes can cause false detections and impede resolution of adjacent pulses. For example, if the transmitted pulse is a 13-bit Barker code and a matched filter receiver is used, the output is compressed to 1/13 of the original pulse length, improving range resolution with no detection loss. Unfortunately, the output pulse is accompanied by 12 sidelobes, where the amplitude of each of these sidelobes is 1/13 of that of the mainlobe peak. With a pulse modulation system operating in an environment of severely degraded signal to noise ratio [76], the sidelobes often lead to false detection and erroneous delay estimation. Thus, it is necessary in certain cases to remove or at least restrict the height of these sidelobes to an acceptable level.

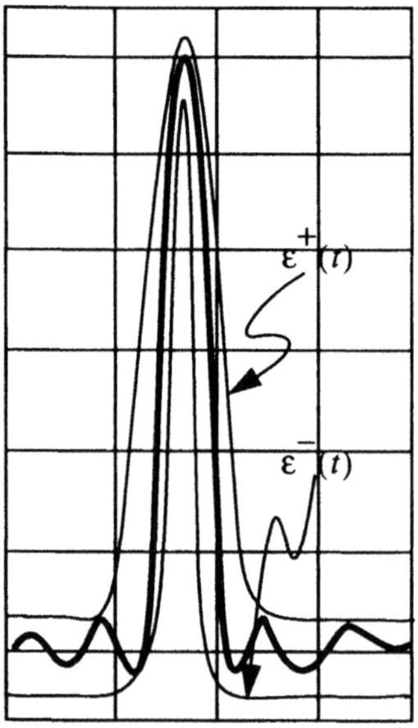

Figure 1.1.2. Pulse shape constraints for radar/sonar problem

Problems of this type are often treated by using a least square approach, where the mean squared difference between the output and some desired pulse shape is minimized. If the sidelobes are reduced using a least square cost, there is no guarantee that the appearance of low-energy high-peak sidelobes (which can cause false detections) can be avoided. Moreover, the solution can be sensitive to the detailed structure of the desired pulse, and it is usually not obvious how the shape of the desired pulse should be altered in order to achieve better performance in terms of low level sidelobes.

The reduction of sidelobes can be formulated in terms of an envelope constrained (EC) problem: find the filter (if it exists) that causes least detection loss while reducing the output sidelobes to a specified fraction of the mainlobe peak. The envelope or pulse shape mask takes on the form shown in Figure 1.1.2. This was first considered in [75], where the mask consists of a finite number of points. Since the probability of detection is proportional to the signal to noise ratio, which, in turn, is inversely proportional to square of the norm of the filter, we are faced with the task of finding a minimum norm filter subject to the constraint that the sidelobe peaks staying within the specified values.

TV waveform equalization

An interesting application of the envelope-constrained filtering problem is the equalization of television transmission channel. Here, a test pulse, for example a $\sin^2$ pulse, is sent through a dispersive channel and becomes distorted. According to the psycho-physical Lewis criteria [7], [103], if this distorted pulse fits into a prescribed envelope called a K-mask (see Figure 1.1.3) and satisfies a specified K-rating, the channel will not subjectively distort transmitted video information, that is, the quality of the picture is acceptable to human viewers. If the distorted test pulse does not satisfy the K-criteria, then it is necessary to insert an equalizing filter. The function of this equalizing filter is to take the distorted pulse as input and produces an output that satisfies the K-criteria (for this reason, it is sometimes known as a pulse-shaping problem). In this way, the channel has been compensated or equalized so that TV video data does not appear distorted to a human observer. Notice that detailed knowledge of the dispersive channel is not required in the design of the equalizer, only the distorted pulse is required. Since any broadband white noise at the input to the filter is multiplied by the square of the norm of the filter, the envelope constrained formulation appears to be tailored for this application.

Most researchers investigating pulse shaping and equalization problems use least square (LS) error and minimax error approximation techniques [63], [107], [127]. These techniques, although very powerful, are difficult to apply to the K-mask equalization problem. Figure 1.1.3 shows the equalized pulses for the EC and LS approaches, where the channel under consideration is a coaxial cable. Both methods yielded solutions that satisfy the constraints. However, the noise enhancement for the LS approach is 9475 times larger than that of the EC approach. When fitting a signal into an envelope, we are not approximating anything, rather we are trying to satisfy a given continuum of constraints. It is important to note that, for this particular application, approaches using least squares or other "soft" criteria to force the equalized pulse to stay inside the prescribed envelope by approximating some desired pulse shape have generally been unsatisfactory.

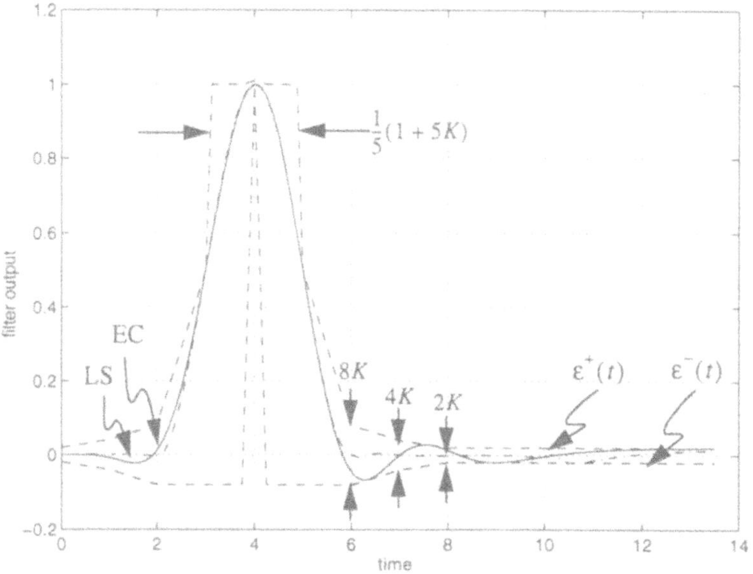

Figure 1.1.3. K-rating mask for equalization of TV channel.

Data channel equalization

In communications, much research has been directed towards the problem of high-speed data transmission over dispersive channels. A wide variety of high-speed data communication systems can be modeled as in Figure 1.1.4.

The input, consisting of a train of M-ary impulses spaced τ time units apart, is transmitted through a channel, where it is subjected to phase and amplitude distortions (resulting in interference) and additive noise contamination. The signal available at the detector is therefore a time spread pulse train corrupted by additive noise. It is often the case that the effect of the resulting intersymbol interference is an order of magnitude worse than the effect of channel noise and the problem is to recover the input sequence with minimum probability of error.

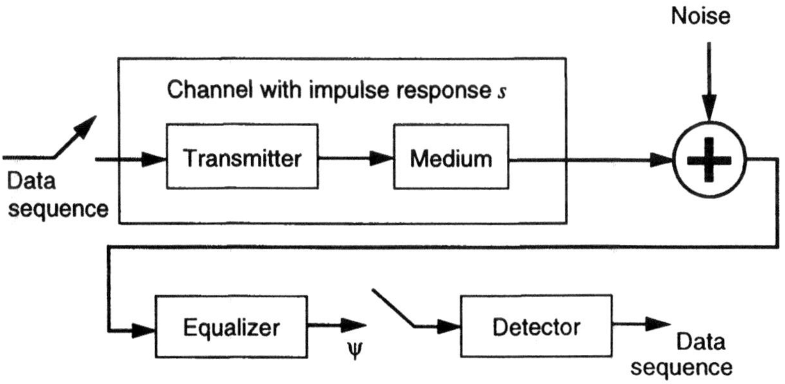

Figure 1.1.4. Model of a data channel

One method of recovering the input sequence is to equalize the channel with a linear prefilter in the receiver and then use symbol by symbol detection. In this case, the equalized channel is required to approximate a channel model with no intersymbol interference. Unfortunately, the linear equalizer, which minimizes the probability of error, is difficult to determine [64]. Thus, a number of suboptimal procedures are used in practice [10], [36], [42], [64], [89], [114]. One such procedure is to minimize the mean square error between the channel input and the equalizer output [10], [36], [89]. This approach clearly attempts to minimize the effects of both noise and intersymbol interference. Other techniques, such as peak distortion [64] and minimax [42], attempt to reduce the intersymbol interference by ignoring the noise enhancement caused by the equalizer. Non-linear equalization techniques have also been investigated in [87], [114].

As an alternative to the least mean square error method, the envelope constrained approach can be used to provide a flexible trade-off between output noise power and intersymbol interference. Since symbol by symbol detection is being used, pulse shape constraints will only be required at the sampling instants, provided that the samplers at the transmitter and receiver are exactly synchronized. This is equivalent

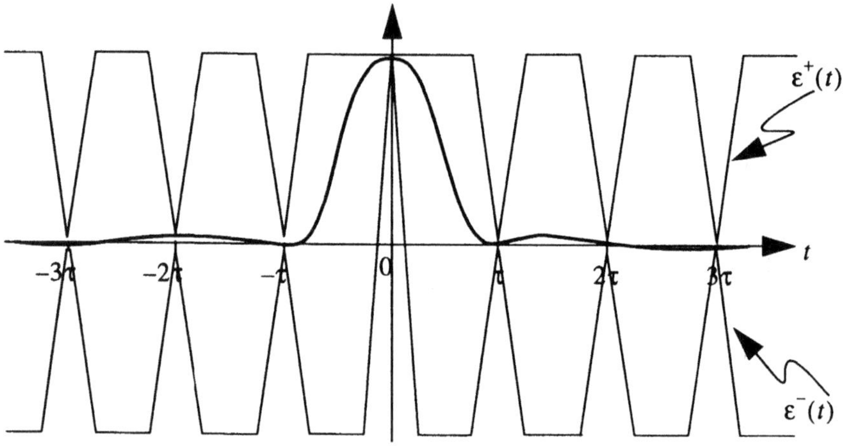

Figure 1.1.5. Mask for constraints at sampling instances

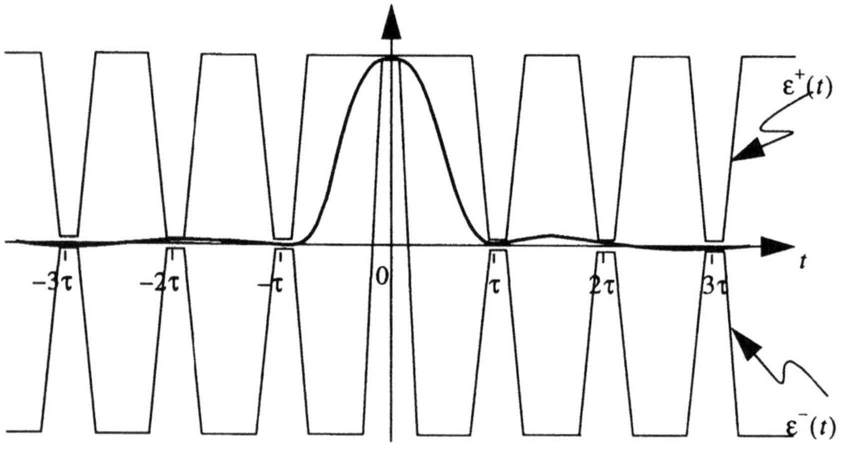

Figure 1.1.6. Mask for handling timing jitter.

to the EC filtering problem, where the equalizer is designed to make the combined channel/equalizer impulse response fit into a mask at the sampling instants, see Figure 1.1.5.

In practice, the samplers are never perfectly synchronized and the error caused by this timing jitter can be severe for systems designed by using the mean square error, peak distortion, or minimax criteria. However, in the envelope constrained context, it is easy to combat timing jitter with a mask of the form shown in Figure 1.1.6.

Design to meet standards

In telecommunication standards, the performance of a digital link is often characterized by pulse masks applied to the test pulses (see, for example, the CCITT recommendations [12]). The signal s would correspond to the test signal specified in the standards. As an example, consider the equalization of an RG59B/U coaxial cable channel whose attenuation follows an approximate $\sqrt{f}$ law (one with a 150m length has approximately 12dB loss at 70MHz). For a coaxial cable of length l, its frequency response is given by

$$H(j\omega) = e^{-Al\sqrt{j\omega}} = e^{-Al(\sqrt{\omega/2}+j\sqrt{\omega/2})}.$$

Let A_0 denote the attenuation of the cable in dB at a frequency f_0 Hz. Then,

$$A_0 = (20Al\log e)\sqrt{\frac{\omega_0}{2}} = (20Al\log e)\sqrt{\pi f_0},$$

where $\omega_0 = 2\pi f_0$. The impulse response of the cable is given by

$$h(t) = \frac{Al}{2\sqrt{\pi t^3}}\exp\left(-\frac{A^2l^2}{4t}\right),\ t > 0.$$

A plot showing the impulse response for several values of Al is given in Figure 1.1.7. Since Al increases with increasing cable length, the peak of the pulse in transit decreases and its base width widens as the cable lengthens. In order to successfully detect these pulses, the width must be compressed by means of pulse shaping networks at the receiving end.

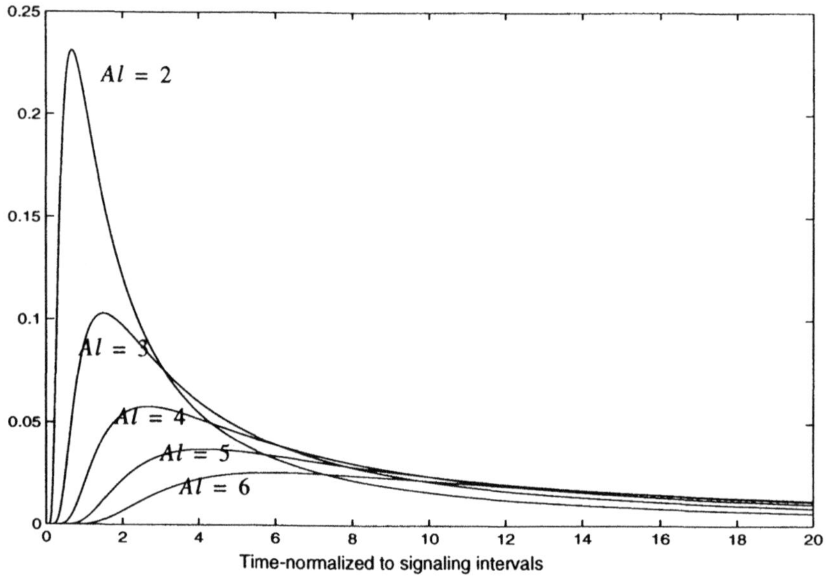

Figure 1.1.7. Impulse response of coaxial cable for various lengths

For a digital transmission channel consisting of a coaxial cable operating at the DX3 rate i.e. 45Mb/s (see [5], [12]), the American National Standards Institute (ANSI) specifies that the pulse received after transmission should fit in the DX3 pulse template illustrated in Figure 1.1.8. From the characteristic of the cable shown in Figure. 1.1.7, it can be seen that for long cables the received pulse will not fit in the mask. An equalizing filter is required to shape the impulse response of the cable so that it fits in the mask given by the DX3 pulse template. In Figure 1.1.8, we have shown the received pulse and pulse generated by an EC equalizing filter designed so that the processed received signal fits in the prescribed mask. The cable has a 30dB attenuation at frequency $\omega_0 = 2\pi/\beta$, where β denotes the Baud period (22.35ns).

Another practical example of designing pulse shapes to meet certain standard specifications involves a variation of the equalization problem. Consider the transmission system shown in Figure 1.1.9, where a pulse pre-shaping filter is placed before the channel. When a rectangular pulse is transmitted, the noiseless output of the channel

$$\psi(t) = \int_0^t s(t-\lambda)u(\lambda)d\lambda$$

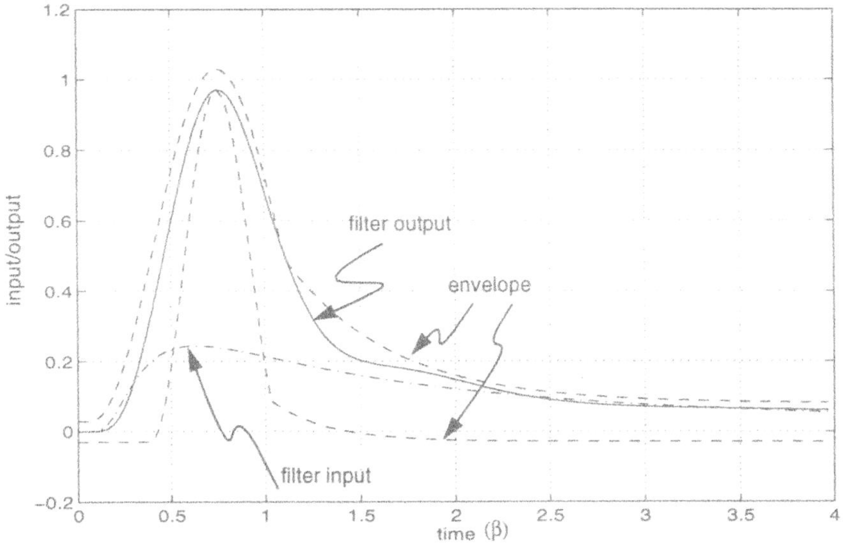

Figure 1.1.8. DSX3 pulse template, coaxial cable response and filter output

is required to fit into a T1 mask specified by the ANSI (shown in Figure 1.1.10). Again the norm of the pulse shaping filter is to be minimized so as to reduce the crosstalk (proportional to the transmit signal power) to other signals.

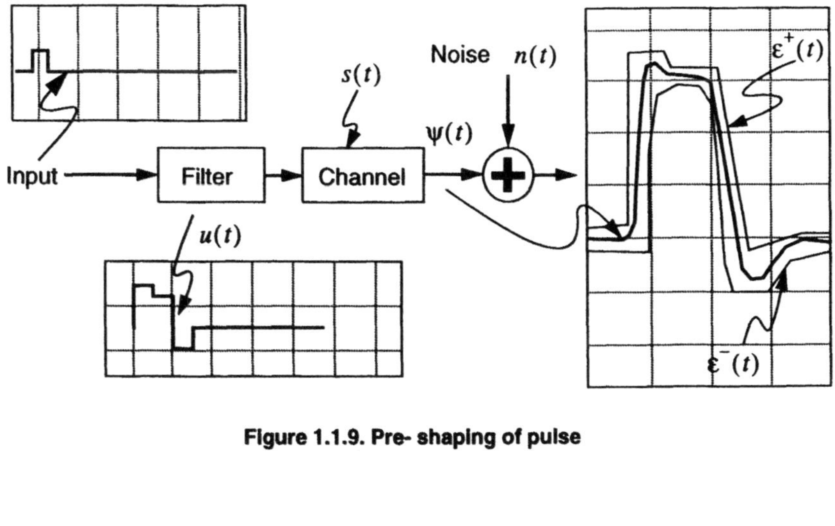

Figure 1.1.9. Pre- shaping of pulse

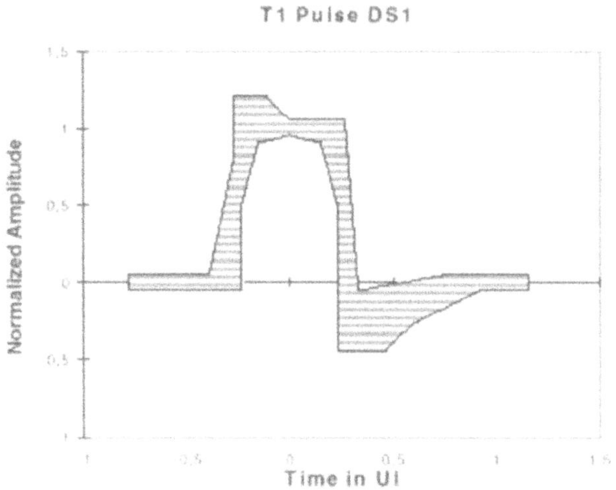

Figure 1.1.10. ANSI T1.403 for T1 -1.544 Mb/s

Deconvolution

Activities such as seismic exploration [129] and ultra-sonic medical imaging [20] involve the transmission of acoustic signals into a medium and analysis of the reflected signals in order to determine the structure of the medium. In many cases, the objective is to estimate the impulse response of the medium. For example, a reflection from a discrete horizon (discontinuity) corresponds to a single impulse whose arrival time determines the location of the horizon and whose magnitude determines the impedance change across it.

Because of the physical limitations, the transmitted signal usually has a complicated shape and a relatively long duration. The received waveform is the convolution of the transmitted signal with the unknown impulse response, and hence the problem is said to be one of deconvolution or inverse filtering. It is identical to the radar/sonar pulse compression problem described earlier.

If a filter u could be designed to produce a single impulse when excited by the transmitted signal s, then the same filter would transform the received waveform back to the impulse response being sought. Of course such a u can never be found in practice because the problem is quite ill-conditioned (the inverse operator is unbounded), and a variety of approximations has been developed. The most popular of these [129] involves selecting a narrow but smooth "desired" pulse shape d, and then choosing u so as to minimize the mean-squared error between the output of the filter when excited by s and the pulse shape d. This approach has been successful in some applications, but the usual deficiencies (such as uncontrolled sidelobes) associated with least square approaches still remain.

The envelope constrained problem is quite appropriate in this context and provides a versatile alternative to least square approaches. The pulse shape constraints may be chosen as in Figure 1.1.1. In fact, they may be altered to suit a particular application. For example, if a large discrete reflection is followed a short time later by a smaller reflection of particular interest, the upper and lower boundaries of the envelope may be chosen so as to force the sidelobes of the first reflection down in the vicinity of the second one. This situation arises in ultrasonic brain scanning, where sidelobes from large skull reflections tend to interfere with reflections from adjacent tissue structures.

Note that the deconvolution problem may also be attacked directly by reinterpreting u as the unknown impulse response of the medium, and choose a narrow impulse-like envelope about the actual recorded data. The problem becomes that of choosing u such that the convolution product $s * u$ fits the data within a given error tolerance. Unfortunately, the duration of u is often too long to render this latter approach intractable.

1.2 Envelope Constrained Filtering

The previous section has outlined some applications that motivated the study of the EC filtering problem without an explicit problem statement. This section presents a more precise statement that covers all the applications previously discussed.

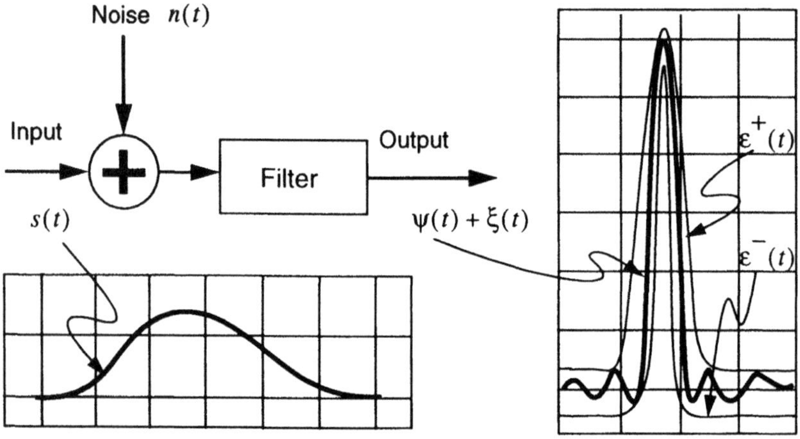

Figure 1.2.1. Receiver model and output mask.

Consider the filtering function shown in Figure 1.2.1 be it for pulse compression, equalization or inverse filtering. The excitation s entering the receiver is corrupted by additive zero-mean, stationary noise n. The impulse response u of the linear time-invariant filter to be determined is in $L_2[0, \infty)$, the space of all square-integrable functions on $[0, \infty)$. The output consists of two components ψ and ξ, which are due to the signal and noise, respectively. The noiseless output ψ is given by the convolution of s with u,

$$\psi(t) = \int_0^{\infty} s(t-\lambda)u(\lambda)d\lambda = \int_0^{t} s(t-\lambda)u(\lambda)d\lambda, \; t \in [0, \infty). \tag{1.2.1}$$

Let ε^- and ε^+ be two piecewise continuous functions of time representing the lower and upper boundaries of the output mask, respectively. Then, the envelope constraints require the filter output ψ to fit into a fully specified envelope as depicted in Figure 1.2.1, i.e.

$$\varepsilon^-(t) \le \psi(t) \le \varepsilon^+(t), \; \forall t \in \Omega \subset [0, \infty), \tag{1.2.2}$$

or alternatively

$$|\psi(t) - d(t)| \le \varepsilon(t), \; \forall t \in \Omega \subset [0, \infty), \tag{1.2.3}$$

where $d = 0.5(\varepsilon^+ + \varepsilon^-)$ and $\varepsilon = 0.5(\varepsilon^+ - \varepsilon^-)$.

Assuming that there are filters that satisfy the envelope constraints and taking into consideration the various applications discussed earlier, we seek those filters that give the least noise enhancement.

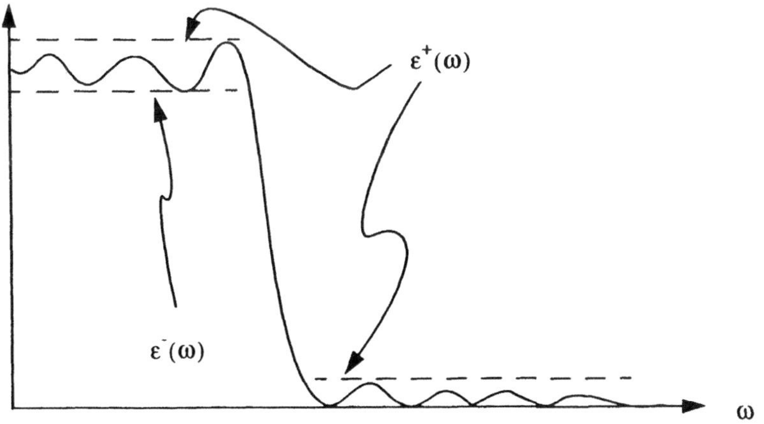

Figure 1.2.2. Magnitude response constraint

In the body of this monograph, a more general formulation that considers other constraints on the signals and a wider range of cost functionals will be presented. The mathematical problem considered here not only applies to EC filters but also to other applications such as the design of a low pass filter to minimize a certain objective function, while meeting a given constraint on stopband attenuation and passband ripple as shown in Figure 1.2.2. The constraint is:

$$|H(\omega) - d(\omega)| \le \varepsilon(\omega), \tag{1.2.4}$$

for all ω on the frequency range of interest. Note that $d(\omega)$ can be complex. For the case of constant ε, this is equivalent to requiring that the Chebyshev error [8]:

$$\max_{\omega \in \Omega} |H(\omega) - d(\omega)|, \tag{1.2.5}$$

does not exceed the tolerance. Similarly, for other types of filters such as highpass, bandpass, notch, bandstop, or filters with multiple stopbands and passbands, appropriate mask specifications can be applied. Another related problem is the design of an optimal pulse shape for use in digital communication systems [82]. Modern digital signal processing (DSP) hardware enables almost any pulse shape to be used for digital data transmission. Hence, there is a need for algorithms capable of constructing pulse shapes to match the properties of a given channel by specifying both time and frequency domain constraints.

There are several useful methods for synthesizing antennas with a prescribed radiation pattern [14]. These methods generally approximate the desired pattern using either a least square error criterion [14], [93] or a minimax error criterion [14], [107]. These algorithms are also used to design highly directive patterns. However, they often suffer from sidelobe and supergain problems. It is possible to overcome some of these difficulties by considering a constrained antenna pattern shaping problem, where we seek to minimize the aperture excitation energy, while constraining the shape of the far field radiation pattern. Mathematically, this constrained antenna pattern shaping problem may be stated as:

$$\begin{aligned} &\min \|A\|^2 \\ &\text{subject to } \varepsilon^-(\phi) \le E(\phi) \le \varepsilon^+(\phi), \ \forall \phi \in [-\pi/2, \pi/2], \end{aligned} \tag{1.2.6}$$

where $A(z)$, $z \in [-L/2, L/2]$, is the uniform phase current distribution over the aperture of length L, and

$$E(\phi) = \int_{-L/2}^{L/2} A(z) e^{j(2\pi/\lambda) z \sin\phi} dz, \quad \phi \in [-\pi/2, \pi/2]$$

is the electric field intensity in the far field region at an angle ϕ from the aperture excited at frequency λ.

The envelope constrained approach can also be used to design receiving arrays of the form shown in Figure 1.2.3. We consider the problem of choosing the weights $\boldsymbol{h} = [h_1, \ldots, h_n]^T$ such that the array has a given receiving gain pattern. If a test (pilot) signal is moved over the angular region, where the antenna gain is to be shaped, and the output $\psi(t)$ is observed, then $\psi(t)$ will represent the antenna gain pattern. Thus, we can select the weights so that

$$\varepsilon^-(t) \le \sum_{i=1}^{n} h_i s_i(t) \le \varepsilon^+(t) \qquad (1.2.7)$$

and the antenna pattern will be shaped accordingly. The test signal can be inserted, and by delay adjustments, it can simulate a test signal at different points in space [127].

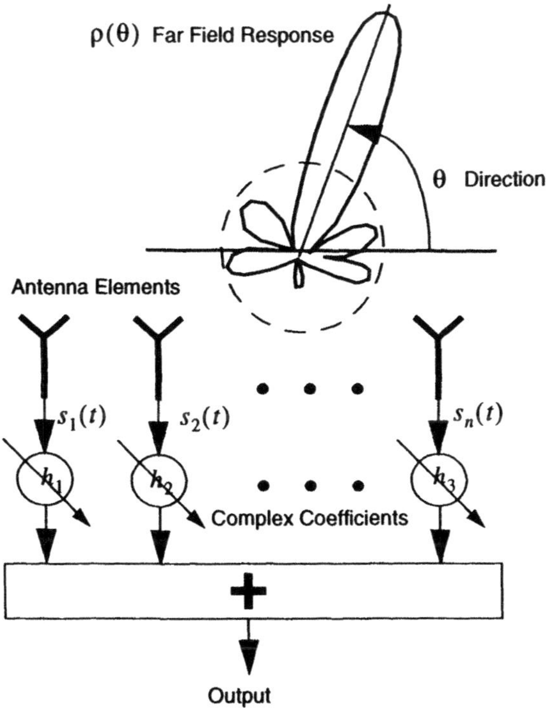

Figure 1.2.3. Antenna receiver

For a transmitting array, we seek to minimize the excitation energy (norm of the coefficients), while constraining the shape of the far field radiation pattern (e.g. low sidelobe levels). For a receiving array, we seek to minimize the isotropic noise gain (norm of the coefficients), while constraining the response of the array in certain directions (e.g. a degree of flatness in the look direction and low sidelobe levels in certain other directions).

1.3 Historical Notes

Arguably, McAulay [73], [75] appears to be the first to consider the design of signals and filters subject to sidelobe constraints. The sidelobes are constrained approximately with inequalities at a finite number of points. This early formulation of the EC filtering problem assumes finite support signals and filters.

Around the same time, Fortmann and Athans [28] investigated similar signal and filter design problems with sidelobes constrained at an uncountable number of points. This approach was extended by Fortmann and Evans to allow for time varying sidelobe constraints in [21], [28], [29]. This problem was tackled by the primal-dual method, which involves transforming the constrained optimization problem into a non-smooth unconstrained dual problem [21] [29]. The unconstrained problem is then discretized into a finite dimensional problem to be solved by a steepest ascent type algorithm [21], [29]. The approximate solutions obtained by this technique are not guaranteed to satisfy the envelope constraints. Only in the limiting case, i.e. with infinitesimal discretization, would the constraints be satisfied.

A discrete-time version of the EC filtering problem was also formulated for finite impulse response (FIR) filters [23], and the primal-dual algorithm was also used to solve this problem [22]. The iterative nature of this method allows it to be implemented on-line. However, convergence can be quite slow. In [138], [116], [117], penalty based methods were developed to solve the discrete-time EC filtering problem iteratively. These new methods are shown to be faster than the primal-dual algorithm in both theory and practice. These techniques are extended to include finite structure infinite impulse response (IIR) filters in [132], [133].

Also for FIR filters, Evans et. al. [24] generalized the EC filtering problem to cover the case where the input is not specified exactly, but known to stay within an input mask. This is known as the envelope constrained with uncertain input (ECUI) filtering problem and EC filtering is the special case where the input uncertainty is zero. The primal-dual algorithm was used to solve this problem. A more efficient technique was developed in [108] by transforming the non-smooth problem into a smooth QP problem, which can then be easily solved by available QP techniques.

1.4 Road Map

In Chapter 2, we introduce a convex programming problem in Hilbert space that covers all digital, analog and hybrid filter design problems of interest in this monograph. This abstraction is introduced to avoid repetition of key ideas and results while addressing a wide range of problems which may appear to be different at first glance. For this broad class of problems, we address the issue of existence and uniqueness and finite dimensional approximations. It is shown that feasible finite dimensional solutions converge to the optimal solution as the dimension tends to infinity. This allows optimum filters to be approximated by realizable filters that satisfy the constraints. The EC filtering problem and its variations, which form the main theme of this monograph, are studied as special cases of this convex program.

Chapter 3 redirects our attention back to the EC filtering problem. In particular, generalized versions of the Kuhn-Tucker Theorem and Duality Theorem (Appendix B.5) are used to convert the quadratic programming problem in Hilbert space into its dual in the space of regular Borel measures. Careful examination of the relationship between the primal and dual solutions reveals an interesting interpretation of the structure of optimal EC filters in terms of matched filters and whitening filters. For the case where the filter impulse response can be characterized by a finite number of parameters, interesting results are obtained through the use of Carathéodory's Dimensionality Theorem and duality (Appendices B.3 and B.5). In fact, the dual problem can be reduced to one in the space of finite support atomic measures. This means that primal problem with a continuum of constraints can be written as one with a finite number of constraints. Unfortunately, this problem is non convex.

In Chapter 4, we further narrow our scope down to the discrete-time EC filtering problem to address numerical methods for finding the optimal solution. From a numerical perspective, this is a good starting point since the discrete-time problem is the simplest in the sense that mathematical details are relatively easy to follow. Furthermore, it can be extended to the continuous time counterpart. Three classes of algorithms are presented, namely active set strategy, primal-dual and penalty methods. We provide convergence analysis and discussion on the relative merits of each of these classes of algorithms. The active set strategy is by far the most efficient, however it is not as well suited to on-line implementation as the primal-dual method or penalty method.

The approaches of Chapter 4 are extended in Chapter 5 to solve the continuous-time EC filtering problem, which can be formulated for purely analog filter structure or a hybrid between analog and digital components. The continuous-time problem contains a continuum of constraints. By discretizing the continuum of constraints to a finite set, the active set method can then be used. The primal-dual method also requires discretization of the dual variable. On the other hand, the penalty method can be extended to handle the case involving continuums of constraints without resorting to discretization.

Observe that for this type of problems, the response of the optimum filter to the prescribed input always touches the output envelope at some points. Consequently, any perturbation of the prescribed input or implementation errors could cause the optimal response to violate the output constraints. The problem of designing filters which are robust to such perturbations is essential for the implementation of EC filters. Chapter 6 addresses these robustness issues by introducing three robustness formulations. The first incorporates input uncertainty into the constraints. This problem has been formulated as a variation of the EC filtering problem earlier in Chapter 2, and is treated in detail in Chapter 6 including solution methods for solving special cases. The second approach incorporates implementation uncertainty into the constraints. This problem takes on the same form as the EC filtering problem and so the algorithms of Chapter 5 can be applied. The third approach does not consider the source of the disturbances, the key idea is to force the filter output to stay away from the envelope boundaries.

CHAPTER 2 FILTERING WITH CONVEX RESPONSE CONSTRAINTS

The envelope constrained (EC) filtering problem has been outlined in Chapter 1 as a constrained optimization problem in Hilbert space, where the filter's response to a prescribed signal is required to stay inside a given envelope. In this chapter, we introduce a convex programming problem in Hilbert space, which covers all digital, analog and hybrid filter design problems of interest in this monograph. The formulation considers the general case where all filter responses to a bounded set of excitations are constrained to a closed and convex set. This abstraction is developed to address a wide range of EC filtering problems in a unified manner. An important result is the convergence of the costs of feasible and finite dimensional solutions to the optimal cost. This allows optimum filters to be approximated by realizable filters that satisfy the constraints.

This chapter is organized as follows: Section 2.1 formulates the *convexly constrained responses* (*CCR*) filtering problem for analog filters as an infinite dimensional convex optimization problem. In Section 2.2, we digress from this abstraction and present specific cases of CCR filtering that form the main theme of this monograph, namely envelope constrained (EC) filtering and its variation EC filtering with uncertain input (ECUI). In Section 2.3, we direct our attention to general convex programming problems. In particular, we address the issues of existence and uniqueness and finite dimensional approximations. Formulation of analog CCR filtering problem for realizable circuit structures is addressed in Section 2.4. All results concerning analog CCR filtering carry over to discrete-time cases, and Section 2.5 presents a summary of discrete-time CCR filtering. Apart from analog filters, continuous-time filters can be realized with digital components via a hybrid configuration. In Sections 2.6 and 2.7, we present formulations of CCR filtering for hybrid structures. Proofs of results are provided in the appendix of the chapter unless otherwise stated.

2.1 Analog Filtering with Convexly Constrained Responses

Instead of imposing envelope constraints on the filter response to one prescribed excitation as described in Chapter 1, this section considers a more general case where all filter responses to a bounded set of excitations are constrained to a closed and convex set. We consider various types of convex cost functionals and discuss their physical implications. This problem is referred to as the convexly constrained response (CCR) filtering problem. This is a convex programming problem in Hilbert space as the properties of the cost functional and the feasible set reveal.

To formulate the CCR filtering problem for analog filters, it is necessary to consider appropriate spaces for the excitations and filter impulse responses. Let the intervals Ω_s, Ω_u, and Ω_ψ denote the supports for the input, the filter impulse response, and the filter output, respectively. By "the support" of a function f, we mean the complement of the largest open set on which f is zero and so the support is always a closed subset of the domain. Note that Ω_ψ is dependent on the choice of Ω_s and Ω_u. For example, if $\Omega_s = [0, T_s]$, and $\Omega_u = [0, T_u]$, then $\Omega_\psi = [0, T_s + T_u]$.

For any interval $\Omega \subseteq \boldsymbol{R}$, let $L_2(\Omega)$ denote the Hilbert space of real square integrable functions on Ω with inner product

$$\langle x, y\rangle = \int_{\Omega} x(\lambda)y(\lambda)d\lambda;$$

$C(\Omega)$ the Banach space of bounded real continuous functions on Ω with norm

$$\|x\|_{\infty} = \sup_{t \in \Omega}|x(t)|$$

and $L_{\infty}(\Omega)$ the Banach space of essentially bounded measurable functions on Ω with norm

$$\|x\|_{\infty} = \operatorname*{ess\,sup}_{t \in \Omega}|x(t)|\,.$$

Corresponding to an input x, we define, for any $t \in \Omega_{\psi}$, a function x_t in $L_2(\Omega_u)$ as

$$x_t(\lambda) = \begin{cases} x(t-\lambda), \; t-\lambda \geq 0 \\ 0, \qquad \text{otherwise.} \end{cases} \tag{2.1.1}$$

Then, the convolution $x * u$ of an impulse response u with x can be expressed as the inner product $\langle x_t, u\rangle$, i.e.

$$(x * u)(t) = \int_{\Omega} x(t-\lambda)y(\lambda)d\lambda = \langle x_t, u\rangle \tag{2.1.2}$$

The notation Ξ_x is adopted to denote the mapping of u to the filter response

$$\Xi_x u = x * u. \tag{2.1.3}$$

Lemma 2.1.1. *If $x \in L_2(\Omega_s)$, then the mapping $\Xi_x : L_2(\Omega_u) \to L_{\infty}(\Omega_{\psi})$ defined by (2.1.3) is a bounded (hence continuous) linear operator.*

The output constraints alone cannot guarantee that the output is bounded everywhere. For stability, the output should be bounded at all times. The restriction that the filter impulse responses are square integrable is not sufficient to guarantee bounded outputs. Boundedness of outputs also depends on the characteristics of the inputs. The following result presents a sufficient condition on the excitation x to yield a bounded output for any filter impulse responses in $L_2(\Omega_u)$. While only a sufficient condition, it proves adequate for the problems of interest in this monograph.

Lemma 2.1.2. *Suppose that* $x \in L_2(\Omega_s)$ *and* $u \in L_2(\Omega_u)$. *Then* $x * u$ *is continuous and bounded on* Ω_ψ, *i.e.* $x * u \in C(\Omega_\psi)$.

This lemma states that if the excitation x has finite energy, then the response ψ is continuous and bounded at all times, i.e. $\psi \in C(\Omega_\psi)$. The proof can be found in pp. 951 of [19]. This justifies the choice of the excitation space $L_2(\Omega_s)$ and the response space $L_\infty(\Omega_\psi)$ since $C(\Omega_\psi)$ is a subset of $L_\infty(\Omega_\psi)$.

If the input signal is not an energy signal but a power signal, i.e. a periodic signal, then we only need to consider the output during one period. Let x, a signal with support $[0, \tau]$, denote the input during one period. Then, the periodic extension $\tilde{x}$ of x is given by

$$\tilde{x}(t) = \sum_{h=-\infty}^{\infty} x(t - i\tau),$$

and the filter output is the convolution of the periodic extension $\tilde{x}$ with u. The convolution $\tilde{x} * u$ defines the circular convolution $x \otimes u$ of x with u, i.e.

$$(x \otimes u)(t) \equiv (\tilde{x} * u)(t), \forall t \in [0, \tau].$$

In effect, the convolution operator is replaced by the circular convolution operator.

The cost functional

In many applications, it is desirable to minimize the noise enhancement of the filter. With this rationale, we defined our cost function f to be the output noise power. Let ξ denote the noise component of the filter output. Then,

$$\xi(t) = \int_{\Omega_u} n(t - \lambda) u(\lambda) d\lambda .$$

Assuming that the additive noise n at the receiver input is zero-mean, stationary, with auto-correlation $R_{nn}(\tau)$, the output noise power is

$$f(u) = E[\xi^2(t)] = \int_{\Omega_u}\int_{\Omega_u} R_{nn}(\kappa - \lambda) u(\lambda) u(\kappa) d\kappa d\lambda . \quad (2.1.4)$$

By defining a linear operator L based on the autocorrelation function R_{nn} as

$$(Lu)(\lambda) = \int_{\Omega_u} R_{nn}(\kappa - \lambda)u(\kappa)d\kappa \tag{2.1.5}$$

and using the usual inner product, the cost functional defined by (2.1.3) can be expressed in a more convenient and intuitive form as a quadratic function of the filter impulse response

$$f(u) = \langle u, Lu \rangle . \tag{2.1.6}$$

Lemma 2.1.3.

(i) $\sqrt{f(u)} \geq 0, \forall u \in L_2(\Omega_u)$,

(ii) $\sqrt{f(\alpha u)} = |\alpha| \sqrt{f(u)}, \forall u \in L_2(\Omega_u), \forall \alpha \in \boldsymbol{R}$,

(iii) $\sqrt{f(u+v)} \leq \sqrt{f(u)} + \sqrt{f(v)}, \forall u, v \in L_2(\Omega_u)$.

Clearly, $\sqrt{f(u)}$ satisfies all the properties of a norm except that $\sqrt{f(u)} = 0$ does not necessarily imply $u = 0$. Hence, it is only a semi-norm or pseudo-norm on $L_2(\Omega_u)$. Part (i) of Lemma 2.1.3 is equivalent to saying that the linear operator L defined by (2.1.5) is positive semi-definite. Consequently, f is convex with respect to $u \in L_2(\Omega_u)$, although not necessarily strictly convex.

The cost functional is strictly convex when L is positive definite. In this case,

$$\|u\|_L \equiv \sqrt{\langle u, Lu \rangle}$$

is a norm, and

$$\langle v, u \rangle_L \equiv \langle v, Lu \rangle$$

generates an inner product. Hence, $\sqrt{f(u)}$ is the norm induced by this inner product.

Suppose that in addition to being additive, zero-mean and stationary, the input noise n is also *white*, with autocorrelation

$$R_{nn}(\tau) = N_0 \delta(\tau),$$

where δ is a unit impulse. Then, the output noise power $E[\xi^2(t)]$, defined by (2.1.4), is proportional to the square of the L_2-norm of the filter,

$$\begin{aligned} E[\xi^2(t)] &= N_0 \int_{\Omega_u} \int_{\Omega_u} \delta(\kappa - \lambda) u(\lambda) u(\kappa) d\kappa d\lambda \\ &= N_0 \int_{\Omega_u} u^2(\lambda) d\lambda = N_0 \|u\|_2^2. \end{aligned} \tag{2.1.7}$$

The cost functional in this case is strictly convex (not all norms are strictly convex), and L, defined by (2.1.5), is the identity operator.

In the derivation of the cost functionals (2.1.4) and (2.1.7), it is assumed that the statistics of the noise is known. In practice, information about the power spectral density of the noise is often obtained experimentally (see e.g. [54]). Suppose that the noise spectral density is $\Phi_N(\omega)$. Then, it can be verified that the output noise power due to the input noise n is given by

$$P_N = \frac{1}{2\pi} \int_{-\infty}^{\infty} \Phi_N(\omega) |U(j\omega)|^2 d\omega, \tag{2.1.8}$$

where U denotes the Laplace transform of u, i.e.

$$U(s) = \int_0^{\infty} u(t) e^{-st} dt.$$

Let $\mathcal{H}_2$ be the space of all complex-valued functions that are analytic in the open right half-plane with square integrable boundary functions. Then, U belongs to $\mathcal{H}_2$ since u is square integrable on $[0, \infty)$.

Assume that Φ_N is bounded in the sense that its noise power satisfies

$$\|\Phi_N\| = \frac{1}{2\pi} \int_{-\infty}^{\infty} \Phi_N(\omega) d\omega \le 1, \tag{2.1.9}$$

but otherwise unknown. It makes sense to consider the output noise power for the worst case input noise, i.e.

$$P_N = \max_{\|\Phi_N\| \le 1} \frac{1}{2\pi}\int_{-\infty}^{\infty} \Phi_N(\omega)|U(j\omega)|^2 d\omega. \tag{2.1.10}$$

It can be shown [30] that

$$\|U\|_\infty^2 = \max_{\|\Phi_N\| \le 1} \frac{1}{2\pi}\int_{-\infty}^{\infty} \Phi_N(\omega)|U(j\omega)|^2 d\omega, \tag{2.1.11}$$

where $\|U\|_\infty$ denotes the H_∞-norm of U, defined by

$$\|U\|_\infty = \sup_{\omega \in \boldsymbol{R}} |U(j\omega)|. \tag{2.1.12}$$

Hence, minimizing the output noise power for the worst case input noise is equivalent to finding a filter of transfer function $U(s)$ with minimum H_∞-norm. In this case, the corresponding cost functional can be expressed explicitly in terms of u as

$$f(u) = \left\|\int_0^\infty u(t)e^{-st}dt\right\|_\infty^2. \tag{2.1.13}$$

This cost functional is convex (but not strictly convex) since the Laplace transform is a linear isomorphism (see Appendix A.3) that maps the space $L_2[0, \infty)$ onto $\mathcal{H}_2$ and the H_∞-norm is convex. For cases in which the input signal is subjected to random disturbances with unknown but bounded power spectrum, the H_∞ optimization approach may offer a more robust design.

The use of these convex cost functionals can also be motivated by consideration of the crosstalk power, the filter's sensitivity to implementation errors, and uncertainties in filter parameters [13].

The feasible region

Let $\mathcal{S}$ be a bounded set of inputs in $L_2(\Omega_s)$ and Ψ be the set of permissible responses which is closed and convex in $L_\infty(\Omega_\psi)$. Then, the feasible region $\mathcal{F}$ is defined as the set of filters with square integrable impulse responses such that when excited by any $x \in \mathcal{S}$, the corresponding output is in Ψ, i.e.

$$\mathcal{F} \equiv \{u \in L_2(\Omega_u) \colon \forall x \in \mathcal{S},\ x * u \in \Psi\}. \tag{2.1.14}$$

***Lemma 2.1.4**. Let H be a Hilbert space, Y be a Banach space, and $B(H, Y)$ denote the (Banach) space of bounded linear operators from H to Y. Suppose that the set $\Psi \subset Y$ is closed and convex, and that the set $\Phi \subset B(H, Y)$ is bounded. Then, the set*

$$\{u \in H \colon \forall X \in \Phi,\ Xu \in \Psi\}$$

is closed and convex.

Using the notation Ξ_x to denote the linear operator associated with x as defined in (2.1.3), the feasible region $\mathcal{F}$ can also be written as

$$\mathcal{F} = \{u \in L_2(\Omega_u) \colon \forall \Xi_x \in \Phi, \Xi_x u \in \Psi\},$$

where

$$\Phi = \{\Xi_x : x \in \mathcal{S}\}.$$

From Lemma 2.1.1, Φ is a bounded set in $B(L_2(\Omega_u), L_\infty(\Omega_\psi))$. Hence, it follows from Lemma 2.1.4 that the feasible region $\mathcal{F}$ is closed and convex.

2.2 Filter Design with Envelope Constraints

In line with the main theme of the monograph, this section is devoted to formal aspects of the design of filters with envelope constraints. One of the simplest instance of the CCR filtering problem is the EC filtering problem discussed in Chapter 1.

When the filter responses on an interval, say $\Omega_c \subseteq \Omega_\psi$, are constrained to lie within an envelope described by the upper and lower boundaries ε^+, $\varepsilon^- \in C(\Omega_\psi)$, respectively, the set Ψ of permissible responses is given by

$$\Psi = \{\psi \in L_\infty(\Omega_\psi) \colon \varepsilon^- \preceq \psi \preceq \varepsilon^+\},$$

where the partial ordering $\preceq$ (see Appendix B.3 for definition) on the response space $L_\infty(\Omega_\psi)$ is defined by

$$x, y \in L_\infty(\Omega_\psi),\ x \preceq y \Leftrightarrow \forall t \in \Omega_c,\ x(t) \le y(t). \tag{2.2.1}$$

Alternatively, by defining

$$\varepsilon \equiv 0.5(\varepsilon^+ - \varepsilon^-) \text{ and } d \equiv 0.5(\varepsilon^+ + \varepsilon^-),$$

the set of permissible responses can be written as

$$\Psi(d, \varepsilon) = \{\psi \in L_\infty(\Omega_\psi) : |\psi - d| \preceq \varepsilon\}, \tag{2.2.2}$$

where $|\bullet|$ denotes the pointwise absolute value of a function, i.e.

$$|x|(t) = |x(t)|, \forall t.$$

Intuitively, the waveform d can be interpreted as the desired output, and ε is the time-dependent tolerance about d. The upper and lower envelope boundaries are recovered from d and ε by

$$\varepsilon^+ = d + \varepsilon \text{ and } \varepsilon^- = d - \varepsilon.$$

It is a straightforward task to show that the set $\Psi(d, \varepsilon)$ of permissible responses is closed and convex. Furthermore, since Ξ_s is bounded, it is clear that the EC filtering problem

$$\begin{gathered} \min f(u) \\ \text{subject to } \varepsilon^- \preceq \Xi_s u \preceq \varepsilon^+ \end{gathered} \tag{2.2.3}$$

is, in fact, a CCR problem.

Another class of CCR problems considered in this monograph involves the design of EC filters, where the excitation s is perturbed so that it is only known to stay within an input envelope described by the upper and lower boundaries s^- and s^+. This filtering problem is said to be envelope constrained with uncertain input (ECUI) [24]. Note that the EC filtering problem is a special case of the ECUI problem.

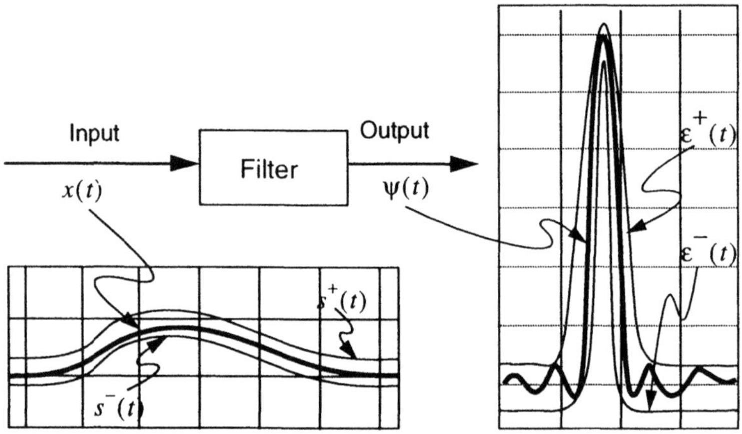

Figure 2.2.1. EC filtering with uncertain input.

When the excitations are also constrained to stay within an envelope described by the upper and lower boundaries s^+, $s^- \in L_2(\Omega_s)$, the set $\mathcal{S}$ of all possible excitations is given by

$$\mathcal{S} = \{x \in L_2(\Omega_s)\colon s^- \preceq x \preceq s^+\},$$

where the partial ordering $\preceq$ on the excitation space $L_2(\Omega_s)$ is defined by

$$x, y \in L_2(\Omega_s),\ x \preceq y \Leftrightarrow \forall t \in \Omega_s,\ x(t) \le y(t). \qquad (2.2.4)$$

It is assumed that $s^- \preceq s^+$ and s^+, s^- are finite energy signals so that $\mathcal{S}$ is non-empty and bounded. Similar to $\Psi(d, \varepsilon)$ given in (2.2.2), by defining

$$s \equiv 0.5(s^+ + s^-) \text{ and } \theta \equiv 0.5(s^+ - s^-),$$

an alternative expression for the set of all possible excitations is

$$\mathcal{S}(s, \theta) = \{x \in L_2(\Omega_s)\colon |x - s| \preceq \theta\}, \qquad (2.2.5)$$

Thus, the feasible set of ECUI filters is

$$\begin{aligned}\mathcal{F} &= \{u \in L_2(\Omega_u)\colon x * u \in \Psi(d, \varepsilon),\ \forall x \in \mathcal{S}(s, \theta)\} \\ &= \{u \in L_2(\Omega_u)\colon |x * u - d| \preceq \varepsilon,\ \forall x\colon |x - s| \preceq \theta\}.\end{aligned} \qquad (2.2.6)$$

Remark 2.2.1: Although the same notation is used for both partial orderings, there should be no ambiguity if the specific signal space is taken into consideration. The major difference is that in the partial ordering for the excitation space $L_2(\Omega_s)$, defined in (2.2.4), the inequality is required to hold for the entire support Ω_s, while in the partial ordering for the response space $L_\infty(\Omega_\psi)$, defined in (2.2.1), the inequality only holds on the set Ω_c rather than the support Ω_ψ.

Since the set $\Psi(d, \varepsilon)$ of permissible responses is closed and convex, the ECUI filtering problem

$$\begin{aligned} &\min\ f(u) \\ &\text{subject to } |x * u - d| \preceq \varepsilon,\ \forall x\colon |x - s| \preceq \theta \end{aligned} \tag{2.2.7}$$

fits neatly into the framework of the optimization problem presented in Section 2.1. Clearly, the feasible region $\mathcal{F}$ given by (2.2.6) is closed and convex, since the set S of all possible excitations is bounded (see Lemma 2.1.4). For the choice of an appropriate cost functional, any of the criteria discussed in Section 2.1 can be a candidate, depending on the specific emphasis of the problem concerned.

To eliminate output envelopes that permit the trivial solution $u = 0$, we make the following assumptions:

Assumption A-2.2.1. *There exists an open subset I of Ω_c such that*

$$\varepsilon^+(t)\varepsilon^-(t) > 0,\ \forall t \in I.$$

Assumption A-2.2.2. $\varepsilon^-(t) \le \varepsilon^+(t)$ *for all* $t \in \Omega_c$.

Assumption A-2.2.1 means that there is an open interval where both the upper and lower boundaries have the same sign. Assumption A-2.2.2 is an obvious necessary condition for the feasible region $\mathcal{F}$ to have a non-empty interior. Because of the nature of the envelope constraints, there may be no feasible solutions at all, if the tolerance ε is too small. However, if there is a feasible solution, then an infinitesimal increase in ε will allow the constraints to be satisfied as strict inequality constraints.

2.3 Convex Programming

The problems considered in this monograph involve the minimization of a continuous and convex cost functional f on a Hilbert space H over a convex subset $\mathcal{F}$. In this section, we address the issue of uniqueness and finite dimensional approximations for this class of problems.

Theorem 2.3.1. *Let $\mathcal{F}$ be a closed and convex subset in a Hilbert space H, and f be a continuous and convex functional on H. If $\mathcal{F}$ is non-empty, then there exists a $u^o \in \mathcal{F}$ such that*

$$f(u^o) \leq f(u), \forall u \in \mathcal{F}.$$

Moreover, if f is strictly convex, then u^o is unique.

If H is infinite dimensional, one approach to solving the convex programming problem is to introduce finite dimensional approximations. This can be done by fixing a basis $\{v_i\}_{i=0}^{\infty}$ and determine a sub-optimum solution for each finite span $[\{v_i\}_{i=0}^{n-1}]$ of this basis, assuming H is separable. The n-dimensional approximation problem can be written as

$$\begin{gathered} \min\ f(u) \\ \text{subject to } u \in \mathcal{F} \cap [\{v_i\}_{i=0}^{n-1}]. \end{gathered} \tag{2.3.1}$$

Notice that $\mathcal{F} \cap [\{v_i\}_{i=0}^{n-1}]$ is also closed and convex. Solutions to this finite dimensional problem exist if the span $[\{v_i\}_{i=0}^{n-1}]$ is large enough to intersect the original feasible set $\mathcal{F}$.

Intuitively, the sub-optimum costs converge to the optimum cost as the number of basis vectors used in the finite span tends to infinity.

Theorem 2.3.2. *Let $\mathcal{F}$ be a closed and convex subset of a separable Hilbert space H with non-empty interior $\mathcal{F}^o$ (see Appendix A.1 for definition), and f be a continuous and convex functional on H. Let $u^o \in \mathcal{F}$ be such that*

$$f(u^o) \le f(u), \forall u \in \mathcal{F},$$

and $u_n^o \in \mathcal{F} \cap [\{v_i\}_{i=0}^n]$ *be such that*

$$f(u_n^o) \le f(u_n), \ \forall u_n \in \mathcal{F} \cap [\{v_i\}_{i=0}^n].$$

If $\{v_i\}_{i=0}^{\infty}$ *is total in* H, *then* $f(u_n^o)$ *monotonically decreases to* $f(u^o)$ *as* n *tends to infinity. Moreover, if* f *is an inner-product-induced norm, then*

$$\lim_{n \to \infty} \|u_n^o - u^o\| = 0.$$

To write the finite dimensional problem, i.e. Problem (2.3.1), in an equivalent form in $\boldsymbol{R}^n$, let $K : \boldsymbol{R}^n \to H$ be a bounded linear operator defined by

$$Ka = \sum_{i=0}^{n-1} a_i v_i, \tag{2.3.2}$$

where $\boldsymbol{a} \in \boldsymbol{R}^n$. Then, Problem (2.3.1) can be written as

$$\min_{\boldsymbol{a} \in K^{-1}(\mathcal{F})} (f \circ K)(\boldsymbol{a}), \tag{2.3.3}$$

where $(f \circ K)(\boldsymbol{a}) = f(K(\boldsymbol{a}))$. From the boundedness and linearity of K, it can be easily verified that $f \circ K$ is a continuous and convex function, and the set

$$K^{-1}(\mathcal{F}) \equiv \{\boldsymbol{a} \in \boldsymbol{R}^n: K\boldsymbol{a} \in \mathcal{F}\}$$

is closed and convex in $\boldsymbol{R}^n$.

Theorem 2.3.2 allows us to approximate the optimum solution to any degree of accuracy (and satisfy the constraints) by using enough elements from any total sequence in H. This type of finite dimensional approximation is crucial for filter design applications since it is unlikely that the optimum filter can be realized without approximation. Suppose that one seeks to realize such a filter by designing a network whose impulse response approximates that of the optimum filter in some sense, e.g. least squares, Páde, orthogonal approximations. The important question that arises is: Would this approximation still satisfy the convex constraints? Thus, it would seem more appropriate to choose a particular filter structure and then impose the constraints. To this end, the sub-optimum solutions of Theorem 2.3.2 are more useful from a practical viewpoint. With an appropriate choice of basis, these sub-optimum solutions correspond to finite-structured filters that can be readily realized as shown in Section 2.4.

2.4 Finite Dimensional Analog CCR Filters

In Section 2.1, we have shown how a filtering problem with time domain response constraints can be formulated as a convex optimization problem. A large class of filtering problems including frequency domain constraints and/or time-frequency domain constraints can be similarly formulated. By virtue of Theorem 2.3.2, we show, in this section, how these filtering problems can be approximated by practical filters that have a finite structure.

The optimum CCR filter in $L_2(\Omega_u)$ is only of theoretical interest since it is unlikely that this filter can be realized with finite circuitry without violating the constraints. It is of great practical importance that finite filter structures for the CCR filtering problem is addressed. The aim is to provide a firm theoretical basis so that implementable algorithms and suitable circuit structures can be developed.

Let $\{v_i\}_{i=0}^{\infty}$ be a total sequence in $L_2(\Omega_u)$. Consider finite dimensional filters with impulse responses of the form $\boldsymbol{Ka} \in L_2(\Omega_u)$, where $K : \boldsymbol{R}^n \to L_2(\Omega_u)$ is a bounded linear operator defined by

$$(\boldsymbol{Ka})(t) = \sum_{i=0}^{n-1} a_i v_i(t) = \boldsymbol{v}^T(t)\boldsymbol{a}, \qquad (2.4.1)$$

in which $\boldsymbol{v}^T = [v_0, v_1, \ldots, v_{n-1}]$, and the superscript T denotes transpose.

The filters considered in this problem are n-dimensional. The set of feasible filters is embedded in an n-dimensional sub-space of $L_2(\Omega_u)$ and can be written as

$$\mathcal{F} \cap [\{v_i\}_{i=0}^{n-1}].$$

Since all n-dimensional vector space is isomorphic to $\boldsymbol{R}^n$, this feasible set can be characterized by the set of feasible filter coefficients which is defined in $\boldsymbol{R}^n$ as

$$\mathcal{F}^n = \{\boldsymbol{a} \in \boldsymbol{R}^n : \forall x \in \mathcal{S},\ \Xi_x K\boldsymbol{a} \in \Psi\}. \tag{2.4.2}$$

It is clear that the set $\mathcal{F}^n$ of feasible filter coefficients is convex by Lemma 2.1.4, since the linearity of K implies the linearity of $\Xi_x K$. To obtain an explicit expression for the cost functional in terms of the vector of filter parameters $\boldsymbol{a}$, we have

$$f(K(\boldsymbol{a})) = \langle K\boldsymbol{a}, LK\boldsymbol{a}\rangle = \boldsymbol{a}^T K^{\dagger} LK\boldsymbol{a}, \tag{2.4.3}$$

where $K^{\dagger} : L_2(\Omega_u) \to \boldsymbol{R}^n$ is the adjoint (operator) of K defined by

$$K^{\dagger}u = [\langle v_0, u\rangle, \langle v_1, u\rangle, \ldots, \langle v_{n-1}, u\rangle]^T. \tag{2.4.4}$$

The operator $K^{\dagger}LK : \boldsymbol{R}^n \to \boldsymbol{R}^n$ can be represented by the following $n \times n$ matrix, (often called a Gram matrix):

$$K^{\dagger}LK = \begin{bmatrix} \langle v_0, Lv_0\rangle & \langle v_0, Lv_1\rangle & \ldots & \langle v_0, Lv_{n-1}\rangle \\ \langle v_1, Lv_0\rangle & \langle v_1, Lv_1\rangle & \ldots & \langle v_1, Lv_{n-1}\rangle \\ \vdots & & \ddots & \vdots \\ \langle v_{n-1}, Lv_0\rangle & \langle v_{n-1}, Lv_1\rangle & \ldots & \langle v_{n-1}, Lv_{n-1}\rangle \end{bmatrix}. \tag{2.4.5}$$

The Gram matrix $K^{\dagger}LK$ is positive semi-definite since L is positive semi-definite. Furthermore, if L is positive definite and the v_i's are linearly independent, then $K^{\dagger}LK$ is also positive definite, i.e. $f \circ K$ is strictly convex. In most practical situations, L is the identity operator and K is specified by a finite subset of a complete set of orthonormal v_i's over the interval Ω_u. In this case, it is easily seen that the Gram matrix $K^{\dagger}K$ is the identity matrix I. The main reason for considering filter realization by orthonormal sets is that it is possible to generate and combine certain set of vectors using simple finite filter structures. The most notable examples are Laguerre and Legendre functions [55], [61], [67], [68] [101] [106], [126].

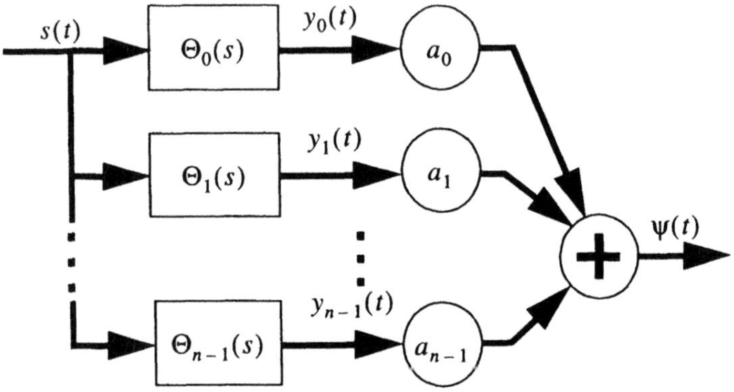

Figure 2.4.1. Parallel filter structure

There are two main techniques for building filters. The parallel structure is the simplest and most direct method (see Figure 2.4.1). The transfer function for each block is denoted by $\Theta_i(s), i = 0, 1, \ldots, n-1$, and the transfer function of the filter is immediately given by

$$U(s) = \sum_{i=0}^{n-1} a_i \Theta_i(s) .$$

In this case, $v_i = \theta_i$, where θ_i is the inverse Laplace transform of Θ_i. The parallel structure is widely used for many filters employing orthonormal functions.

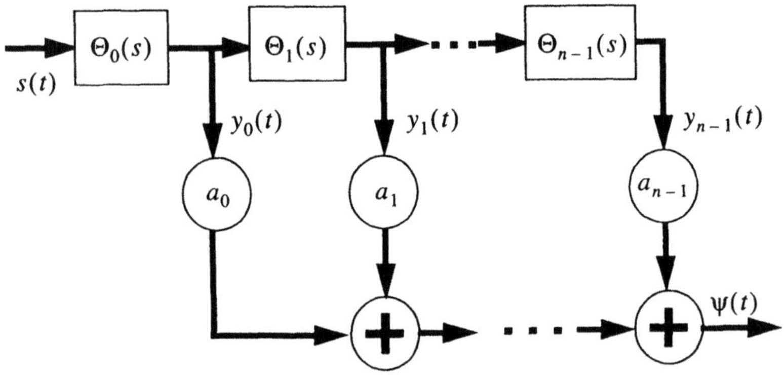

Figure 2.4.2. Transversal filter structure

The second method is the transversal structure, (see Figure 2.4.2). The transfer function for each block is denoted by $\Theta_j(s), j = 0, 1, \ldots, n-1$. Let V_i denote the Laplace transform of v_i. Then,

$$V_i(s) = \prod_{j=0}^{i} \Theta_j(s)$$

and the transfer function of the transversal filter is given by

$$U(s) = \sum_{i=0}^{n-1} a_i \prod_{j=0}^{i} \Theta_j(s).$$

The best known examples of filters employing the transversal structure are Legendre or Laguerre filters or when the Θ_i's are pure delay elements, which is a limiting case of the latter.

These finite-structured filters restrict the space of filter impulse responses to a finite dimensional subspace of $L_2(\Omega_u)$. That is, the optimization is carried out on this subspace rather than on the whole of $L_2(\Omega_u)$. The subspace spanned by $\{v_i\}_{i=0}^{n-1}$ (denoted as $[\{v_i\}_{i=0}^{n-1}]$) is closed and convex. Hence, the feasible region $\mathcal{F} \cap [\{v_i\}_{i=0}^{n-1}]$ is also closed and convex. Strict convexity of the cost functional ensures a unique optimum solution if an interior point exists. For a norm cost, it follows from Theorem 2.3.2 that the sub-optimum CCR filters for the finite filter structures converge to the optimum CCR filter as the filter structure grows. For a general convex cost, the sub-optimum costs converge to the optimum cost.

2.5 Discrete-time CCR Filter Design

Results from the continuous-time case apply to the discrete-time case if the functions of time are replaced by sequences and the Lebesque measure is replaced by the usual counting measure. This section provides a summary of these results.

To distinguish between discrete-time and continuous-time, let $\tilde{\Omega}_s$, $\tilde{\Omega}_u$, $\tilde{\Omega}_\psi$ denote the discrete-time support of the input, filter impulse response and output respectively.

Additionally, let $\tilde{\Omega}_c$ be the set of instances where the constraints are imposed on the filter output. The notation $l_2(\tilde{\Omega})$ is used to denote the Hilbert space of square summable sequences on $\tilde{\Omega}$ with inner product

$$\langle x, y \rangle = \sum_{i \in \tilde{\Omega}} x(i) y(i) .$$

The notation $l_\infty(\tilde{\Omega})$ denotes the Banach space of bounded sequences on $\tilde{\Omega}$ with norm:

$$\|x\|_\infty = \sup_{k \in \tilde{\Omega}} |x(k)| .$$

Similar to the continuous-time case, the convolution of x and u can be written as an inner product:

$$(x * u)(k) = \sum_{i \in \tilde{\Omega}_u} x(k-i) u(i) = \langle x_k, u \rangle, \tag{2.5.1}$$

where x_k is a sequence defined for any non-negative integer k by

$$x_k(i) = \begin{cases} x(k-i), & k-i \geq 0 \\ 0 & , \; k-i < 0. \end{cases}$$

Suppose that the excitation x is square-summable on $\tilde{\Omega}_s$, i.e. $x \in l_2$. Then, the response ψ of a filter $u \in l_2(\tilde{\Omega}_u)$ is bounded on $\tilde{\Omega}_\psi$, i.e. $\psi \in l_\infty(\tilde{\Omega}_\psi)$. Moreover, the mapping

$$\Xi_x : l_2(\tilde{\Omega}_u) \to l_\infty(\tilde{\Omega}_\psi)$$

defined by

$$\Xi_x u = x * u \tag{2.5.2}$$

is a continuous linear operator. This linear operator is also invertible if $x \neq 0$ as shown in the following lemma.

***Lemma 2.5.1**. If the excitation $x \in l_2$ is non-zero, then the associated linear operator Ξ_x, defined by (2.5.2), is invertible.*

Let $\mathcal{S}$ be the bounded set of possible excitations, and Ψ be the closed and convex set of permissible responses. Then, the feasible set given by

$$\mathcal{F} = \{u \in l_2(\tilde{\Omega}_u): \forall x \in \mathcal{S},\ x * u \in \Psi\}$$

is closed and convex by Lemma 2.1.4.

Let $R_{nn}(j)$ denote the auto-correlation of the zero-mean, stationary input noise n. Using the stationarity of the input noise n together with the filter's linearity and time invariance, the output noise power is given by

$$f(u) = \sum_{i \in \Omega_u} \sum_{j \in \Omega_u} R_{nn}(i-j)u(i)u(j) = \langle u, Lu \rangle \tag{2.5.3}$$

This cost functional defines a pseudo-norm on $l_2(\tilde{\Omega}_u)$, and hence, is convex although not necessarily strictly convex.

Suppose that in addition to being additive, zero-mean and stationary, the input noise n is also white. Then, it is easy to see that the output noise power (2.5.3) is proportional to the square of the l_2-norm of the filter, i.e.

$$f(u) = N_0 \sum_{i \in \Omega_u} |u(i)|^2 = N_0\|u\|^2.$$

The cost functional in this case is strictly convex.

For input noise with unknown statistics but bounded noise power, it is desirable to minimize the output noise power for the worst case input noise. By the same arguments as for the analog case, the cost functional corresponds to the H-infinity norm of the transfer function $U(z)$ of the filter, i.e.

$$f(u) = \max_{\omega \in [-\pi, \pi]} \left| \sum_{k \in \Omega_u} u(k)e^{-j\omega k} \right|^2.$$

Discrete-time Filters with Finite Structures

For finite filter structures such as transversal and parallel filters, the block diagrams are the same as for the analog case, with time t replaced by k and frequency s replaced by z. A tapped delay line finite impulse response (FIR) filter is an example of a transversal filter. A discrete-time Laguerre network is another example of a transversal filter. However, this is an infinite impulse response (IIR) filter. In general, these filters have impulse response that can be written in the form:

$$Ka = \sum_{k=0}^{n-1} a_k v_k,$$

where $\boldsymbol{a} \in \boldsymbol{R}^n$, and $K : \boldsymbol{R}^n \to l_2(\tilde{\Omega}_u)$.

The set of feasible filter coefficients is given by

$$\mathcal{F} = \{\boldsymbol{a} \in \boldsymbol{R}^n : \forall x \in \mathcal{S},\ x * (K\boldsymbol{a}) \in \Psi\}.$$

The cost for the finite dimensional problem can be expressed as a quadratic function of the vector of filter coefficients, i.e.

$$(f \circ K)(\boldsymbol{a}) = \langle K\boldsymbol{a}, LK\boldsymbol{a} \rangle = \boldsymbol{a}^T K^{\dagger} LK\boldsymbol{a}.$$

As in the continuous-time case, this is a finite dimensional optimization problem. From Theorem 2.3.2, if the sequence $\{v_k\}_{k=0}^{\infty}$ is total in $l_2(\tilde{\Omega}_u)$, then the sequence of sub-optimum solutions approaches the optimum solution as the dimension of filter increases. For a general convex cost, the sub-optimum costs converge to the optimum cost.

2.6 Continuous-time CCR Filtering via DSP Approach

The advances in the development of digital processors motivate the consideration of filter structures realized with digital components. This section presents a formal study of the continuous-time CCR filtering problem for a hybrid filter composed of an analog-to-digital (A/D) converter, a discrete-time filter, a digital-to-analog (D/A) converter and a post-filter. The task is to determine the discrete-time component of such a filter so as to minimize the effect of input noise whilst satisfying the constraint that the responses of the filter to a specified set fit into a prescribed convex set.

Problem formulation for hybrid filter

In this section, the CCR filtering problem for the hybrid filter shown in Figure 2.6.1 is formulated. Given digital signal processing (DSP) technology, this is a likely method of realization. To process a continuous-time signal by DSP techniques, it is necessary to convert the signal into a sequence of numbers. This is usually done by sampling the analog signal. The samples must then be quantized to a discrete set of amplitude levels, and the resulting signal is passed to the digital processor. After processing, the output sequence of numbers is then converted to continuous-time signal via the D/A converter.

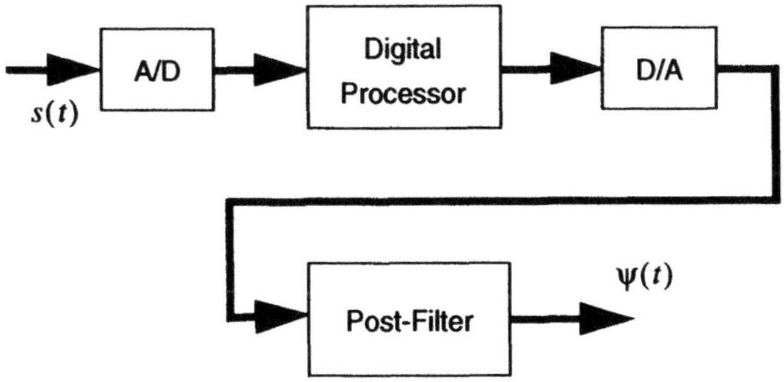

Figure 2.6.1. Configuration for digital processing of continuous-time signal

The post-filter's function is to smooth the output of the D/A block. In other words, the combined function of the D/A block and the post-filter is to interpolate the (discrete-time) output of the digital processor. The absence of a post-filter corresponds to piecewise constant interpolation of the digital output. Next on the ladder of complexity is linear interpolation. For a smoother output, techniques such as Shannon interpolation [70] can be used. The output of the system varies from a staircase to much smoother waveforms depending on the type of interpolation.

In practice, the post-filter is often implemented as a lowpass filter with cut-off frequency being half of the sampling frequency. Examples of these filters include Butterworth, Bessel, Chebyshev and elliptic filters [91].

It is assumed that the incoming signal is sampled at or above the Nyquist rate [88]. To simplify matters, quantization effects inherent in digital processes are neglected. This assumption results in the linearity of the system, since quantization is a non-linear operation. Hence, the term 'digital filters' actually means discrete-time filters.

Feasible region

Suppose that the digital processor has a discrete-time impulse response $u \in l_2(\tilde{\Omega}_u)$. Then, the (discrete-time) output $\hat{\psi}$ of the digital processor, when the system in Figure 2.6.1 is excited by a continuous-time signal x, is given by

$$\hat{\psi}(i) = \sum_{j \in \tilde{\Omega}_u} x[(i-j)T_s]u(j), \tag{2.6.1}$$

where T_s is the sampling period of the A/D converter. If x is square-integrable (i.e. $\{x(kT_s)\}_{k=-\infty}^{\infty}$ is square-summable), then the output sequence $\{\hat{\psi}(i)\}_{i=-\infty}^{\infty}$ is bounded, from the discrete-time filtering results presented in Section 2.5. This discrete-time signal enters the D/A block, which then produces the following waveform:

$$\sum_{i=-\infty}^{\infty} \hat{\psi}(i)\Pi(t-iT_s), \tag{2.6.2}$$

where Π is the (continuous-time) response of the D/A block to a (discrete-time) unit impulse. In practice, Π assumes a shape such that the sum in (2.6.2) converges for all $t \in [0, \infty)$. Hence, the output of the D/A block is bounded and piecewise continuous. An example of such a pulse shape is a rectangular pulse of length T_s, i.e.

$$\Pi(t) = \begin{cases} 1, t \in [0, T_s] \\ 0, \text{otherwise} \end{cases}.$$

Let h denote the impulse response of the post-filter, and $\Lambda = \Pi * h$ the response of the post-filter to the pulse Π. Then, the response ψ of the hybrid filter to a continuous-time excitation x is given by

$$\psi(t) = \sum_{i=-\infty}^{\infty} \hat{\psi}(i)\Lambda(t - iT_s), \quad t \in [0, \infty).$$

Using (2.6.1) and (2.6.2), the output ψ can be expressed in terms of u as

$$\psi(t) = \sum_{i=-\infty}^{\infty} \sum_{j \in \tilde{\Omega}_u} x[(i-j)T_s]u(j)\Lambda(t - iT_s), \quad t \in [0, \infty). \tag{2.6.3}$$

Since we are only dealing with excitation with support in $[0, \infty)$, the index i in (2.6.3) can be taken from zero to infinity. Assuming appropriate post-filtering such that (2.6.3) converges for all $t \in [0, \infty)$ (e.g. bounded input bounded output stability), ψ is bounded and continuous on $[0, \infty)$. Moreover, the mapping

$$\Xi_x : l_2(\tilde{\Omega}_u) \to C(\Omega_\psi)$$

defined by

$$(\Xi_x u)(t) = \sum_{j \in \tilde{\Omega}_u} \sum_{i=0}^{\infty} x[(i-j)T_s]\Lambda(t - iT_s)u(j) \tag{2.6.4}$$

is a continuous linear operator. This operator is analogous to the convolution operator defined for continuous-time and discrete-time filters in (2.1.3) and (2.5.2), respectively. Making the index change $k = i - j$, for all $j \in \tilde{\Omega}_u$ (hence $k \in \tilde{\Omega}_s$), we can rewrite $\Xi_x u$ in the following form:

$$\begin{aligned}(\Xi_x u)(t) &= \sum_{k \in \tilde{\Omega}_s} \sum_{j \in \tilde{\Omega}_u} \Lambda(t - kT_s - jT_s)u(j)x(kT_s) \\ &= \sum_{k \in \tilde{\Omega}_s} (\Psi u)(t - kT_s)x(kT_s) \\ &= (\tilde{\Xi}_x \Psi u)(t)\end{aligned} \tag{2.6.5}$$

where $\Psi : l_2(\tilde{\Omega}_u) \to L_\infty(\Omega_\psi)$ and $\tilde{\Xi}_x : L_\infty(\Omega_\psi) \to L_\infty(\Omega_\psi)$ are linear mappings given, respectively, by

$$(\Psi u)(\varsigma) \equiv \sum_{l = \tilde{\Omega}_u} \Lambda(\varsigma - lT_s)u(l), \tag{2.6.6}$$

and

$$(\tilde{\Xi}_x v)(t) \equiv \sum_{k \in \tilde{\Omega}_s} v(t - kT_s)x(kT_s). \tag{2.6.7}$$

Lemma 2.6.1*. The linear operator* Ψ*, defined by (2.6.6), is a one-to-one mapping if the pulse response* Λ *is non-zero.*

Let $\mathcal{S}$ denote the bounded set of excitations, and Ψ denote the closed and convex set of permissible responses. The feasible region $\mathcal{F}$ for this problem is defined as the set of discrete-time filters with square-summable impulse responses such that when evoked by any excitation in $\mathcal{S}$, the corresponding output lies in Ψ, i.e.

$$\mathcal{F} = \{u \in l_2(\tilde{\Omega}_u) \colon \forall x \in \mathcal{S},\ \Xi_x u \in \Psi\}. \tag{2.6.8}$$

This feasible region is the same as that for the all-analog case, apart from being a subset of $l_2(\tilde{\Omega}_u)$ instead of $L_2(\Omega_u)$. Since Ψ is closed and convex, it follows from Lemma 2.1.4 that $\mathcal{F}$ is also a closed and convex set.

Cost Functional

A simple cost functional one could use is the norm of the discrete-time filter. However, the output noise power of this system is not proportional to this norm. In this section, an expression for the output noise power is derived, assuming that the sequence of input noise samples is a stationary process as for the discrete-time case.

Similar to the expression for the noiseless output, the output noise can be written as

$$\xi(t) = \sum_{i = -\infty}^{\infty} n(iT_s)[(\Psi u)(t - iT_s)]$$

Since input noise sample sequence is a stationary process with mean μ_n and autocorrelation $R_{nn}((j-k)T_s)$, it is easily seen that the mean $\mu_\xi(t)$ and autocorrelation $R_{\xi,\xi}(t+\lambda, t)$ of the output noise are given, respectively, by

$$\mu_\xi(t) \equiv E[\xi(t)]$$
$$= \sum_{i=-\infty}^{\infty} \mu_n[(\Psi u)(t - iT_s)],$$

and

$$R_{\xi,\xi}(t+\lambda, t) \equiv E[\xi(t+\lambda)\xi(t)]$$
$$= \sum_{j=-\infty}^{\infty} \sum_{k=-\infty}^{\infty} R_{nn}((j-k)T_s)[(\Psi u)(t+\lambda - jT_s)][(\Psi u)(t - kT_s)].$$

It is clear from these expressions that the output noise is no longer stationary as for the analog filters. However, it can be easily verified that $\mu_\xi(t)$ and $R_{\xi,\xi}(t+\lambda, t)$ are periodic functions of t, with period T_s. The noise process ξ is said to be wide sense cyclo-stationary (see [34]). The averaged output noise power over one period,

$$\bar{R}_{\xi,\xi}(0) \equiv \frac{1}{T_s}\int_0^{T_s} E[\xi^2(t)]dt,$$

is independent of time, and hence, $f(u) = \bar{R}_{\xi,\xi}(0)$ is a well-defined cost. Thus, substituting for $E[\xi^2(t)]$, the cost functional is obtained as follows:

$$f(u) = \frac{1}{T_s}\int_0^{T_s} E\left\{\sum_{j=-\infty}^{\infty} n(jT_s)[(\Psi u)(t - jT_s)] \sum_{k=-\infty}^{\infty} n(kT_s)[(\Psi u)(t - kT_s)]\right\}dt$$
$$= \frac{1}{T_s}\int_0^{T_s} \sum_{j=-\infty}^{\infty}\sum_{k=-\infty}^{\infty} R_{nn}((j-k)T_s)[(\Psi u)(t - jT_s)][(\Psi u)(t - kT_s)]dt \qquad (2.6.9)$$
$$= \sum_{l \in \Omega_u}\sum_{m \in \Omega_u} u(l)u(m)\Gamma_{l,m}$$

where (using (2.6.6) for Ψu)

$$\Gamma_{l,m} = \frac{1}{T_s}\sum_{j=-\infty}^{\infty}\sum_{k=-\infty}^{\infty} R_{nn}((j-k)T_s)\int_0^{T_s} \Lambda[t - (j+l)T_s]\Lambda[t - (k+m)T_s]dt$$
$$= \frac{1}{T_s}\sum_{j=-\infty}^{\infty}\sum_{k=-\infty}^{\infty} R_{nn}((j-k)T_s)\int_0^{T_s} \Lambda[t - (l-m+j)T_s]\Lambda[t - kT_s]dt \qquad (2.6.10)$$

Note that the coefficient $\Gamma_{l,m}$ is symmetric, i.e. $\Gamma_{l,m} = \Gamma_{m,l}$. Moreover, it depends only on $l - m$. Thus, we can write $\Gamma_{l,m}$ as $\Gamma_{|l-m|}$.

As in the analog and digital cases, the cost functional can be written in a more intuitive and compact form as the following quadratic:

$$f(u) = \langle u, Lu \rangle, \tag{2.6.11}$$

where $L : l_2(\tilde{\Omega}_u) \to l_2(\tilde{\Omega}_u)$ is a positive semi-definite linear operator defined by

$$(Lu)(l) = \sum_{m \in \tilde{\Omega}_u} u(m)\Gamma_{l-m} \tag{2.6.12}$$

Lemma 2.6.2.

(i) $\sqrt{f(u)} \geq 0, \ \forall u \in l_2(\tilde{\Omega}_u)$,

(ii) $\sqrt{f(\alpha u)} = |\alpha|\sqrt{f(u)}, \ \forall u \in l_2(\tilde{\Omega}_u), \forall \alpha \in \boldsymbol{R}$,

(iii) $\sqrt{f(u+v)} \leq \sqrt{f(u)} + \sqrt{f(v)}, \ \forall u, v \in l_2(\tilde{\Omega}_u)$.

Proof. Use the same procedure as for the analog filter given for Lemma 2.1.3.

Similar to the continuous-time and discrete-time cases, $\sqrt{f(u)}$ is a pseudo-norm as the operator L is positive semi-definite. Clearly, $f(u)$ is convex with respect to u, although not strictly convex in general. Consequently, the CCR filtering problem for hybrid filter fits neatly in the frame work of Section 2.3 as the minimization of a convex cost f over a convex set $\mathcal{F}$.

If the sequence of noise samples is statistically independent, i.e.

$$R_{nn}((j-k)T_s) = N_0 \delta_{j-k},$$

then, making the change of variable $\zeta = t - jT_s$, the cost functional f becomes:

$$
\begin{aligned}
f(u) &= \frac{N_0}{T_s}\int_0^{T_s}\sum_{j=-\infty}^{\infty}[(\Psi u)(t-jT_s)]^2 dt = \frac{N_0}{T_s}\sum_{j=-\infty}^{\infty}\int_0^{T_s}[(\Psi u)(t-jT_s)]^2 dt \\
&= \frac{N_0}{T_s}\sum_{j=-\infty}^{\infty}\int_{-jT_s}^{(1-j)T_s}[(\Psi u)(\zeta)]^2 d\zeta \\
&= \frac{N_0}{T_s}\int_{-\infty}^{\infty}[(\Psi u)(\zeta)]^2 d\zeta.
\end{aligned}
\tag{2.6.13}
$$

In this case, the cost functional is strictly convex. To show this, let $N{:}L_\infty(\Omega_\psi) \to \boldsymbol{R}$ be a strictly convex mapping defined by

$$N(v) = \frac{N_0}{T_s}\int_{-\infty}^{\infty}[v(\zeta)]^2 d\zeta,$$

so that the cost functional f can be expressed as

$$f(u) = N(\Psi u).$$

For any $\alpha \in (0, 1)$ and any distinct $u, v \in l_2(\tilde{\Omega}_u)$, it follows from the linearity of Ψ that

$$
\begin{aligned}
\Psi(\alpha u + (1-\alpha)v) &= \alpha\Psi u + (1-\alpha)\Psi v \\
N(\Psi(\alpha u + (1-\alpha)v)) &= N(\alpha\Psi u + (1-\alpha)\Psi v)
\end{aligned}
$$

From Lemma 2.6.1, Ψ is one-to-one, and hence, $\Psi u \neq \Psi v$. Moreover, since N is strictly convex,

$$N(\alpha\Psi u + (1-\alpha)\Psi v) < \alpha N(\Psi u) + (1-\alpha)N(\Psi v)$$

Therefore, the composite mapping $N \circ \Psi : l_2(\tilde{\Omega}_u) \to \boldsymbol{R}$ is strictly convex. In fact, we can use the same procedure to show that, if the autocorrelation of the noise sample give rise to a positive definite mapping, i.e. the mapping of ζ to

$$\sum_{j=-\infty}^{\infty}\sum_{k=-\infty}^{\infty} R_{nn}((j-k)T_s)\zeta(j)\zeta(k)$$

is positive definite, then L is also positive definite.

Note that since a strictly convex pseudo-norm is a norm, $\sqrt{f(u)}$ defines a norm on $l_2(\tilde{\Omega}_u)$. Moreover, it is clear that $\langle u, v\rangle_L \equiv \langle u, Lv\rangle$ is an inner product on $l_2(\tilde{\Omega}_u)$. Thus, $\sqrt{f(u)}$ is the norm induced by this inner product.

By the same variable change $\zeta = t - jT_s$, used earlier for the cost functional in (2.6.13), the expression for the coefficient Γ_{l-m} of the operator L also reduces to:

$$\begin{aligned}\Gamma_{l-m} &= \frac{N_0}{T_s}\sum_{j=-\infty}^{\infty}\int_0^{T_s}\Lambda[t-(j+l-m)T_s]\Lambda(t-jT_s)dt \\ &= \frac{N_0}{T_s}\sum_{j=-\infty}^{\infty}\int_{-jT_s}^{(1-j)T_s}\Lambda[\zeta-(l-m)T_s]\Lambda(\zeta)d\zeta \\ &= \frac{N_0}{T_s}\int_{-\infty}^{\infty}\Lambda[\zeta-(l-m)T_s]\Lambda(\zeta)d\zeta.\end{aligned} \tag{2.6.14}$$

In this case, the coefficient Γ_{m-l} can be viewed as the continuous-time autocorrelation of the pulse response Λ. If $\Lambda(t)$ is time limited to qT_s, where q is a positive integer, then $\Gamma_k = 0$ for $|k| \geq q$.

Similar to the analog case, suppose that the power spectral density of the noise samples is denoted as $\tilde{\Phi}_N(\omega)$. Then, it can be verified that the output noise power due to the input noise n is given by

$$P_N = \frac{1}{2\pi}\int_{-\infty}^{\infty}\tilde{\Phi}_N(\omega)|H_{zoh}(j\omega)|^2|U(e^{j\omega})H(j\omega)|^2 d\omega, \tag{2.6.15}$$

where H is the transfer function of the post-filter, U is the Z-transform of u, i.e.

$$U(z) = \sum_{k=0}^{\infty} u(k)z^{-k},$$

and H_{zoh} is the transfer function of a zero-order-hold given by

$$H_{zoh}(s) = \frac{1-e^{-sT_s}}{s}.$$

Clearly, with $s = j\omega$, we have

$$H_{zoh}(j\omega) = T_s e^{-j\omega T_s/2}\mathrm{sinc}(\omega T_s/2).$$

Thus,

$$P_N = \frac{1}{2\pi}\int_{-\infty}^{\infty}\tilde{\Phi}_N(\omega)\left|T_s\mathrm{sinc}\left(\omega\frac{T_s}{2}\right)\right|^2\left|U(e^{j\omega})H(j\omega)\right|^2 d\omega \tag{2.6.16}$$

Note that since u is square summable on $\tilde{\Omega}_u$, U belongs to $\mathcal{H}_2$, the Hardy space of all complex-valued functions that are analytic in the open unit disk with square integrable boundary functions on $\{|z| = 1\}$.

We assume that $\tilde{\Phi}_N$ is bounded in the sense that its noise power satisfies:

$$\|\tilde{\Phi}_N\| = \frac{1}{2\pi}\int_{-\omega_B}^{\omega_B}\tilde{\Phi}_N(\omega)d\omega \le 1,$$

but otherwise unknown. Furthermore, assuming the post-filter is also bandlimited to ω_B, we have

$$P_N = \frac{1}{2\pi}\int_{-\omega_B}^{\omega_B}\tilde{\Phi}_N(\omega)\left|T_s\mathrm{sinc}(\omega T_s/2)U(e^{j\omega})H(j\omega)\right|^2 d\omega.$$

It makes sense to consider the output noise power for the worst case input noise, i.e.

$$P_N = \max_{\|\tilde{\Phi}_N\| \le 1}\frac{1}{2\pi}\int_{-\omega_B}^{\omega_B}\tilde{\Phi}_N(\omega)\left|T_s\mathrm{sinc}(\omega T_s/2)U(e^{j\omega})H(j\omega)\right|^2 d\omega. \tag{2.6.17}$$

Then, it can be shown in the same way as in the analog and discrete-time cases that

$$\sup_{\omega \in [-\omega_B,\,\omega_B]}\left|T_s\mathrm{sinc}(\omega T_s/2)U(e^{j\omega})H(j\omega)\right|^2 = P_N. \tag{2.6.18}$$

In this case, the cost functional can be expressed explicitly in terms of u as

$$f(u) = \sup_{\omega \in [-\omega_B,\,\omega_B]}\left|T_s\mathrm{sinc}(\omega T_s/2)H(j\omega)\sum_{k=0}^{\infty}u(k)e^{-j\omega k T_s}\right|^2.$$

This cost functional is convex (but not strictly convex).

2.7 Finite Dimensional Hybrid Filter for CCR Filtering

The analysis in the previous section is for filters with impulse responses in $l_2(\tilde{\Omega}_u)$. This section narrows down the discrete-time filters to those with finite structure. The resulting CCR filtering problem is a finite dimensional optimization problem because the digital processor is finite in structure. Note that the filters can be FIR or IIR.

Let $\{v_i\}_{i=0}^{\infty}$ be a total sequence in $l_2(\tilde{\Omega}_u)$, and let the impulse response of finite filter structures be described by

$$u = Ka = \sum_{k=0}^{n-1} a_k v_k, \tag{2.7.1}$$

where $K : \boldsymbol{R}^n \to l_2(\tilde{\Omega}_u)$ is a bounded linear operator.

Such filters can be realized as FIR filters or discrete-time Laguerre networks where, incidentally, the corresponding v_k's are also orthonormal. Similar to the all-analog case, the set $\mathcal{F}^n$ of feasible filter coefficients is given by

$$\mathcal{F}^n = \{\boldsymbol{a} \in \boldsymbol{R}^n : \forall x \in \mathcal{S},\ \Xi_x K\boldsymbol{a} \in \Psi\}, \tag{2.7.2}$$

and the general form of the cost functional is

$$(f \circ K)(\boldsymbol{a}) = \langle K\boldsymbol{a}, LK\boldsymbol{a}\rangle = \boldsymbol{a}^T K^{\dagger} LK\boldsymbol{a}, \tag{2.7.3}$$

where $K^{\dagger}$ is the adjoint of K (see (2.4.4)). The composite operator $K^{\dagger}LK$, takes $\boldsymbol{R}^n$ to $\boldsymbol{R}^n$, can be represented by an $n \times n$ matrix. To obtain an explicit expression for this matrix, observe that

$$\langle K\boldsymbol{a}, LK\boldsymbol{a}\rangle = \sum_{j=0}^{n-1}\sum_{k=0}^{n-1} a_j a_k \langle v_j, Lv_k\rangle$$

Comparing this with (2.7.3), we see that the term $\langle v_j, Lv_k\rangle$ corresponds to $[K^{\dagger}LK]_{j,k}$, the (j, k)th entry of the matrix $K^{\dagger}LK$. Thus, using (2.6.12) gives

$$[K^{\dagger}LK]_{j,k} = \langle v_j, Lv_k\rangle = \sum_{l \in \tilde{\Omega}_u}\sum_{m \in \tilde{\Omega}_u} v_j(l)v_k(m)\Gamma_{l,m}, \tag{2.7.4}$$

where the general expression for $\Gamma_{l,m}$ is given by (2.6.10). For independent noise sample, $\Gamma_{l,m}$ simplifies to (2.6.14) and so

$$[K^{\dagger}LK]_{j,k} = \frac{N_0}{T_s}\sum_{l \in \Omega_u}\sum_{m \in \Omega_u} v_j(l)v_k(m)\int_{-\infty}^{\infty}\Lambda(\zeta - lT_s)\Lambda(\zeta - mT_s)d\zeta. \tag{2.7.5}$$

By the same argument as in Section 2.4, it can be shown that the cost functional $f \circ K$, given in (2.7.3), is a strictly convex function of $\boldsymbol{a}$ when L is positive definite and the v_j's are linear independent. This cost functional thus generates a norm, and the matrix $K^{\dagger}LK$ is positive definite. It follows from Theorem 2.3.2 that the optimum finite-structured hybrid CCR filters converge to the optimum hybrid CCR filter. In the general case where $f \circ K$ is a pseudo-norm cost, the optimum costs of the finite-structured filters decrease to the optimum cost as the dimension of filter grows.

FIR digital processor

The best known discrete-time filter is probably the FIR filter for its simplicity and wide range of applicability. With an FIR filter as the digital processor, the problem reduces to a simpler form. The linear operator K becomes the identity operator. Thus, substituting $v_k(m) = \delta_{k,m}$ into (2.7.4) yields

$$[K^{\dagger}LK]_{j,k} = \Gamma_{j,k} \tag{2.7.6}$$

The (j, k)th entry of $K^{\dagger}LK$ is now simply the coefficient $\Gamma_{j,k}$. Since $\Gamma_{j,k}$ is a function of $|j-k|$, the notation $\Gamma_{|j-k|}$ can be used to denote $[K^{\dagger}LK]_{j,k}$. The matrix $[K^{\dagger}LK]$ takes the form of the following Toeplitz persymmetric matrix:

$$\begin{bmatrix} \Gamma_0 & \Gamma_1 & \cdots & \Gamma_{n-2} & \Gamma_{n-1} \\ \Gamma & \Gamma_0 & & & \Gamma_{n-2} \\ \vdots & & \ddots & & \vdots \\ \Gamma_{n-2} & & & \Gamma_0 & \Gamma_1 \\ \Gamma_{n-1} & \Gamma_{n-2} & \cdots & \Gamma_1 & \Gamma_0 \end{bmatrix}.$$

Suppose that $\Lambda(t)$ is time limited to qT_s, where q is a positive integer less than n. Then, $\Gamma_k = 0$ for $|k| \geq q$. Thus, only the coefficients Γ_k, $k = 0, \ldots, q-1$, of $[K^{\dagger}LK]$ are required to be computed.

2.8 Appendix

Lemma 2.1.1. *If $x \in L_2(\Omega_s)$, then the mapping $\Xi_x : L_2(\Omega_u) \to L_\infty(\Omega_\psi)$ defined by (2.1.3) is a bounded (hence continuous) linear operator.*

Proof. It is straightforward to see that the operator Ξ_x is linear. Since x is square-integrable on its support Ω_u, it is clear that $x_t \in L_2(\Omega_u)$, for any $t \in \Omega_\psi$, i.e.

$$\|x_t\|^2 = \int_{\Omega_u} |x_t(\lambda)|^2 d\lambda \le \|x\|^2,$$

where x_t is defined by (2.1.1).

For any $t \in \Omega_\psi$, it follows from the Cauchy-Schwartz inequality that

$$|\langle x_t, u\rangle| \le \|x_t\| \|u\| \le \|x\| \|u\|.$$

Hence, for any $u \in L_2(\Omega_u)$,

$$\|\Xi_x u\|_\infty = \sup_{t \in \Omega_\psi} |\langle x_t, u\rangle| \le \|x\| \|u\|.$$

Since x is square integrable, $\|x\| < \infty$. Thus, Ξ_x is bounded, and hence continuous. ❐

Lemma 2.1.3.

(i) $\sqrt{f(u)} \ge 0, \forall u \in L_2(\Omega_u)$,

(ii) $\sqrt{f(\alpha u)} = |\alpha| \sqrt{f(u)}, \forall u \in L_2(\Omega_u), \forall \alpha \in \boldsymbol{R}$,

(iii) $\sqrt{f(u+v)} \le \sqrt{f(u)} + \sqrt{f(v)}, \forall u, v \in L_2(\Omega_u)$.

Proof. (i) and (ii) are straightforward from the definition of f. To prove (iii), consider

$$\begin{aligned} f(u+v) &= \langle u+v, L(u+v)\rangle \\ &= \langle u, Lu\rangle + \langle v, Lv\rangle + \langle u, Lv\rangle + \langle v, Lu\rangle. \end{aligned}$$

Since the autocorrelation kernel R_{nn} is symmetric, $\langle u, Lv\rangle = \langle v, Lu\rangle$. Thus,

$$f(u+v) = f(u) + f(v) + 2\langle u, Lv\rangle. \tag{2.8.1}$$

Now let, $\alpha = \langle u, Lv\rangle / f(v)$. Then, $\alpha^2 f(v) = \alpha\langle u, Lv\rangle$, and hence,

$$\begin{aligned} 0 \le f(u-\alpha v) &= f(u) + \alpha^2 f(v) - 2\alpha\langle u, Lv\rangle \\ &= f(u) - \frac{[\langle u, Lv\rangle]^2}{f(v)} \end{aligned}$$

That is,

$$\langle u, Lv\rangle \le \sqrt{f(u)f(v)}.$$

Thus, from (2.8.1),

$$f(u+v) \le f(u) + f(v) + 2\sqrt{f(u)f(v)} = (\sqrt{f(u)} + \sqrt{f(v)})^2. \square$$

Lemma 2.1.4*. Let H be a Hilbert space, Y be a Banach space, and $B(H, Y)$ denote the (Banach) space of bounded linear operators from H to Y. Suppose that the set $\Psi \subset Y$ is closed and convex, and that the set $\Phi \subset B(H, Y)$ is bounded. Then, the set*

$$\{u \in H\colon \forall X \in \Phi,\ Xu \in \Psi\}$$

is closed and convex.

Proof. Define $\mathcal{U} \equiv \{u \in H: \forall X \in \Phi, Xu \in \Psi\}$. Then, $\mathcal{U}$ can also be expressed as

$$\mathcal{U} = \bigcap_{X \in \Phi} X^{-1}(\Psi).$$

Since, the linear operator $X \in \Phi$ is bounded, it is continuous. Moreover, since Ψ is closed, it follows that $X^{-1}(\Psi)$ is also closed. Hence, $\mathcal{U}$ is closed since it is the intersection of closed sets. To show convexity, let u and v be any points in $\mathcal{U}$, i.e. for all $X \in \Phi$, $Xu, Xv \in \Psi$. Since Ψ is convex, we have, for any $\alpha \in [0, 1]$,

$$X[\alpha u + (1-\alpha)v] = \alpha Xu + (1-\alpha)Xv \in \Psi.$$

Thus, $\alpha u + (1-\alpha)v \in \mathcal{F}$, and hence $\mathcal{U}$ is convex. $\square$

Theorem 2.3.1. *Let $\mathcal{F}$ be a closed and convex subset in a Hilbert space H, and f be a continuous and convex functional on H. If $\mathcal{F}$ is non-empty, then there exists a $u^o \in \mathcal{F}$ such that*

$$f(u^o) \le f(u), \ \forall u \in \mathcal{F}.$$

Moreover, if f is strictly convex, then u^o is unique.

Proof. Since the feasible region $\mathcal{F}$ is non-empty, convexity of $\mathcal{F}$ and the cost f implies that a minimum of f exists and the optimum cost is unique (see Theorem B.3.7, Appendix B.3). For the case where f is strictly convex, let u and v be two distinct minima of f on $\mathcal{F}$. Convexity of $\mathcal{F}$ implies that, for any $\alpha \in (0, 1)$, $\alpha u + (1-\alpha)v \in \mathcal{F}$. Since $f(v) = f(u)$, it follows from the strict convexity of f that

$$f(\alpha u + (1-\alpha)v) < \alpha f(u) + (1-\alpha)f(v) = f(u).$$

This contradicts the minimality of u. Hence, the minimum solution is unique. ❐

Theorem 2.3.2. *Let $\mathcal{F}$ be a closed and convex subset of a separable Hilbert space H with non-empty interior $\mathcal{F}^o$ (see Appendix A.1 for definition), and f be a continuous and convex functional on H. Let $u^o \in \mathcal{F}$ be such that*

$$f(u^o) \le f(u), \forall u \in \mathcal{F},$$

and $u_n^o \in \mathcal{F} \cap [\{v_i\}_{i=0}^n]$ be such that

$$f(u_n^o) \le f(u_n), \ \forall u_n \in \mathcal{F} \cap [\{v_i\}_{i=0}^n].$$

If $\{v_i\}_{i=0}^\infty$ is total in H, then $f(u_n^o)$ monotonically decreases to $f(u^o)$ as n tends to infinity. Moreover, if f is an inner-product-induced norm, then

$$\lim_{n \to \infty} \left\| u_n^o - u^o \right\| = 0.$$

Proof. (i) Suppose not, i.e. there is an $\hat{\varepsilon} > 0$ such that for any integer N

$$\exists n > N : f(u_n^o) - f(u^o) > \hat{\varepsilon}.$$

Let L denote the least n for which u_n^o exists. Then,

$$n \geq N \geq L \Rightarrow f(u_n^o) \leq f(u_N^o) \leq f(u_L^o),$$

i.e. the sequence $\{f(u_n^o)\}_{n=L}^{\infty}$ is decreasing. Hence

$$\forall N \geq L, \hat{\varepsilon} < f(u_N^o) - f(u^o).$$

Using the continuity of f, let $\delta > 0$ be such that

$$x \in B(u^o, \delta) \Rightarrow |f(x) - f(u^o)| < \hat{\varepsilon}.$$

Since the feasible set has a non-empty interior and u^o is an accumulation point of $\mathcal{F}^o$, it follows that $B(u^o, \delta) \cap \mathcal{F}^o$ is non-empty. Hence, we can choose a point $x \in B(u^o, \delta) \cap \mathcal{F}^o$ and an $\alpha > 0$ such that

$$B(x, \alpha) \subset B(u^o, \delta) \cap \mathcal{F}^o.$$

From the totality of $\{v_i\}_{i=0}^{\infty}$, there exists an M and $\hat{x}_M \in [\{v_i\}_{i=0}^{M}]$ such that

$$\|\hat{x}_M - x\| < \alpha,$$

that is, $\hat{x}_M \in B(u^o, \delta) \cap \mathcal{F}$. This implies

$$f(\hat{x}_M) - f(u^o) \leq \hat{\varepsilon} < f(u_M^o) - f(u^o).$$

That is, there exists an M and a feasible $\hat{x}_M \in [\{v_i\}_{i=0}^{M}]$ such that $f(\hat{x}_M) < f(u_M^o)$. This is clearly a contradiction. Hence,

$$\lim_{n \to \infty} f(u_n^o) = f(u^o).$$

(ii) Since f is strictly convex, closeness and convexity of $\mathcal{F}$ and $\mathcal{F} \cap [\{v_i\}_{i=0}^{n}]$ imply uniqueness of u^o and u_n^o.

Suppose $\lim_{n \to \infty} \|u_n^o - u^o\| \neq 0$, i.e. there is an $\hat{\varepsilon} > 0$ such that for any integer N,

$$\exists n > N : \left\| u_n^o - u^o \right\| > \hat{\varepsilon}. \tag{2.8.2}$$

Substituting $\left\| u_n^o - u^o \right\|^2 = \left\| u_n^o \right\|^2 + \left\| u^o \right\|^2 - 2\langle u_n^o, u^o \rangle$ into (2.8.2), we have

$$\forall N, \exists n > N : \left\| u_n^o \right\|^2 + \left\| u^o \right\|^2 - 2\langle u_n^o, u^o \rangle > \hat{\varepsilon}^2.$$

Assuming $0 \notin \mathcal{F}$, the projection theorem (Theorem B.1.2, Appendix B.1) asserts that for all $u \in \mathcal{F}$,

$$\langle 0 - u^o, u - u^o \rangle \leq 0.$$

Hence, expanding the inner product and rearranging yields

$$-\left\| u^o \right\|^2 \geq \left\| u^o \right\|^2 - 2\langle u^o, u \rangle.$$

This is also true for $0 \in \mathcal{F}$, because, in this case $u^o = 0$.

Hence,

$$\forall N, \exists n > N : \left\| u_n^o \right\|^2 - \left\| u^o \right\|^2 > \hat{\varepsilon}^2.$$

Let L denote the least n for which u_n^o exists. Since $f(u) = \left\| u \right\|^2$, we have

$$\forall N \geq L, f(u_N^o) - f(u^o) > \hat{\varepsilon}^2.$$

This leads to a contradiction by using the same arguments as in part (i). Hence, the conclusion follows. ❒

Lemma 2.5.1*. If the excitation $x \in l_2$ is non-zero, then the associated linear operator Ξ_x defined by (2.5.2) is invertible.*

Proof. Since Ξ_x is a linear operator, it is sufficient to show that $\Xi_x u = 0 \Rightarrow u = 0$. We do this by induction, that is, first show that $u(0) = 0$ and that for any k in the support $\tilde{\Omega}_u$, $u(j) = 0$ for all non-negative $j < k$ implies $u(k) = 0$.

Let $x(l)$ be the first non-zero element of the sequence x (since x is non-zero), i.e. $x(j) = 0$ for all non-negative $j < l$ and $x(l) \neq 0$. Then,

$$(\Xi_x u)(l) = \sum_{i \in \tilde{\Omega}_u} x(l-i)u(i) = x(l)u(0).$$

Furthermore, by the assertion $(\Xi_x u)(j) = 0$ for all $j \in \tilde{\Omega}_\psi$, we have $u(0) = 0$.

Now given any $k \in \tilde{\Omega}_u$, suppose that $u(j) = 0$ for all non-negative $j < k$. Since $x(j) = 0$ for $j < l$ and $u(j) = 0$ for $j < k$, it follows that

$$\begin{aligned}(\Xi_x u)(l+k) &= \sum_{i \in \tilde{\Omega}_u} x(l+k-i)u(i) \\ &= x(l)u(k) + \sum_{i<k,\, i \in \tilde{\Omega}_u} x(l+k-i)u(i) + \sum_{i>k,\, i \in \tilde{\Omega}_u} x(l+k-i)u(i) \\ &= x(l)u(k)\end{aligned}$$

By the assertion, $(\Xi_x u)(l+k) = 0$ and hence, $u(k) = 0$. Therefore, $u = 0$ by the principle of induction. ❐

Lemma 2.6.1*. The linear operator* Ψ*, defined by (2.6.6), is a one-to-one mapping if the pulse response* Λ *is non-zero.*

Proof. Since Ψ is linear, we only need to show that $\Psi u = 0$ implies $u = 0$. Let $t_0 \geq 0$ be such that $\Lambda(t) = 0, \forall t \leq t_0$ and $\Lambda(t) \neq 0$ for some $t \in [t_0, t_0 + T_s]$. Such t_0 exists since the post-filter is causal (no response before input is applied). Hence,

$$\Lambda(t - jT_s) = 0, \quad \forall t \leq t_0 + jT_s$$

and

$$\Lambda(t - jT_s) \neq 0 \text{ for some } t \in [t_0 + jT_s, t_0 + (j+1)T_s].$$

For any $t \leq t_0 + T_s$, we have $\Lambda(t - jT_s) = 0, \ \forall j \geq 1$, and hence,

$$(\Psi u)(t) = \sum_{j \in \tilde{\Omega}_u} \Lambda(t - jT_s)u(j) = \Lambda(t)u(0),$$

By the assertion $\Psi u = 0$, we have $\Lambda(t)u(0) = 0$, for all $t \leq t_0 + T_s$. But $\Lambda(t) \neq 0$ for some $t \in [t_0, t_0 + T_s]$, hence, we can conclude that $u(0) = 0$.

Now, for any $t \leq t_0 + (m+1)T_s$, we have $\Lambda(t - jT_s) = 0$, $\forall j \geq m+1$. Suppose that $u(k) = 0$, for all $k \leq m-1$. Then,

$$(\Psi u)(t) = \sum_{j \in \bar{\Omega}_u} \Lambda(t - jT_s)u(j) = \Lambda(t - mT_s)u(m).$$

Since $\Lambda(t - mT_s) \neq 0$, for some $t \in [t_0 + mT_s, t_0 + (m+1)T_s]$, and $\Psi u = 0$, we have

$$u(m) = 0.$$

In summary, we have shown that $u(0) = 0$ and that $u(k) = 0, \forall k \leq m-1$ implies $u(m) = 0$. Hence, by the principle of induction, $u = 0$. ❒

CHAPTER 3 ANALYSIS AND PROBLEM CHARACTERIZATION

The previous chapter addresses the envelope constrained (EC) filtering problem from the general view point of a convex programming problem by examining properties of the cost functional and the feasible region. This chapter examines the primal-dual nature of the EC filtering problem. In particular, generalized versions of the Kuhn-Tucker Theorem and Duality Theorem (Appendix B.5) are used to convert the quadratic programming problem in Hilbert space into its dual in the space of regular Borel measures. Furthermore, by exploiting the mutual singularity of the dual variables, we can rewrite the dual problem as an unconstrained one. This form, however, does not retain the desirable smooth feature of the original dual. The relationship between the primal and dual solutions suggests an interesting interpretation of the structure of optimal EC filters. For the case in which the decision variable is finite dimensional, we have a semi-infinite programming (SIP) problem, i.e. finite number of degrees of freedom but infinite number of constraints. In this case, Carathéodory's Dimensionality Theorem (Appendix B.3) can be applied to convert the convex but infinite dimensional dual problem into a non-convex finite dimensional problem.

In Section 3.1, we present a brief background on the primal-dual nature of quadratic programming with affine inequality constraints. This is then applied to the EC filtering problem in Section 3.2, yielding a smooth dual problem. We also exploit the nature of the constraints to derive an unconstrained dual problem which turns out to be non-smooth. In Section 3.3, we derive constrained and unconstrained dual problems for the finite dimensional EC filter problem. Section 3.4 presents a brief background of semi-infinite programming (SIP). In particular, we apply Carathéodory's Dimensionality Theorem (Theorem B.3.6, Appendix B.3) to reduce the dual problem in the space of regular Borel measures to one in the space of finite support atomic measures. This leads to a non-convex finite dimensional problem. Section 3.5 focuses on linearly constrained SIP and its application to EC filtering.

3.1 Duality of Quadratic Program

We begin this section by studying a general quadratic programming (QP) problem of the form

$$\begin{aligned} &\min_u \ \langle u, Lu\rangle + \langle b, u\rangle \\ &\text{subject to } \ (A_i u)(t) - c_i(t) \le 0, \ \ \forall t \in \Omega_c, \, i = 1, \ldots, m, \end{aligned} \tag{3.1.1}$$

where $u, b \in L_2(\Omega_u)$; Ω_u is the support of u; $c_i \in C(\Omega_\psi)$; Ω_ψ is the support of c_i; $\Omega_c \subseteq \Omega_\psi$ is a compact set; L is a positive semi-definite linear operator on $L_2(\Omega_u)$; and each A_i is a bounded linear operator from $L_2(\Omega_u)$ to $C(\Omega_\psi)$ defined by

$$(A_i u)(t) = \int_{\Omega_u} a_i(t, \tau) u(\tau) d\tau . \tag{3.1.2}$$

Recall from Chapter 2 that the EC filtering problem (2.2.3) takes on this form if $f(u) = \langle u, Lu\rangle$.

It is assumed that there exists $u \in L_2(\Omega_u)$ satisfying

$$[G_i(u)](t) \equiv (A_i u)(t) - c_i(t) < 0, \ \forall t \in \Omega_c, \, i = 1, \ldots, m. \tag{3.1.3}$$

The dual space $C^*(\Omega_\psi)$ of $C(\Omega_\psi)$ is well known [66] to be represented by $M(\Omega_\psi)$, the space of regular Borel measures on Ω_ψ, where a linear functional $\phi \in C^*(\Omega_\psi)$ corresponding to an $m \in M(\Omega_\psi)$ is given by

$$\phi \bullet x = \int_{\Omega_\psi} x dm, \quad x \in C(\Omega_\psi). \tag{3.1.4}$$

For notational convenience, we use m in place of ϕ, and $\phi \bullet x$ is written as $m \bullet x$. The ε-weighted norm of a measure $m \in M(\Omega_\psi)$ is given by the total variation

$$\|m\|_\varepsilon = \sup \sum_{i=1}^{n} \left\{ |m(E_i)| \int_{E_i} \varepsilon(t) dt \right\},$$

where the supremum is taken over all Borel partitions $\{E_i\}_{i=1}^{n}$ of Ω_ψ. An equivalent definition of the ε-weighted norm on $M(\Omega_\psi)$ is

$$\|m\|_\varepsilon = (m_1 + m_2) \bullet \varepsilon = \int_{\Omega_\psi} \varepsilon(t) dm_1(t) + \int_{\Omega_\psi} \varepsilon(t) dm_2(t), \tag{3.1.5}$$

where m_1 and m_2 are the positive and negative variations of m, respectively (see [53], [95]). For a positive measure p, i.e. $p(E) \geq 0$ for all $E \subseteq \Omega_\psi$, we have

$$\|p\|_\varepsilon = p \bullet \varepsilon.$$

The partial ordering $\preceq$ on $C(\Omega_\psi)$ is generated by the positive cone P_C^+ (for definitions of cones and partial ordering, see Appendix B.3) consisting of functions in $C(\Omega_\psi)$ that are non-negative on Ω_c, i.e.

$$P_C^+ = \{x \in C(\Omega_\psi) \colon \ 0 \leq x(t), \ \forall t \in \Omega_c\}. \tag{3.1.6}$$

The interior of P_C^+ is

$$\{x \in C(\Omega_\psi) \colon \ 0 < x(t), \ \forall t \in \Omega_c\}.$$

The boundary of P_C^+ consists of $x \in P_C^+$ such that $x(t) = 0$ for some $t \in \Omega_c$. It is easy to verify that P_C^+ is a closed convex cone with non-empty interior. A non-empty interior is essential for the application of the Kuhn-Tucker theorem (i.e. Theorem B.5.1, Appendix B.5). Note that corresponding positive cones in $L_p(\Omega_\psi)$ have no interior points (see [28], [66]).

Lemma 3.1.1. The dual positive cone for P_C^+ is given by

$$P_M^{\oplus} = \{m \in M(\Omega_\psi):\ m(E) \geq 0, \forall E \subseteq \Omega_\psi,\ m(F) = 0, \forall F \subseteq \Omega_\psi \backslash \Omega_c\}, \quad (3.1.7)$$

where $\Omega_\psi \backslash \Omega_c$ denotes the difference between the sets Ω_ψ and Ω_c.

The above lemma shows that $P_M^{\oplus}$, the dual positive cone of P_C^+, generates the partial ordering $\preceq$ on $M(\Omega_\psi)$, consists of non-negative measures on Ω_ψ that are zero outside of Ω_c. For example, in the simplest case where $\Omega_\psi = \Omega_c$,

$$P_M^{\oplus} = \{m \in M(\Omega_\psi):\ m(E) \geq 0, \forall E \subseteq \Omega_\psi\}.$$

Typically, $\Omega_\psi = [0, \infty)$ and the constraints are on $\Omega_c = [0, T]$. In this case, the dual positive cone is given by

$$P_M^{\oplus} = \{m \in M(\Omega_\psi):\ m(E) \geq 0, \forall E \subseteq [0, T]\}.$$

As a consequence, we can regard the dual positive cone $P_M^{\oplus}$ as the set of non-negative measures of $M[0, T]$. It it easily shown that the dual positive cone $P_M^{\oplus}$ is closed (even if the positive cone P_C^+ is not).

The necessary and sufficient conditions for optimality follow a straightforward application of the generalized Kuhn-Tucker theorem (see Appendix B.5).

Making the identifications:

$$\begin{aligned} X &= L_2(\Omega_u) \\ Z_i &= C(\Omega_\psi) \\ P_i^+ &= P_C^+ \\ P_i^{\oplus} &= P_M^{\oplus},\ i = 1, \ldots, m, \end{aligned}$$

the Lagrangian for Problem (3.1.1) is

$$\begin{aligned} \mathcal{L}(u, p_1, \ldots, p_m) &= \langle Lu, u\rangle + \langle b, u\rangle + \sum_{i=1}^{m} p_i \bullet [A_i u - c_i] \\ &= \langle Lu, u\rangle + \langle b, u\rangle + \langle \alpha(p_1, \ldots, p_m), u\rangle - \sum_{i=1}^{m} p_i \bullet c_i, \end{aligned} \quad (3.1.8)$$

where $\alpha(p_1, \ldots, p_m)$ is the Riesz representation (Theorem A.6.5, Appendix A.6) of the linear functional that maps u to the real number $\sum_{i=1}^{m} p_i \bullet A_i u$. In other words, $\alpha(p_1, \ldots, p_m)$ is defined such that

$$\langle \alpha(p_1, \ldots, p_m), u \rangle = \sum_{i=1}^{m} p_i \bullet A_i u .$$

By expanding $p_i \bullet A_i u$ in terms of the kernel a_i and applying Fubini's Theorem (see Theorem A.8.10, Appendix A.8), it follows that

$$[\alpha(p_1, \ldots, p_m)](\tau) = \sum_{i=1}^{m} \int_{\Omega_\psi} a_i(t, \tau) dp_i(t). \tag{3.1.9}$$

To check that all the conditions of the Kuhn-Tucker theorem are satisfied, observe that

(i) P_C^+ *is a closed, convex cone with non-empty interior*

(ii) *The mappings* $f : X \to \boldsymbol{R}$, *and* $G_i : X \to Z_i$ *are all convex by virtue of the convexity of the cost and the linearity of the constraints*

The existence of a solution u^o (satisfying the constraints) and the Kuhn-Tucker Theorem (Theorem B.5.1, Appendix B.5) now say that there exist non-negative Lagrange multipliers

$$p_i^o \in P_M^{\oplus} \subseteq M(\Omega_\psi), \; i = 1, \ldots, m,$$

such that

$$\mathcal{L}(u^o, p_1^o, \ldots, p_m^o) \le \mathcal{L}(u, p_1^o, \ldots, p_m^o), \forall u \in L_2(\Omega_u), \tag{3.1.10}$$

$$p_i^o \bullet G_i(u^o) = p_1^o \bullet [A_i u^o - c_i] = \int_{\Omega_c} [A_i u^o - c_i] dp_i^o = 0. \tag{3.1.11}$$

Note that the region of integration is now Ω_c instead of Ω_ψ because $p_i^o \in P_M^{\oplus}$ and the dual positive cone $P_M^{\oplus}$ lies in the subspace $M(\Omega_c)$ of $M(\Omega_\psi)$ (from Lemma 3.1.1). Now according to the Generalized Duality Theorem (Theorem B.5.2, Appendix B.5), such Lagrange multipliers $p_i^o \in P_M^{\oplus}$ for the primal problem may be found by solving the dual problem.

One method of finding the dual functional

$$\phi(p_1, \ldots, p_m) \equiv \inf_u \mathcal{L}(u, p_1, \ldots, p_m)$$

is to find the Fréchet derivative of $\mathcal{L}$ with respect to u and equate it to zero. This yields a Fredholm integral equation of the first kind for u^o (see [48]):

$$2Lu + b + \alpha(p_1, \ldots, p_m) = 0. \qquad (3.1.12)$$

Since the Lagrangian (3.1.8) is convex in u, the solution set of $\inf_u \mathcal{L}(u, p_1, \ldots, p_m)$ is completely characterized by (3.1.12) for any given $(p_1, \ldots, p_m)$. Substituting (3.1.12) into (3.1.8) gives

$$\mathcal{L}(u, p_1, \ldots, p_m) = -\langle Lu, u\rangle - \sum_{i=1}^{m} p_i \bullet c_i.$$

This observation leads to an equivalent dual problem known as Dorn's dual form:

$$\begin{aligned} &\min_{(u, p_1, \ldots, p_m)} \langle Lu, u\rangle + \sum_{i=1}^{m} p_i \bullet c_i \\ &\text{subject to } 2Lu + b + \alpha(p_1, \ldots, p_m) = 0, \\ &\qquad p_i \in P_M^{\oplus},\ i = 1, \ldots, m. \end{aligned} \qquad (3.1.13)$$

Our primary intention is to find an optimal solution to the primal problem. For a given dual solution $(p_1^o, \ldots, p_m^o)$, the duality theorem (Theorem B.5.2, Appendix B.5) asserts that any primal solution minimizes the Lagrangian $\mathcal{L}(\bullet, p_1^o, \ldots, p_m^o)$. But not every minimizer of the Lagrangian $\mathcal{L}(\bullet, p_1^o, \ldots, p_m^o)$ solves the primal problem unless it is primal feasible and satisfies the complementarity condition, i.e.

$$G_i(u^o) \preceq 0,$$

and

$$p_i^o \bullet G_i(u^o) = 0, \text{ for each } i = 1, \ldots, m.$$

In the case of quadratic programming, the following result enables us to obtain the primal solution.

Theorem 3.1.1. *Let $(u^o, p_1^o, \ldots, p_m^o)$ be a solution of Dorn's dual Problem (3.1.13), and z^o be a multiplier corresponding to its equality constraint. Then, z^o gives a solution to the primal Problem (3.1.1).*

Note that for strictly convex quadratic programs, i.e. L is positive definite and hence invertible. The solution to (3.1.12) given by

$$u^o = -\frac{1}{2}L^{-1}[\alpha(p_1, \ldots, p_m) + b],$$

minimizes the Lagrangian (3.1.8) over all u. Hence, the dual functional is

$$\phi(p_1, \ldots, p_m) = -\frac{1}{4}\langle L^{-1}[\alpha(p_1, \ldots, p_m) + b], \alpha(p_1, \ldots, p_m) + b\rangle - \sum_{i=1}^{m} p_i \bullet c_i . \quad (3.1.14)$$

The invertiblity of L also reduces Dorn's dual Problem (3.1.13) to the standard dual problem

$$\max_{p_1, \ldots, p_m \in P_M^{\oplus}} \phi(p_1, \ldots, p_m). \quad (3.1.15)$$

Hence, if $(u^o, p_1^o, \ldots, p_m^o)$ is a solution of Problem (3.1.13), then

$$u^o = -\frac{1}{2}L^{-1}[\alpha(p_1^o, \ldots, p_m^o) + b]. \quad (3.1.16)$$

Suppose that z^o is a multiplier corresponding to the equality constraint of Problem (3.1.13). Then, from the invertibility of L, it can be shown that $z^o = u^o$ (see the first condition of (3.6.1) appeared in the proof of Theorem 3.1.1, Section 3.6). Thus, by Theorem 3.1.1, the primal solution is given by (3.1.16).

3.2 Dual Problems for EC Filtering

In this section, we present theoretical results regarding the dual of the EC filtering problem. By exploiting the nature of the constraints, we arrive at an important theoretical result that allows the dual problem to be expressed as an unconstrained problem. Although the analysis in this section is based on an analog setting, it still carries over to the discrete-time and hybrid cases. Whenever there is a difference, we will point it out.

For analog EC filtering, we have the following QP problem:

$$\begin{aligned} &\min \ \langle u, Lu \rangle, \ u \in L_2(\Omega_u) \\ &\text{subject to } \varepsilon^- \preceq \Xi_s u \preceq \varepsilon^+, \end{aligned} \tag{3.2.1}$$

in which L is a linear operator on $L_2(\Omega_u)$ defined by a stationary (time-invariant) autocorrelation kernel R_{nn} (see (2.1.5)), $\Xi_x : L_2(\Omega_u) \to C(\Omega_\psi)$ is the linear operator defined by the convolution with the input s (see (2.1.3)), Ω_u is the support for filter impulse response, and Ω_c is the set where the envelope constraints are imposed on the filter output which has support Ω_ψ. In the discrete-time setting, we have $l_2(\tilde{\Omega}_u)$ and $l_\infty(\tilde{\Omega}_\psi)$ (the dual of $l_\infty(\tilde{\Omega}_\psi)$ is $l_1(\tilde{\Omega}_\psi)$) in place of $L_2(\Omega_u)$ and $C(\Omega_\psi)$, respectively. The linear operators L and Ξ_x are defined in (2.5.2) and (2.5.3), respectively. For the hybrid setting, we have $l_2(\tilde{\Omega}_u)$ instead of $L_2(\Omega_u)$, L and Ξ_x are defined in (2.6.4) and (2.6.12), respectively.

The QP Problem (3.2.1) is, in fact, Problem (3.1.1) with $A_1 = \Xi_s$, $A_2 = -\Xi_s$, $c_1 = \varepsilon^+$ and $c_2 = -\varepsilon^-$. Thus, Dorn's dual for Problem (3.2.1) can be stated as

$$\begin{aligned} &\min_u \ \langle Lu, u \rangle + p_1 \bullet \varepsilon^+ - p_2 \bullet \varepsilon^- \\ &\text{subject to } 2Lu + \alpha(p_1, p_2) = 0, \\ &\qquad p_1, p_2 \in P_M^{\oplus} \end{aligned} \tag{3.2.2}$$

where $\alpha(p_1, p_2) \in L_2(\Omega_u)$ defined by (3.1.9) reduces to the following form for analog filters:

$$[\alpha(p_1, p_2)](\tau) = \int_{\Omega_\psi} s(t-\tau) d(p_1 - p_2)(t). \tag{3.2.3}$$

In the case of hybrid filters, $\alpha(p_1, p_2) \in l_2(\tilde{\Omega}_u)$ is given by

$$[\alpha(p_1, p_2)](j) = \int_{\Omega_\psi} \left\{ \sum_{i=0}^{\infty} s[(i-j)T_s] \Lambda(t - iT_s) \right\} d(p_1 - p_2)(t). \tag{3.2.4}$$

Observe from these equations that $\alpha(p_1, p_2)$ is a function of $p_1 - p_2$. Thus, we can define

$$\hat{\alpha}(p_1 - p_2) \equiv \alpha(p_1, p_2). \tag{3.2.5}$$

Since the operator L is defined by a stationary (time-invariant) autocorrelation kernel R_{nn} (see (2.1.5)), the constraints of (3.2.2) can be expressed in the terms of the autocorrelation R_{nn} and the input s as follows:

$$2\int_{\Omega_u} R_{nn}(\kappa - \tau)u(\kappa)d\kappa + \int_{\Omega_v} s(t-\tau)d(p_1 - p_2)(t) = 0. \tag{3.2.6}$$

This equation is a Weiner-Hopf equation [112], similar to those found in least square detection and estimation theory. In order to solve (3.2.6) and obtain a unique closed form solution, certain assumptions are necessary. For (3.2.6) to have a unique solution, L must be a one-to-one and onto mapping, i.e. L must be non-singular so that the inverse L^{-1} exists. Further, since the operator L is defined by a stationary autocorrelation kernel, we will assume that this autocorrelation is non-singular. This assumption also turns out to be important for filter realizations. The following result can now be applied.

***Lemma** 3.2.1. Suppose that a continuous linear operator L on $L_2(\Omega_u)$ can be expressed in the form $L = I + H$, where I is the identity operator, and that H is a positive operator, i.e. $\langle Hu, u\rangle \geq 0, \forall u \in L_2(\Omega_u)$. Then, a continuous linear inverse for L exists.*

Let J_{nn} be the kernel of the inverse L^{-1}, i.e.

$$(L^{-1}u)(t) = \int_{\Omega_u} J_{nn}(\tau - t)u(\tau)d\tau. \tag{3.2.7}$$

Then, J_{nn} is defined by

$$\int_{\Omega_u} J_{nn}(\tau - t)R_{nn}(\theta - \tau)d\tau = \delta(t - \theta). \tag{3.2.8}$$

It can be easily verified that $L^{-1}Lu = LL^{-1}u = u$.

For the discrete-time setting, the inverse operator (if it exists) is defined by (3.2.7), but with integrals replaced by sums. The hybrid setting is a little different. In the general case where the kernel of L is given by a symmetric kernel Γ_{l-m}, the inverse (if it exists) is given by the reciprocal kernel J_{nn} defined by

$$\sum_{m \in \Omega_u} J_{nn}(m-k)\Gamma_{l-m} = \delta_{k-l}. \tag{3.2.9}$$

In terms of L^{-1}, the solution to the constraint of Problem (3.2.2) is

$$u^o = -\frac{1}{2}L^{-1}\hat{\alpha}(p_1 - p_2). \tag{3.2.10}$$

Using (3.2.3) and (3.2.7), we can rewrite the above equation as

$$u^o(\tau) = -\frac{1}{2}\int_{\Omega_u} J_{nn}(\sigma-\tau)\left[\int_{\Omega_\psi} s(t-\sigma)d(p_1-p_2)(t)\right]d\sigma.$$

The dual problem to Problem (3.2.1) is

$$\max_{p_1, p_2 \in P_M^{\oplus}} \phi(p_1, p_2), \tag{3.2.11}$$

where the dual functional ϕ is given by

$$\phi(p_1, p_2) = -\frac{1}{4}\langle L^{-1}\hat{\alpha}(p_1-p_2), \hat{\alpha}(p_1-p_2)\rangle - p_1 \bullet \varepsilon^+ + p_2 \bullet \varepsilon^-. \tag{3.2.12}$$

Since L is a positive semi-definite operator, by Theorem 3.1.1, once p_1^o and p_2^o are found, the primal solution u^o can be obtained from (3.2.10), i.e.

$$u^o = -\frac{1}{2}L^{-1}\hat{\alpha}(p_1^o - p_2^o),$$

or in terms of the reciprocal kernel J_{nn}:

$$u^o(\tau) = -\frac{1}{2}\int_{\Omega_u} J_{nn}(\sigma-\tau)\left[\int_{\Omega_c} s(t-\sigma)d(p_1^o-p_2^o)(t)\right]d\sigma. \tag{3.2.13}$$

The output of the optimum filter is

$$\Xi_s u^o = -\frac{1}{2}\Xi_s L^{-1}\hat{\alpha}(p_1^o - p_2^o).$$

Unconstrained dual problem

The Kuhn-Tucker necessary and sufficient conditions for optimality for Problem (3.2.1) are as follows: u^o is an optimal filter if and only if $\varepsilon^- \preceq \Xi_s u^o \preceq \varepsilon^+$ and there exist Lagrange multipliers $p_1^o, p_2^o \in P_M^{\oplus} \subseteq M(\Omega_\psi)$ such that

(i) u^o *minimizes the Lagrangian* $\mathcal{L}(u, p_1^o, p_2^o) = \langle Lu, u\rangle + p_1^o \bullet \varepsilon^+ - p_2^o \bullet \varepsilon^-$,

(ii) $p_1^o \bullet [\Xi_s u^o - \varepsilon^+] = 0,\ p_2^o \bullet [\varepsilon^- - \Xi_s u^o] = 0,$

where $P_M^{\oplus}$ is defined by (3.1.7). Observe that the Lagrange multipliers p_1^o and p_2^o are positive measures. Thus, it follows from the Kuhn-Tucker condition (ii) that p_1^o and p_2^o can be non-zero only where their respective constraints are active. Moreover, from physical considerations, we know that both upper and lower pulse shape constraints cannot be simultaneously active. Hence, we conclude that the Lagrange multipliers p_1^o and p_2^o must be mutually singular [4], i.e. each must be zero whenever the other is not. Because of the mutual singularity of p_1^o and p_2^o, the real measure $p_1^o - p_2^o$, which belongs to the space of regular Borel measures $M(\Omega_\psi)$, has an ε-weighted norm (see (3.1.5)):

$$\|p_1^o - p_2^o\|_\varepsilon = \|p_1^o\|_\varepsilon + \|p_2^o\|_\varepsilon = \|p_1^o + p_2^o\|_\varepsilon$$

In view of (3.1.7), we note that the measure $p_1^o - p_2^o$ is in $M(\Omega_c)$, where $M(\Omega_c)$ is the subspace of $M(\Omega_\psi)$ containing measures that are zero outside of Ω_c.

Recall that $\varepsilon^+ = d + \varepsilon$ and $\varepsilon^- = d - \varepsilon$, hence

$$\begin{aligned} p_1^o \bullet \varepsilon^+ - p_2^o \bullet \varepsilon^- &= (p_1^o - p_2^o) \bullet d + (p_1^o + p_2^o) \bullet \varepsilon \\ &= (p_1^o - p_2^o) \bullet d + \|p_1^o - p_2^o\|_\varepsilon. \end{aligned}$$

This suggests that, by combining the positive measures into a single signed measure, i.e. setting $p = p_1 - p_2$, we can define the dual functional (3.2.12) as

$$\eta(p) \equiv \phi(p_1, p_2) = -\frac{1}{4}\langle L^{-1}\hat{\alpha}(p), \hat{\alpha}(p)\rangle - p \bullet d - \|p\|_\varepsilon . \tag{3.2.14}$$

It follows from the Hahn-Jordan decomposition theorem (Theorem A.8.9, Appendix A.8) that every element p in $M(\Omega_c)$ is decomposable into two such mutually singular measures known as the positive and negative variations of p. Therefore, maximizing $\phi(p_1, p_2)$ over $p_1, p_2 \in P_M^{\oplus}$ is completely equivalent to maximizing $\eta(p)$ over $M(\Omega_c)$ in the following sense.

Theorem 3.2.1. *If p^o maximizes η over $M(\Omega_c)$, then the positive and negative measures (p^{o+}, p^{o-}) of p^o maximizes ϕ over $p_1, p_2 \in P_M^{\oplus}$. Conversely, if (p_1^o, p_2^o) maximizes ϕ over $p_1, p_2 \in P_M^{\oplus}$, then $p_1^o - p_2^o$ maximizes η over $M(\Omega_c)$.*

The unconstrained dual of the EC filtering Problem (3.2.1) is given by

$$\max_{p \in M(\Omega_c)} \eta(p). \tag{3.2.15}$$

Once p^o is found to maximize η, if L is a positive semi-definite operator, then by Theorem 3.1.1, the solution to Problem (3.2.1) is

$$u^o = -\frac{1}{2}L^{-1}\hat{\alpha}(p^o), \tag{3.2.16}$$

where

$$[\hat{\alpha}(p^o)](\tau) = \int_{\Omega_c} s(t-\tau)dp^o(t).$$

Observe that $\hat{\alpha}(p^o)$ is essentially a weighted sum of shifted versions of the matched filter where the weights are determined by the Lagrange multiplier p^o. Hence, $\hat{\alpha}(p^o)$ can be interpreted as a filter matched to the input signal s, followed by a pulse shaping filter determined by p^o. In fact, the impulse response of this pulse shaping filter is given by the distribution function defining p^o. Since p^o can only be non-zero where the constraints are active, the shifts in the matched filter correspond to the intances when the filter output touches the envelope.

There are many practical situations where coloured noise can be passed through an invertible (reversible) time-invariant linear filter such that the output is white noise. In fact, if the noise covariance function can be decomposed into a sum of delta functions at the origin (white noise) plus a remaining H_{nn} which is square-integrable and satisfies the Paley-Wiener [32] condition:

$$\int_{-\infty}^{\infty} \frac{|\log \Phi_{nn}(f)|}{1+f^2} df < \infty,$$

where $\Phi_{nn}(f)$ is the Fourier transform of H_{nn}, then a whitening filter can be found. Suppose that the impulse response of the whitening filter is q. Then, q satisfies the spectral factorization equation [112]

$$\int_{\Omega_u}\int_{\Omega_u} R_{nn}(\theta-\tau)q(\lambda)q(\tau-\sigma+\lambda)d\tau d\lambda = \delta(\sigma-\theta).$$

Comparing this with the defining equation for L^{-1} given by (3.2.8), we have

$$J_{nn}(\zeta) = \int_{\Omega_u} q(\zeta+\lambda)q(\lambda)d\lambda. \tag{3.2.17}$$

Let $\hat{s}$ be the filter matched to the whitened input $s * q$, i.e.

$$\hat{s}(\varsigma) = (s * q)(-\varsigma). \tag{3.2.18}$$

Then, it follows from (3.2.13) and (3.2.17) that

$$\begin{aligned} u^o(\tau) &= -\frac{1}{2}\int_{\Omega_u}\int_{\Omega_u}\int_{\Omega_c} q(\sigma-\tau+\lambda)q(\lambda)s(t-\sigma)dp^o(t)d\lambda d\sigma \\ &= -\frac{1}{2}\int_{\Omega_c}\int_{\Omega_u}\int_{\Omega_u} q(\lambda-\tau+t-\zeta)s(\zeta)d\zeta q(\lambda)d\lambda dp^o(t) \\ &= -\frac{1}{2}\int_{\Omega_c}\int_{\Omega_u} \hat{s}(\tau-t-\lambda)q(\lambda)d\lambda dp^o(t) \\ &= -\frac{1}{2}\int_{\Omega_c} (\hat{s} * q)(\tau-t)dp^o(t). \end{aligned} \tag{3.2.19}$$

Hence, if the additive noise is coloured and a whitening filter q can be found, then the optimum filter consists of a whitening filter q, followed by a filter matched to the whitened input pulse $\hat{s}$, and then followed by an equalizing or pulse shaping filter determined by p^o as shown in Figure 3.2.1. The same interpretation follows in the discrete-time setting but not for hybrid filters.

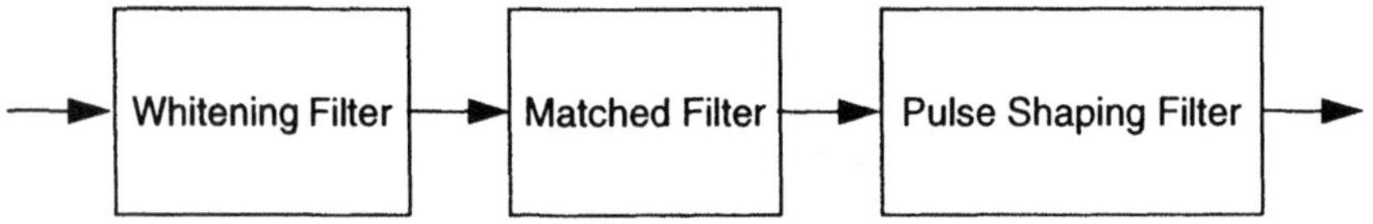

Figure 3.2.1. Configuration of optimal analog EC filter

For the hybrid configuration, we have

$$\begin{aligned}[\hat{\alpha}(p^o)](j) &= \int_{\Omega_c} \sum_{i=0}^{\infty} s[(i-j)T_s]\Lambda(t-iT_s)dp^o(t) \\ &= \sum_{i=0}^{\infty} s[(i-j)T_s]\int_{\Omega_c} \Lambda(t-iT_s)dp^o(t) \qquad (3.2.20) \\ &= \sum_{i=0}^{\infty} s[(i-j)T_s][\hat{\beta}(p^o)](iT_s),\end{aligned}$$

where

$$[\hat{\beta}(p^o)](\tau) \equiv \int_{\Omega_c} \Lambda(t-\tau)dp^o(t).$$

Again, $\hat{\alpha}(p^o)$ is a weighted sum of shifted versions of the discrete-time filter matched to the sampled input s. It can be interpreted as a discrete-time matched filter followed by a pulse shaping filter whose impulse response is a sampled version of $\hat{\beta}(p^o)$. However, unlike the analog case, the shifts are in multiples of the sampling period T_s, and do not correspond to instances where the constraints are active. The filter $\hat{\beta}(p^o)$ consists of a filter matched to the pulse response Λ, followed by a pulse shaping filter determined by p^o. In this case, the shifts in the filter matched to Λ do correspond to the instances where the constraints are active as in the case for the analog $\hat{\alpha}(p^o)$.

Since $J_{nn}(k)$ is the kernel of L^{-1}, recall from (3.2.9) that,

$$\sum_{\tau=0}^{\infty} J_{nn}(\tau-k)\Gamma_{l-\tau} = \delta_{k-l}.$$

Suppose that the impulse response of the whitening filter is q. Then, q satisfies the spectral factorization equation [112]

$$\sum_{\tau=0}^{\infty}\sum_{m=0}^{\infty} \Gamma_{l-\tau}q(m)q(\tau-k+m) = \delta_{k-l}.$$

Hence,

$$J_{nn}(k) = \sum_{m=0}^{\infty} q(k+m)q(m). \tag{3.2.21}$$

Again, let $\hat{s}$ be the filter matched to the whitened input, $s * q$, i.e.

$$\hat{s}(\varsigma) = (s * q)(-\varsigma).$$

Then, from (3.2.10), (3.2.20), and (3.2.21), it follows that

$$\begin{aligned}
u^o(k) &= -\frac{1}{2}\sum_{j=0}^{\infty}\sum_{m=0}^{\infty} q(j-k+m)q(m)\sum_{i=0}^{\infty} s[(i-j)\tau][\hat{\beta}(p^o)](i\tau) \\
&= -\frac{1}{2}\sum_{m=0}^{\infty}\sum_{\zeta=0}^{\infty}\sum_{i=0}^{\infty} q(m-k+i-\zeta)s(\zeta\tau)q(m)[\hat{\beta}(p^o)](i\tau) \\
&= -\frac{1}{2}\sum_{i=0}^{\infty}\sum_{m=0}^{\infty} \hat{s}(k-i-m)q(m)[\hat{\beta}(p^o)](i\tau) \\
&= -\frac{1}{2}\sum_{i=0}^{\infty} (\hat{s} * q)(k-i)[\hat{\beta}(p^o)](i\tau),
\end{aligned} \tag{3.2.22}$$

Thus, the overall optimum hybrid EC filter consists of a whitening filter q, a filter $\hat{s}$ which is matched to the whitened input, followed by a filter matched to the pulse response Λ, which is finally followed by a pulse shaping filter determined by p^o as shown in Figure 3.2.2.

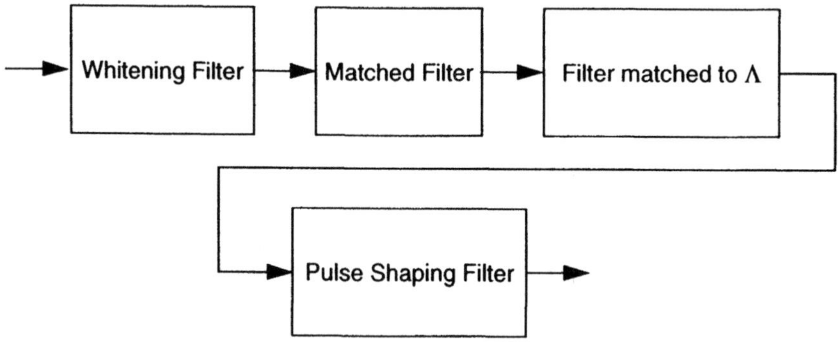

Figure 3.2.2. Configuration of optimal hybrid EC filter

3.3 Dual Problem for Finite Dimensional Filter

It is clear that the results of Section 3.2 carry over to finite-dimensional EC filters by replacing the impulse response u with

$$Ka = v^T a,$$

where $v^T = [v_0, v_1, \ldots, v_{n-1}]$ is a vector of n linearly independent elements of $L_2(\Omega_u)$ and $a \in R^n$ is the vector of filter parameters (see Section 2.4). The EC filtering problem for finite filter structures is stated as follows:

$$\begin{gathered} \min \ \langle Ka, LKa \rangle, u \in L_2(\Omega_u) \\ \text{subject to } \varepsilon^- \preceq \Xi_s Ka \preceq \varepsilon^+. \end{gathered} \tag{3.3.1}$$

This section presents the dual of Problem (3.3.1) involving the vector a instead of the filter impulse response u. Although results from Section 3.2 can be applied, we have to take precaution since the derivatives involved are taken with respect to $u = Ka$ rather than the vector a itself.

Let us examine the Lagrangian for the finite dimensional EC filtering Problem (3.3.1). By substituting $u = Ka$, $A_1 = \Xi_s$, $A_2 = -\Xi_s$, $c_1 = \varepsilon^+$ and $c_2 = -\varepsilon^-$ into (3.1.8), we have

$$\begin{aligned}\mathcal{L}(Ka, p_1, p_2) &= \langle LKa, Ka\rangle + \langle \hat{\alpha}(p_1 - p_2), Ka\rangle - p_1 \bullet \varepsilon^+ + p_2 \bullet \varepsilon^- \\ &= \langle K^\dagger LKa, a\rangle + \langle K^\dagger \hat{\alpha}(p_1 - p_2), a\rangle - p_1 \bullet \varepsilon^+ + p_2 \bullet \varepsilon^-,\end{aligned} \tag{3.3.2}$$

where $K^\dagger$ is the adjoint of K. The operator $K^\dagger LK$ can be represented by the $n \times n$ Gram matrix (see (2.4.5) of Section 2.4). We can derive an expression for the term $K^\dagger \hat{\alpha}(p_1 - p_2) \in \boldsymbol{R}^n$ by considering (3.1.9), which is the defining equation of $\hat{\alpha}(p_1 - p_2)$, and the inner product $\langle K^\dagger \hat{\alpha}(p_1 - p_2), a\rangle$ in (3.3.2),

$$\begin{aligned}\langle K^\dagger \hat{\alpha}(p_1 - p_2), a\rangle &= \langle \hat{\alpha}(p_1 - p_2), Ka\rangle \\ &= (p_1 - p_2) \bullet \Xi_s Ka \\ &= [(p_1 - p_2) \bullet y_s]^T a.\end{aligned} \tag{3.3.3}$$

where y_s is defined so that $\Xi_s Ka = y_s^T a$, i.e.

$$y_s = [\Xi_s v_0, \ldots, \Xi_s v_{n-1}]^T. \tag{3.3.4}$$

Since a is arbitrary in (3.3.3), it follows that

$$K^\dagger \hat{\alpha}(p_1 - p_2) = [(p_1 - p_2) \bullet y_s] = \int_{\Omega_v} y_s(dp_1 - dp_2). \tag{3.3.5}$$

Differentiating the convex Lagrangian (3.3.2) with respect to a, and then equating the result to zero, we obtain the equivalent of (3.1.12) for the finite dimensional EC filtering Problem (3.3.1):

$$2K^\dagger LKa + K^\dagger \hat{\alpha}(p_1 - p_2) = 0. \tag{3.3.6}$$

Dorn's dual for Problem (3.3.1) is

$$\begin{aligned}&\min_{a \in R^n} \; a^T K^\dagger LKa + p_1 \bullet \varepsilon^+ - p_2 \bullet \varepsilon^- \\ &\text{subject to } 2K^\dagger LKa + \alpha(p_1, p_2) = 0. \\ &\qquad p_1, p_2 \in P_M^\oplus\end{aligned} \tag{3.3.7}$$

Assuming L is invertible, the Gram matrix $K^\dagger LK$ (see Section 2.4) is also invertible since K and $K^\dagger$ are one-to-one mappings. Thus, the vector a^o, which minimizes the Lagrangian (3.3.2), is obtained from (3.3.5) and (3.3.6) as

$$a^o = -\frac{1}{2}(K^\dagger LK)^{-1}K^\dagger\hat{\alpha}(p_1 - p_2) = -\frac{1}{2}(K^\dagger LK)^{-1}(p_1 - p_2)\bullet y_s,$$

and the dual functional of Problem (3.3.1) takes the form:

$$\begin{aligned}\phi(p_1, p_2) &= -\frac{1}{4}\langle K^\dagger\hat{\alpha}(p_1 - p_2), (K^\dagger LK)^{-1}K^\dagger\hat{\alpha}(p_1 - p_2)\rangle - p_1 \bullet \varepsilon^+ + p_2 \bullet \varepsilon^- \\ &= -\frac{1}{4}\langle (p_1 - p_2)\bullet y_s, (K^\dagger LK)^{-1}(p_1 - p_2)\bullet y_s\rangle - p_1 \bullet \varepsilon^+ + p_2 \bullet \varepsilon^-.\end{aligned} \tag{3.3.8}$$

Let (p_1^o, p_2^o) be a solution of the dual Problem (3.3.8). Then, by Theorem 3.1.1, the optimal filter parameter vector a^o is given by

$$a^o = -\frac{1}{2}(K^\dagger LK)^{-1}K^\dagger\hat{\alpha}(p_1^o - p_2^o) = -\frac{1}{2}(K^\dagger LK)^{-1}\int_{\Omega_c} y_s d(p_1^o - p_2^o).$$

Combining the positive measures into a single signed measure, i.e. setting $p = p_1 - p_2$, it follows from (3.2.14) and (3.3.8) that

$$\begin{aligned}\eta(p) &= -\frac{1}{4}\langle K^\dagger\hat{\alpha}(p), (K^\dagger LK)^{-1}K^\dagger\hat{\alpha}(p)\rangle - p \bullet d - \|p\|_\varepsilon \\ &= -\frac{1}{4}\langle p \bullet y_s, (K^\dagger LK)^{-1}p \bullet y_s\rangle - p \bullet d - \|p\|_\varepsilon.\end{aligned} \tag{3.3.9}$$

Let p^o be a minimizer of the dual functional (3.3.9). Then, by Theorem 3.2.1, $p^o = p_1^o - p_2^o$, and the optimal filter parameter vector a^o is

$$a^o = -\frac{1}{2}(K^\dagger LK)^{-1}K^\dagger\hat{\alpha}(p^o) = -\frac{1}{2}(K^\dagger LK)^{-1}\int_{\Omega_c} y_s dp^o. \tag{3.3.10}$$

Although there is no apparent interpretation of the optimum vector of filter parameter a^o, the optimal impulse response Ka^o contains the term $\hat{\alpha}(p^o)$, which can be interpreted as a filter matched to the input signal s, followed by a pulse shaping filter determined by the multiplier p^o as for the infinite dimensional case.

Note that all results in this section hold for the discrete-time and hybrid settings discussed in Section 3.2.

3.4 Semi-Infinite Programming

So far we have presented the dual of the EC filtering problem for both the finite and infinite dimensional cases. In particular, it has been established that the dual of EC filtering can be posed as an unconstrained optimization problem when the cost functional is strictly convex. As pointed out earlier in Sections 2.4, the study of finite dimensional system is more important. In this section, further properties of the finite dimensional problem, which takes on the form of a semi-infinite programming (SIP) problem are investigated. In particular, we provide some background on SIP and explore the primal-dual structure of this class of problems, which can be applied to EC filtering. An important and interesting result of this analysis allows the (SIP) problem to be expressed as a problem with a finite number of constraints. This consequently reveals some insight to the structure of the optimal EC filter.

Semi-infinite programming, optimization in finite-dimensional spaces with infinitely many constraints, arises in various fields of engineering. Extensive survey of semi-infinite programming can be found in [1], [3], [26], [38], [46], [47]. We present, first, a general convex semi-infinite programming problem in a functional analytic setting and explore the primal structure underlying this problem by investigating its global duality and local differentiable properties. The specialized case of semi-infinite quadratic programming is then examined and applied to EC filtering.

Consider the following convex *semi-infinite programming* problem:

$$\begin{aligned} &\min \; f(\boldsymbol{x}) \\ &\text{subject to } \; \boldsymbol{g}(\boldsymbol{x}, \omega) \le 0, \;\; \forall \omega \in \Omega_c, \end{aligned} \tag{3.4.1}$$

where $\boldsymbol{x} \in \boldsymbol{R}^n$ is the decision variable, $f : \boldsymbol{R}^n \to \boldsymbol{R}$ is a convex and differentiable function, $\boldsymbol{g} : \boldsymbol{R}^n \times \Omega \to \boldsymbol{R}^m$ is convex, continuously differentiable in $\boldsymbol{x}$ for each ω in the compact Hausdorff space $\Omega_c \subseteq \Omega$, and continuous in ω for each $\boldsymbol{x} \in \boldsymbol{R}^n$.

Defining an operator $\boldsymbol{G} : \boldsymbol{R}^n \to C^m(\Omega)$, by

$$[\boldsymbol{G}(\boldsymbol{x})](\omega) = \boldsymbol{g}(\boldsymbol{x}, \omega), \tag{3.4.2}$$

we can write the constraints as

$$G(x) \preceq 0,$$

where the partial $\preceq$ ordering on $C^m(\Omega)$ is determined by a cone $P^+_{C^m}$ consisting of m-vectors of functions on Ω that are non-negative on Ω_c. This cone is convex, closed and has non-empty interior (see Section 3.1).

It is well known that strong duality holds for the convex programming Problem (3.4.1) under Slater's constraint qualification, i.e. there exists an $x \in R^n$ satisfying $g(x, \omega) < 0$, $\forall \omega \in \Omega_c$, which we assume from now on.

The Lagrangian $\mathcal{L} : R^n \times M^m(\Omega) \to R$ is defined by

$$\mathcal{L}(x, p) \equiv f(x) + p^T \bullet G(x), \tag{3.4.3}$$

where $M^m(\Omega)$ denotes the dual space of $C^m(\Omega)$ consisting of all m-vectors of regular Borel measures on Ω, and $\bullet$ denotes a duality pair between $M^m(\Omega)$ and $C^m(\Omega)$ given by

$$p^T \bullet G(x) = \sum_{i=1}^{m} \int_\Omega G_i(x) dp_i .$$

The dual of Problem (3.4.1) can then be stated as

$$\begin{gathered} \max_p \min_x \mathcal{L}(x, p) \\ \text{subject to } p \in P^{\oplus}_{M^m} \end{gathered} \tag{3.4.4}$$

where $P^{\oplus}_{M^m}$ is the dual cone of $P^+_{C^m}$ consisting of all m-vectors of non-negative regular Borel measures on Ω that are zero outside Ω_c (see (3.1.7)).

The duality theorem (Theorem B.5.2, Appendix B.5) states that if the minimum of the primal Problem (3.4.1) is achieved by some $x^o \in R^n$, then there exists a solution $p^o \in P^{\oplus}_{M^m} \subseteq M^m(\Omega)$ to the dual problem and

$$\begin{aligned} f(x^o) &= \min_x \mathcal{L}(x, p^o) \\ x^o &= \arg \min_x \mathcal{L}(x, p^o) \\ p^o \bullet G(x^o) &= 0. \end{aligned} \tag{3.4.5}$$

Since the Lagrangian (3.4.3) is convex in x, if we assume differentiability of f and g, the solutions of $\min_x \mathcal{L}(x, p)$ are completely characterized by $(d\mathcal{L}(\bullet, p))_x = 0$ for any given p, where $(d\mathcal{L}(\bullet, p))_x$ denotes the partial derivative of the Lagrangian $\mathcal{L}$ with respect to x. This observation leads to Wolfe's dual [69], [128],

$$\max_{x,p} \mathcal{L}(x, p)$$
$$\text{subject to } (d\mathcal{L}(\bullet, p))_x = 0, \qquad (3.4.6)$$
$$p \in P^{\oplus}_{M^m}.$$

Usually it is not wise to consider the dual problem in this form, since, except for a few typical applications which will be shown next, this form generally does not retain the desirable concavity that the original dual problem has.

Optimality conditions

While the previous section gives a global duality analysis of convex programming, this section investigates its local behaviour to examine some property of the Lagrange multipliers for semi-infinite programming.

Lemma 3.4.1. *The operator* $G : R^n \to C^m(\Omega)$ *given by*

$$[G(x)](\omega) = g(x, \omega),$$

is continuously Fréchet differentiable and its derivative $(dG)_x \in B(R^n, C^m(\Omega))$ *is given by*

$$[(dG)_x h](\omega) = (dg(\bullet, \omega))_x h,$$

where

$$(dg(\bullet, \omega))_x \equiv g_x(x, \omega) = \left.\frac{\partial g}{\partial x}\right|_{(x,\omega)} = \begin{bmatrix} \frac{\partial g_1}{\partial x_1}(x, \omega) & \dots & \frac{\partial g_1}{\partial x_n}(x, \omega) \\ \vdots & & \vdots \\ \frac{\partial g_m}{\partial x_1}(x, \omega) & \dots & \frac{\partial g_m}{\partial x_n}(x, \omega) \end{bmatrix}. \qquad (3.4.7)$$

The Karush-Kuhn-Tucker (*KKT*) conditions (see Theorem B.5.1, Appendix B.5) are given as follows:

Theorem 3.4.1. *The minimum of Problem (3.4.1) is achieved at* $\boldsymbol{x}^o \in \boldsymbol{R}^n$ *if and only if* $\boldsymbol{x}^o$ *is feasible and there exists a* $\boldsymbol{p}^o \in M^m(\Omega)$ *such that*

$$\begin{aligned} (d\mathcal{L}(\bullet, \boldsymbol{p}^o))_{\boldsymbol{x}^o} = (df)_{\boldsymbol{x}^o} + \boldsymbol{p}^{oT} \bullet (d\boldsymbol{G})_{\boldsymbol{x}^o} = 0 \\ \boldsymbol{p}^{oT} \bullet \boldsymbol{G}(\boldsymbol{x}^o) = 0 \\ \boldsymbol{p}^o \in P_{M^m}^{\oplus} \end{aligned} \tag{3.4.8}$$

or, equivalently,

$$\begin{aligned} (df)_{\boldsymbol{x}^o} + \sum_{i=1}^{m} \int_{\Omega} g_{i,\boldsymbol{x}}(\boldsymbol{x}^o, \omega) dp_i^o(\omega) = 0 \\ \sum_{i=1}^{m} \int_{\Omega} g_i(\boldsymbol{x}^o, \omega) dp_i^o(\omega) = 0 \\ \boldsymbol{p}^o \in P_{M^m}^{\oplus} \end{aligned} \tag{3.4.9}$$

where $g_{i,\boldsymbol{x}}(\boldsymbol{x}^o, \omega)$ *denotes the ith row of* $\boldsymbol{g}_{\boldsymbol{x}}(\boldsymbol{x}^o, \omega)$, *i.e.* $\left.\frac{\partial g_i}{\partial \boldsymbol{x}}\right|_{(\boldsymbol{x}^o, \omega)}$.

The multiplier $\boldsymbol{p}^o \in M^m(\Omega)$ satisfying the KKT conditions is not unique in general. The set of such multipliers (if not empty) contains an m-vector of atomic measures with at most $m \times n$ support points as the following result shows.

Theorem 3.4.2. *Suppose that* $\boldsymbol{x}^o$ *is feasible, and that the set of Lagrange multipliers satisfying the KKT conditions is not empty. Then, there exists a measure* $\boldsymbol{p}^o = [p_1^o, \ldots, p_m^o]^T \in M^m(\Omega)$ *which satisfies the KKT conditions, and each* p_i *is given by*

$$p_i^o \bullet z = \sum_{j=1}^{k} \lambda_{i,j}^o z(\omega_{i,j}^o),$$

for any $z \in C(\Omega)$, *where* $\lambda_{i,j}^o \geq 0$, *and* $\omega_{i,j}^o \in \Omega_c$, $j = 1, \ldots, k$, $k \leq n$.

This observation justifies the following classical result [52].

Corollary 3.4.1. The minimum of the primal Problem (3.4.1) is achieved at $x^o \in R^n$ *if and only if* x^o *is feasible and there exists* $\lambda^o_{i,j} \in R$, *and* $\omega^o_{i,j} \in \Omega_c$, $i = 1, \ldots, m$, $j = 1, \ldots, k$, $k \le n$, *such that*

$$(df)_{x^o} + \sum_{i=1}^{m}\sum_{j=1}^{k}\lambda^o_{i,j}g_{i,x}(x^o, \omega^o_{i,j}) = 0,$$

$$\sum_{i=1}^{m}\sum_{j=1}^{k}\lambda^o_{i,j}g_i(x^o, \omega^o_{i,j}) = 0,$$

$$\lambda^o_{i,j} \ge 0, i = 1, \ldots, m, j = 1, \ldots, k.$$

Note that Theorem 3.4.2 does not necessarily imply that for each constraint there are at most n active points. There may be other active points with zero multipliers. Although the non-negativity and the complementarity condition imply that the Lagrange multipliers are zero when the constraints are non-active, the converse does not necessarily hold, i.e. when the constraints are active it does not follow that the corresponding multipliers are non-zero.

Let $S \equiv \bigcup_{i=1}^{m} S_i$ and observe from the key step in the proof of Theorem 3.4.2 (Section 3.6) that $s_i \in \text{cone}S_i$, for each $i = 1, \ldots, m$; and since $\text{cone}S_i \subseteq \text{cone}S$, we have

$$p^{oT} \bullet (dG)_{x^o} = \sum_{i=1}^{m} s_i \in coneS.$$

Carathéodory's Dimensionality Theorem (Theorem B.3.6, Appendix B.3) then ensures the existence of $u_j \in S$ and non-negative λ^o_j for $j = 1, \ldots, k$, $k \le n$, such that

$$\sum_{i=1}^{m} s_i = \sum_{j=1}^{k}\lambda^o_j u_j.$$

Each $u_j \in S$ can be given by

$$u_j = g_{r^o_j, x}(x^o, \omega^o_j),$$

for some $r^o_j \in \{1, \ldots, m\}$ and $\omega^o_j \in \bar{\Omega}(x^o)$. Hence,

$$p^{oT}\bullet(dG)_{x^o} = \sum_{j=1}^{k}\lambda_j^o g_{r_j^o,x}(x^o,\omega_j^o) = \sum_{i=1}^{m}\sum_{j=1}^{k}\lambda_j^o\delta_{ir_j^o}g_{i,x}(x^o,\omega_j^o),$$

where δ_{ij} denotes the Kronecker delta. The corresponding measure p^o is hence a measure with at most n support points satisfying the KKT condition. Thus, not only the set of multipliers satisfying the KKT conditions contains m-vectors of atomic measures with at most $m\times n$ support points, it also contains m-vectors of atomic measures with at most n support points. Hence, we have the following results.

Theorem 3.4.3. *Suppose that x^o is feasible, and that the set of multipliers satisfying the KKT condition is not empty. Then, there exist a measure p^o satisfying the KKT conditions, non-negative λ_j^o, $r_j^o\in\{1,\dots,m\}$, $\omega_j^o\in\Omega_c$ for $j=1,\dots,k$, $k\le n$ such that*

$$p^{oT}\bullet z = \sum_{i=1}^{m}\sum_{j=1}^{k}\lambda_j^o\delta_{ir_j^o}z_i(\omega_j^o) = \sum_{j=1}^{k}\lambda_j^o z_{r_j^o}(\omega_j^o).$$

Corollary 3.4.2. *The minimum of the primal Problem (3.4.1) is achieved at $x^o\in R^n$ if and only if x^o is feasible and there exist $\lambda_j^o\in R$, $\omega_j^o\in\Omega_c$ and $r_j^o\in\{1,\dots,m\}$, $j=1,\dots,k$, $k\le n$ such that*

$$(df)_{x^o} + \sum_{j=1}^{k}\lambda_j^o g_{r_j^o,x}(x^o,\omega_j^o) = 0,$$

$$\sum_{j=1}^{k}\lambda_j^o g_{r_j^o}(x^o,\omega_j^o) = 0,$$

$$\lambda_j^o \ge 0, j = 1,\dots,k.$$

Transformation technique via dual parametrization

Lemma 3.4.2. *Suppose the minimum of the primal Problem (3.4.1) is achieved at $x^o\in R^n$. Then, the set of multipliers satisfying the KKT conditions coincides with the set of solutions to the dual Problem (3.4.4).*

An immediate consequence of Lemma 3.4.2 and Corollary 3.4.1 is the following theorem:

Theorem 3.4.4. *Suppose the minimum of the primal Problem (3.4.1) is achieved at* $x^o \in R^n$. *Then the set of solutions to the dual Problem (3.4.4) always includes a measure* $p^o = [p_1^o, \ldots, p_m^o]^T \in M^m(\Omega)$ *such that each* p_i^o *has at most* n *support points.*

This theorem suggests the search for a dual solution in a subset of $M^m(\Omega_c)$ whose elements are characterized by the location of a finite number of supporting points and the measures of these points. For each $i = 1, \ldots, m$, let us fix the number of supporting points of p_i to k, where $k \le n$ and denote these supporting points and their respective measures by $\omega_{i,j} \in \Omega_c$ and $\lambda_{i,j}$ for $j = 1, \ldots, k$. Then,

$$p^T \bullet G(x) = \sum_{i=1}^{m}\sum_{j=1}^{k} \lambda_{i,j} g_i(x, \omega_{i,j}) = l^T q(x, w),$$

$$p^T \bullet (dG)_x = \sum_{i=1}^{m}\sum_{j=1}^{k} \lambda_{i,j} g_{i,x}(x, \omega_{i,j}) = l^T Q(x, w),$$

where

$$l = \begin{bmatrix} l_1 \\ \vdots \\ l_m \end{bmatrix}, \quad l_i = \begin{bmatrix} \lambda_{i,1} \\ \vdots \\ \lambda_{i,k} \end{bmatrix}, \quad w = \begin{bmatrix} w_1 \\ \vdots \\ w_m \end{bmatrix}, \quad w_i = \begin{bmatrix} \omega_{i,1} \\ \vdots \\ \omega_{i,k} \end{bmatrix}, \tag{3.4.10}$$

$$q(x, w) = \begin{bmatrix} q_1(x, w_1) \\ \vdots \\ q_m(x, w_m) \end{bmatrix}, \quad q_i(x, w_i) = \begin{bmatrix} g_i(x, \omega_{i,1}) \\ \vdots \\ g_i(x, \omega_{i,k}) \end{bmatrix}, \tag{3.4.11}$$

$$Q(x, w) = \begin{bmatrix} Q_1(x, w_1) \\ \vdots \\ Q_m(x, w_m) \end{bmatrix}, \quad Q_i(x, w_i) = \begin{bmatrix} g_{i,x}(x, \omega_{i,1}) \\ \vdots \\ g_{i,x}(x, \omega_{i,k}) \end{bmatrix}. \tag{3.4.12}$$

The dual Problem (3.4.6) can be written as

$$\max_{x,\, \omega_{i,j},\, \lambda_{i,j},\, i = 1, \ldots m, j = 1, \ldots, k} \quad f(x) + \sum_{i=1}^{m} \sum_{j=1}^{k} \lambda_{i,j} g_i(x, \omega_{i,j})$$

$$\text{subject to} \;\; (df)_x + \sum_{i=1}^{m} \sum_{j=1}^{k} \lambda_{i,j} g_{i,x}(x, \omega_{i,j}) = 0, \qquad (3.4.13)$$

$$\omega_{i,j} \in \Omega_c, \lambda_{i,j} \geq 0.$$

In vector notation, we have

$$\max_{x, w, l} \; f(x) + l^T q(x, w)$$

$$\text{subject to} \;\; (df)_x + l^T Q(x, w) = 0,$$

$$w \in \Omega_c^{mk}, l \geq 0.$$

Alternatively using Lemma 3.4.2 and Corollary 3.4.2 instead of Corollary 3.4.1, we can write the dual Problem (3.4.6) as

$$\max_{x,\, \omega_j,\, r_j,\, \lambda_j,\, j = 1, \ldots, k} \quad f(x) + \sum_{j=1}^{k} \lambda_j g_{r_j}(x, \omega_j)$$

$$\text{subject to} \;\; (df)_x + \sum_{j=1}^{k} \lambda_j g_{r_j, x}(x, \omega_j) = 0, \qquad (3.4.14)$$

$$\omega_j \in \Omega_c, \lambda_j \geq 0, r_j \in \{1, \ldots, m\}, j = 1, \ldots, k.$$

In vector notation, we have

$$\max_{x, w, r, l} \; f(x) + l^T q(x, w, r)$$

$$\text{subject to} \;\; (df)_x + l^T Q(x, w, r) = 0,$$

$$w \in \Omega_c^k, r \in \{1, \ldots, m\}^k, l \geq 0.$$

where

$$l = \begin{bmatrix} \lambda_1 \\ \vdots \\ \lambda_k \end{bmatrix}, w = \begin{bmatrix} \omega_1 \\ \vdots \\ \omega_k \end{bmatrix}, r = \begin{bmatrix} r_1 \\ \vdots \\ r_k \end{bmatrix}, \qquad (3.4.15)$$

$$q(x, w, r) = \begin{bmatrix} g_{r_1}(x, \omega_1) \\ \vdots \\ g_{r_k}(x, \omega_k) \end{bmatrix}, \quad Q(x, w, r) = \begin{bmatrix} g_{r_1, x}(x, \omega_1) \\ \vdots \\ g_{r_k, x}(x, \omega_k) \end{bmatrix}. \tag{3.4.16}$$

Getting the primal solution

We have presented a procedure for solving the dual Problem (3.4.4), but our main concern is to find an optimal solution to the primal Problem (3.4.1). When a dual problem is solved yielding a dual solution p^o, the duality theorem (Theorem B.5.2, Appendix B.3) asserts that any primal solution minimizes the Lagrangian $\mathcal{L}(\bullet, p^o)$. But not every minimizer of the Lagrangian $\mathcal{L}(\bullet, p^o)$ solves the primal problem unless it is primal feasible, i.e. $G(x^o) \preceq 0$ and satisfies the complementarity condition, i.e. $p^{oT} \bullet G(x^o) = 0$.

There seems to be no general method of retrieving a primal solution from a dual solution pair that does not satisfy the primal feasibility or the complementarity condition. However, in the case of linearly constrained convex (not necessarily strictly) quadratic programming (including linear programming), Theorem 3.1.1 can be used.

3.5 Linearly Constrained Quadratic SIP and EC Filters

In this section we apply the results of Section 3.4 to a linearly constrained convex quadratic SIP of the form:

$$\begin{aligned} &\min_x \frac{1}{2} x^T Q x + b^T x \\ &\text{subject to } G(x) \equiv Ax - c \preceq 0, \end{aligned} \tag{3.5.1}$$

where $Q \in R^{n \times n}$ is positive semi-definite, $b \in R^n$, the partial ordering $\preceq$ is generated by the positive cone P_C^+ defined by (3.1.6), and $A \in B(R^n, C^m(\Omega))$, i.e. a bounded linear operator from R^n to $C^m(\Omega)$, and $c \in C^m(\Omega)$.

It is assumed that there exists an $\boldsymbol{x} \in \boldsymbol{R}^n$ satisfying $A\boldsymbol{x} - \boldsymbol{c} \prec 0$. Note that the linear operator A can be specified by $a_{i,j} \in C(\Omega)$, $i = 1, \ldots, m$, $j = 1, \ldots, n$, so that

$$(A\boldsymbol{x})_i = \sum_{j=1}^{n} a_{i,j} x_j .$$

The Lagrangian for this problem is

$$\begin{aligned} \mathcal{L}(\boldsymbol{x}, \boldsymbol{p}) &= \frac{1}{2}\boldsymbol{x}^T Q \boldsymbol{x} + \boldsymbol{b}^T \boldsymbol{x} + \boldsymbol{p}^T \bullet (A\boldsymbol{x} - \boldsymbol{c}) \\ &= -\frac{1}{2}\boldsymbol{x}^T Q \boldsymbol{x} - \boldsymbol{p}^T \bullet \boldsymbol{c} + \boldsymbol{x}^T (Q\boldsymbol{x} + \boldsymbol{b} + (\boldsymbol{p}^T \bullet A)^T), \end{aligned} \tag{3.5.2}$$

where $(\boldsymbol{p}^T \bullet A)^T \in \boldsymbol{R}^n$ is given by

$$(\boldsymbol{p}^T \bullet A)_j = \sum_{i=1}^{m} \int_{\Omega} a_{i,j} dp_i .$$

Taking the partial derivative of the Lagrangian (3.5.2) with respect to $\boldsymbol{x}$ yields

$$(d\mathcal{L}(\bullet, \boldsymbol{p}))_x = Q\boldsymbol{x} + \boldsymbol{b} + (\boldsymbol{p}^T \bullet A)^T .$$

Consequently, we obtain Dorn's dual form of the SIP Problem (3.5.1):

$$\begin{aligned} &\min_{\boldsymbol{x}, \boldsymbol{p}} \frac{1}{2}\boldsymbol{x}^T Q \boldsymbol{x} + \boldsymbol{p}^T \bullet \boldsymbol{c} \\ &\text{subject to } Q\boldsymbol{x} + \boldsymbol{b} + (\boldsymbol{p}^T \bullet A)^T = 0, \\ &\boldsymbol{p} \in P^{\oplus}_{M^m} . \end{aligned} \tag{3.5.3}$$

Let $(\boldsymbol{x}^o, \boldsymbol{p}^o)$ be a solution pair of Dorn's dual Problem (3.5.3), and z^o be a multiplier corresponding to its equality constraint. Then, z^o gives a solution to the primal Problem (3.5.1) by Theorem 3.1.1. Moreover, if Q is positive definite, i.e. for strictly convex quadratic programs, Dorn's dual (3.5.3) is reduced to (see(3.1.15)):

$$\begin{aligned} &\min_{\boldsymbol{p}} \frac{1}{2}[\boldsymbol{p}^T \bullet A + \boldsymbol{b}^T] Q^{-1} [(\boldsymbol{p}^T \bullet A)^T + \boldsymbol{b}] + \boldsymbol{p}^T \bullet \boldsymbol{c}, \\ &\text{subject to } \boldsymbol{p} \in P^{\oplus}_{M^m} . \end{aligned} \tag{3.5.4}$$

and the primal solution is given by (see (3.1.16))

$$x^o = -Q^{-1}[(p^{oT} \bullet A)^T + b]. \tag{3.5.5}$$

Finite dimensional dual problem with mxn support points

Restricting the search space to a subset of $M^m(\Omega_c)$, whose elements are characterized by m-vectors of atomic measures with at most $m \times n$ support points, yields a finite dimensional problem. Representing the atomic measures by the location of the support points and the measures of these points as in Section 3.4, we can write

$$\begin{aligned} p^T \bullet c &= l^T \tilde{c}(w), \\ p^T \bullet A &= l^T \tilde{A}(w), \end{aligned} \tag{3.5.6}$$

where l and w are given by (3.4.10), and

$$\tilde{c}(w) = \begin{bmatrix} c_1(w_1) \\ \vdots \\ c_m(w_m) \end{bmatrix}, \quad c_i(w_i) = \begin{bmatrix} c_i(\omega_{i,1}) \\ \vdots \\ c_i(\omega_{i,k}) \end{bmatrix},$$

$$\tilde{A}(w) = \begin{bmatrix} a_{11}(w_1), \ldots, a_{1n}(w_1) \\ \vdots \\ a_{m1}(w_m), \ldots, a_{mn}(w_m) \end{bmatrix}, \quad a_{i,j}(w_i) = \begin{bmatrix} a_{i,j}(\omega_{i,1}) \\ \vdots \\ a_{i,j}(\omega_{i,k}) \end{bmatrix}.$$

We can then write Problem (3.5.3) as

$$\begin{aligned} &\min_{x, w, l} \ \frac{1}{2} x^T Q x + l^T \tilde{c}(w) \\ &\text{subject to } Qx + b + \tilde{A}^T(w) l = 0, \\ &\qquad w \in \Omega_c^{mk}, \ l \geq 0. \end{aligned} \tag{3.5.7}$$

Similarly, Problem (3.5.4) can be written as

$$\begin{aligned} &\min_{w, l} \ \frac{1}{2}[l^T \tilde{A}(w) + b^T] Q^{-1} [\tilde{A}^T(w) l + b] + l^T \tilde{c}(w) \\ &\text{subject to } w \in \Omega_c^{mk}, \ l \geq 0. \end{aligned} \tag{3.5.8}$$

The primal solution is given by

$$x^o = -Q^{-1}[\tilde{A}^T(w^o)l^o + b]. \tag{3.5.9}$$

Although the dual Problems (3.5.3) and (3.5.4) are convex in the dual variables, the equivalent finite dimensional dual Problems (3.5.7) and (3.5.8) are no longer convex in the variables (w, l).

Finite dimensional dual problem with n support points

We now consider the n support points representation as in Problem (3.4.14). Note that

$$\begin{aligned} p^T \bullet c &= l^T\tilde{c}(w, r), \\ p^T \bullet A &= l^T\tilde{A}(w, r), \end{aligned} \tag{3.5.10}$$

where l and w are given by (3.4.15), and

$$\tilde{c}(w, r) = \begin{bmatrix} c_{r_1}(\omega_1) \\ \vdots \\ c_{r_k}(\omega_k) \end{bmatrix}, \tilde{A}(w, r) = \begin{bmatrix} a_{r_1, 1}(\omega_1), & \ldots, a_{r_1, n}(\omega_1) \\ & \vdots \\ a_{r_k, 1}(\omega_k), & \ldots, a_{r_1, n}(\omega_k) \end{bmatrix}$$

We can then write Problem (3.5.3) as

$$\begin{aligned} &\min_{x, w, r, l} \frac{1}{2}x^T Qx + l^T\tilde{c}(w, r) \\ &\text{subject to } Qx + b + \tilde{A}^T(w, r)l = 0, \\ &\qquad w \in \Omega_c^{mk}, r \in \{1, \ldots, m\}^k, \ l \geq 0. \end{aligned} \tag{3.5.11}$$

Similarly, Problem (3.5.4) can be written as

$$\begin{aligned} &\min_{w, r, l} \frac{1}{2}[l^T\tilde{A}(w, r) + b^T]Q^{-1}[\tilde{A}^T(w, r)l + b] + l^T\tilde{c}(w, r) \\ &\text{subject to } w \in \Omega_c^{mk}, r \in \{1, \ldots, m\}^k, \ l \geq 0. \end{aligned} \tag{3.5.12}$$

and the primal solution is given by

$$x^o = -Q^{-1}(\tilde{A}^T(w^o, r^o)l^o + b). \tag{3.5.13}$$

Although the dimension is smaller, these problems involve an extra variable $r \in \{1, \ldots, m\}^k$, which would require integer programming. Again, the equivalent finite dimensional dual Problems (3.5.11) and (3.5.12) are no longer convex.

Finite-dimensional EC filters

For the case of the EC filtering Problem (3.3.1), we have

$$b = 0_n,\ Q = 2K^{\dagger}LK,\ c = \begin{bmatrix} \varepsilon^+ \\ -\varepsilon^- \end{bmatrix},\ A = \begin{bmatrix} y_s^T \\ -y_s^T \end{bmatrix}.$$

Substituting this into the dual Problem (3.5.4), we obtain the dual Problem (3.3.7) derived in Section 3.3. Furthermore, for the finite dimensional dual problem, we have

$$\tilde{c}(t) = \begin{bmatrix} \varepsilon^+(t_1) \\ -\varepsilon^-(t_2) \end{bmatrix},\ \tilde{A}(t) = \begin{bmatrix} y_s^T(t_1) \\ -y_s^T(t_2) \end{bmatrix},$$

$$l^T\tilde{A}(t) = [l_1^T, l_2^T]\begin{bmatrix} y_s^T(t_1) \\ -y_s^T(t_2) \end{bmatrix} = l^T M y_s^T(t)\text{, where } M \equiv \begin{bmatrix} I_k & 0 \\ 0 & -I_k \end{bmatrix}$$

$$\begin{aligned} l^T\tilde{c}(t) &= l_1^T(d(t_1) + \varepsilon(t_1)) - l_2^T(d(t_2) - \varepsilon(t_2)) \\ &= [l_1^T, -l_2^T]\begin{bmatrix} d(t_1) \\ d(t_2) \end{bmatrix} + [l_1^T, l_2^T]\begin{bmatrix} \varepsilon(t_1) \\ \varepsilon(t_2) \end{bmatrix} = l^T M d(t) + l^T \varepsilon(t) \end{aligned}$$

Hence, the finite-dimensional dual problem can be written as the following non convex problem:

$$\begin{aligned} &\min_{t,\,l}\ \frac{1}{2} l^T M y_s^T(t)(K^{\dagger}LK)^{-1} y_s(t) M l + l^T M d(t) + l^T \varepsilon(t) \\ &\text{subject to } t \in \Omega_c^{2k},\ l \geq 0. \end{aligned} \tag{3.5.14}$$

Let (t^o, l^o) be a dual solution. Then, the optimal filter parameter vector, i.e. the primal solution, is given by

$$x^o = -\frac{1}{2}(K^{\dagger}LK)^{-1}[y_s(t_1^o)l_1^o - y_s(t_2^o)l_2] = -\frac{1}{2}(K^{\dagger}LK)^{-1}y_s(t^o)Ml^o. \qquad (3.5.15)$$

The measures p_1^o and p_2^o corresponding to the solution (t^o, l^o) are given by

$$p_i^o \bullet z = \sum_{j=1}^{k} \lambda_{i,j}^o z(t_{i,j}^o),\ i = 1, 2. \qquad (3.5.16)$$

The pair (p_1^o, p_2^o) is a solution to the dual Problem (3.3.8). From Section 3.3, the optimum filter impulse response can be written as

$$Kx^o = -\frac{1}{2}L^{-1}\hat{\alpha}(p_1^o - p_2^o),$$

where

$$[\hat{\alpha}(p_1^o - p_2^o)](\tau) = \int_{\Omega_c} s(t-\tau)d(p_1^o - p_2^o)(t)$$
$$= \sum_{j=1}^{k} [\lambda_{1,j}^o s(t_{1,j}^o - \tau) - \lambda_{2,j}^o s(t_{2,j}^o - \tau)].$$

Thus, not only that the optimum EC filter is a matched filter followed by a pulse shaping filter as shown in Section 3.2, it is a finite sum of at most $2n$ weighted shifts of the filter matched to the input s. The time shifts $t_{1,j}^o$ and $t_{2,j}^o$ correspond to instances, where the filter output touches the upper and lower boundary, respectively. Note that this does not necessarily mean that the output touches each boundary at no more than n points. The corresponding Lagrange multiplier at these points are given by the weights $\lambda_{1,j}^o$ and $\lambda_{2,j}^o$. This is also true for discrete-time filters, where the time shifts are restricted to multiples of the sampling period.

For the hybrid configuration (see (3.2.20)), we have

$$[\hat{\alpha}(p_1^o - p_2^o)](j) = \sum_{i=0}^{\infty} s[(i-j)T_s][\hat{\beta}(p_1^o - p_2^o)](iT_s),$$

where

$$[\hat{\beta}(p_1^o - p_2^o)](\tau) = \int_{\Omega_c} \Lambda(t-\tau)d(p_1^o - p_2^o)(t)$$
$$= \sum_{j=1}^{k} [\lambda_{1,j}^o \Lambda(t_{1,j}^o - \tau) - \lambda_{2,j}^o \Lambda(t_{2,j}^o - \tau)].$$

The filter $\hat{\beta}(p_1^o - p_2^o)$ is a finite sum of weighted shifts in the filter matched to Λ. Thus, the optimum hybrid EC filter consists of a filter matched to the sampled input s, followed by a filter whose impulse response is a finite sum of weighted shifts in the filter matched to Λ.

3.6 Appendix

Lemma 3.1.1. *The dual positive cone for P_C^+ is given by*

$$P_M^{\oplus} = \{m \in M(\Omega_\psi)\colon\ m(E) \geq 0, \forall E \subseteq \Omega_\psi,\ m(F) = 0, \forall F \subseteq \Omega_\psi \backslash \Omega_c\}.$$

Proof. The dual positive cone of P_C^+ is defined (see Appendix B.3) in the dual space $M(\Omega_\psi)$ as

$$P = \{m \in M(\Omega_\psi)\colon m \bullet x \geq 0, \forall x \in P_C^+\}.$$

We now show that $P = P_M^{\oplus}$. Suppose that $x \in P_C^+$ and $m \in P_M^{\oplus}$. Then, it is clear that

$$m \bullet x = \int_{\Omega_\psi} xdm = \int_{\Omega_c} xdm + \int_{\Omega_\psi \backslash \Omega_c} xdm = \int_{\Omega_c} xdm \geq 0$$

so that $m \in P$. Hence, $P_M^{\oplus} \subseteq P$.

Conversely, suppose that $m \in P$. If there is a subset $E \subseteq \Omega_\psi$ with $m(E) < 0$, then, by Urysohn's Lemma (Lemma A.1.1, Appendix A.1) there must exist some $x \in P_C^+$ such that

$$\int_{\Omega_\psi} xdm \leq 0.$$

This contradicts the definition of the dual positive cone, and so $m(E) \geq 0, \forall E \subseteq \Omega_\psi$. Similarly, if there is a subset $F \subseteq \Omega_\psi \backslash \Omega_c$ such that $m(F) < 0$, then, again by Urysohn's lemma there exists a $y \in P_C^+$, with support in $\Omega_\psi \backslash \Omega_c$ such that

$$\int_{\Omega_\psi} ydm \leq 0.$$

This also contradicts the definition of the dual positive cone. Thus, $P \subseteq P_M^{\oplus}$. ❐

Theorem 3.1.1. *Let $(u^o, p_1^o, \ldots, p_m^o)$ be a solution of Dorn's dual Problem (3.1.13), and z^o be a multiplier corresponding to its equality constraint. Then, z^o gives a solution to the primal Problem (3.1.1).*

Proof. The Lagrangian

$$\mathcal{L} : L_2(\Omega_u) \times M^m(\Omega_\psi) \times L_2(\Omega_u) \times M^m(\Omega_\psi)^* \to \mathbf{R}$$

for Problem (3.1.13) is given by

$$\begin{aligned} \mathcal{L}(u, p_1, \ldots, p_m, z, M_1, \ldots, M_m) &= \langle Lu, u\rangle + \sum_{i=1}^{m} p_i \bullet c_i \\ &\quad + \langle z, 2Lu + b + \alpha(p_1, \ldots, p_m)\rangle - \sum_{i=1}^{m} M_i \bullet p_i \\ &= \langle Lu, u\rangle + \sum_{i=1}^{m} p_i \bullet c_i + \langle z, 2Lu + b\rangle \\ &\quad + \sum_{i=1}^{m} p_i \bullet A_i z - \sum_{i=1}^{m} M_i \bullet p_i, \end{aligned}$$

where $M_i \bullet p_i$ denotes the duality pair between $M(\Omega_\psi)^*$ and $M(\Omega_\psi)$. The KKT Theorem (Theorem B.5.1, Appendix B.5) then ensures the existence of z^o and $(M_1^o, \ldots, M_m^o)$ satisfying

$$\begin{aligned} (d\mathcal{L}(\bullet, p_1^o, \ldots, p_m^o, z^o, M_1^o, \ldots, M_m^o))_{u^o} &= Lu^o - Lz^o = 0, \\ (d\mathcal{L}(u^o, p_1^o, \ldots, \bullet, \ldots, p_m^o, z^o, M_1^o, \ldots, M_m^o))_{p_i^o} &= c_i^* - (A_i z^o)^* - M_i^o = 0, \\ M_i^o \bullet p_i^o &= 0, \\ M_i^o &\in P_{M^*}^{\oplus}, \end{aligned} \tag{3.6.1}$$

where $P_{M^*}^{\oplus}$ is the dual positive cone of $P_M^{\oplus}$, c_i^* and $(A_i z^o)^*$ denote, respectively, c_i and $A_i z^o$ canonically imbedded in $M(\Omega_\psi)^* = (C(\Omega_\psi)^*)^*$ by a canonical mapping $c_i^* \bullet p_i = p_i \bullet c_i$ (see Appendix A.7). The second condition implies that each multiplier M_i^o can be identified with $c_i - A_i z^o$ by

$$M_i^o \bullet p_i = p_i \bullet (c_i - A_i z^o), \ \forall p_i \in M(\Omega_\psi).$$

The last two conditions of (3.6.1) can then be replaced by

$$p_i^o \bullet (c_i - A_i z^o) = 0 \text{ and } A_i z^o - c_i \preceq 0,$$

respectively. Note that the positive cone P_C^+, which generates the partial ordering $\preceq$ on $C(\Omega_\psi)$, is closed in its uniform topology. The dual feasibility of $(u^o, p_1^o, \ldots, p_m^o)$ and the first condition of (3.6.1) imply

$$2Lz^o + b + \alpha(p_1^o, \ldots, p_m^o) = 0 \text{ and } p_i^o \in P_M^\oplus.$$

Thus, p_i^o satisfies the KKT conditions for Problem (3.1.1) at z^o. The optimality of z^o then follows from the convexity of the problem. ❐

Lemma 3.2.1. *Suppose that a continuous linear operator L on $L_2(\Omega_u)$ can be expressed in the form $L = I + H$, where I is the identity operator, and that H is a positive operator, i.e. $\langle Hu, u\rangle \geq 0$, $\forall u \in L_2(\Omega_u)$. Then, a continuous linear inverse for L exists.*

Proof (Outline). Since H is a positive operator, it is straightforward to show that $I + H$ is a one-to-one mapping of $L_2(\Omega_u)$ onto itself.

$$(I + H)x = 0 \Rightarrow Hx = -x \Rightarrow \langle Hx, x\rangle = -\|x\|^2 \geq 0 \Rightarrow x = 0$$

Also, it can be shown that the range of $I + H$ is closed and onto. The operator $I + H$ is thus non-singular and has bounded everywhere defined inverse. ❐

Theorem 3.2.1. *If p^o maximizes η over $M(\Omega_c)$, then the positive and negative measures (p^{o+}, p^{o-}) of p^o maximizes ϕ over $p_1, p_2 \in P_M^\oplus$. Conversely, if (p_1^o, p_2^o) maximizes ϕ over $p_1, p_2 \in P_M^\oplus$, then $p_1^o - p_2^o$ maximizes η over $M(\Omega_c)$.*

Proof. By the Hahn-Jordan decomposition theorem (Theorem A.8.9, Appendix A.8), for any $p \in M(\Omega_c)$, let p^+ and p^- be its positive and negative variations. Then, clearly $p^+, p^- \in P_M^\oplus$. Moreover, $p = p^+ - p^-$ and $|p| = p^+ + p^-$. Hence,

$$\begin{aligned}\phi(p^+,p^-) &= -\frac{1}{4}\langle L^{-1}\hat{\alpha}(p^+-p^-), \hat{\alpha}(p^+-p^-)\rangle - (p^+-p^-)\bullet d - (p^++p^-)\bullet \varepsilon \\ &= -\frac{1}{4}\langle L^{-1}\hat{\alpha}(p), \hat{\alpha}(p)\rangle - p\bullet d - \|p\|_\varepsilon \\ &= \eta(p).\end{aligned} \tag{3.6.2}$$

For any $p_1, p_2 \in P_M^{\oplus}$, $\|p_1+p_2\|_\varepsilon \geq \|p_1-p_2\|_\varepsilon$, (as $0 \preceq \varepsilon$)

$$\begin{aligned}\eta(p_1-p_2) &= -\frac{1}{4}\langle L^{-1}\hat{\alpha}(p_1-p_2), \hat{\alpha}(p_1-p_2)\rangle - (p_1-p_2)\bullet d - \|p_1-p_2\|_\varepsilon \\ &\geq -\frac{1}{4}\langle L^{-1}\hat{\alpha}(p_1-p_2), \hat{\alpha}(p_1-p_2)\rangle - (p_1-p_2)\bullet d - \|p_1+p_2\|_\varepsilon \\ &= \phi(p_1,p_2).\end{aligned} \tag{3.6.3}$$

Now, suppose that p^o maximizes η over $M(\Omega_c)$. Then, for any $p_1, p_2 \in P_M^{\oplus}$, it follows from (3.6.2) and (3.6.3) that

$$\phi(p^{o+}, p^{o-}) = \eta(p^o) \geq \eta(p_1-p_2) \geq \phi(p_1, p_2).$$

Therefore, (p^{o+}, p^{o-}) maximizes ϕ over $p_1, p_2 \in P_M^{\oplus}$. Conversely, suppose that (p_1^o, p_2^o) maximizes ϕ over $p_1, p_2 \in P_M^{\oplus}$. Then, for any p, it follows from (3.6.3) and (3.6.2) that

$$\eta(p_1^o - p_2^o) \geq \phi(p_1^o, p_2^o) \geq \phi(p^+, p^-) = \eta(p).$$

Therefore, $p_1^o - p_2^o$ maximizes η over $M(\Omega_c)$. ❐

Note that this proof does not require the mutual singularity of the Lagrange multipliers p_1, p_2. However, the positivity of the tolerance vector ε is necessary.

Lemma 3.4.1. *The operator* $\boldsymbol{G} : \boldsymbol{R}^n \to C^m(\Omega)$ *given by*

$$[\boldsymbol{G}(\boldsymbol{x})](\omega) = \boldsymbol{g}(\boldsymbol{x}, \omega)$$

is continuously Fréchet differentiable and its derivative $(d\boldsymbol{G})_{\boldsymbol{x}} \in B(\boldsymbol{R}^n, C^m(\Omega))$ *is given by*

$$[(dG)_x h](\omega) = (dg(\bullet, \omega))_x h$$

Proof. First, we show that G is Gateaux differentiable and the Gateaux differential $\delta G(x, h) \in C^m(\Omega)$ at x in the direction h is given by

$$\delta G(x, h)(\omega) = g_x(x, \omega)h$$

where

$$g_x(x, \omega) \equiv (dg(\bullet, \omega))_x = \left.\frac{\partial g}{\partial x}\right|_{(x,\omega)} = \begin{bmatrix} \frac{\partial g_1}{\partial x_1}(x, \omega) & \cdots & \frac{\partial g_1}{\partial x_n}(x, \omega) \\ \vdots & & \vdots \\ \frac{\partial g_m}{\partial x_1}(x, \omega) & \cdots & \frac{\partial g_m}{\partial x_n}(x, \omega) \end{bmatrix}. \tag{3.6.4}$$

Since

$$\left\|\frac{G(x+\alpha h)-G(x)}{\alpha} - g_x(x, \bullet)h\right\|_{C^m(\Omega)} \le \max_{\omega \in \Omega}\|g_x(x+\beta(\alpha, \omega)\alpha h, \omega) - g_x(x, \omega)\|\|h\|,$$

where $0 < \beta(\alpha, \omega) < 1$, it follows from the continuity of $(dg(\bullet, \omega))_x$ and the compactness of Ω that the right hand side of the above inequality converges to 0 as $\alpha \to 0$. Hence, we have

$$\delta G(x, h) = \lim_{\alpha \to 0} \frac{G(x + \alpha h) - G(x)}{\alpha} = g_x(x, \bullet)h.$$

The differential $\delta G(x, h)$ is obviously linear and continuous in h, which means that $\delta G(x, \bullet) \in B(R^n, C^m(\Omega))$.

We then show that $\delta G(x^k, \bullet) \to \delta G(x, \bullet)$ in the uniform topology of $B(R^n, C^m(\Omega))$ as $x^k \to x$. Since

$$\|\delta G(x^k, h) - \delta G(x, h)\|_{C^m(\Omega)} \le \max_{\omega \in \Omega}\|g_x(x^k, \omega) - g_x(x, \omega)\|\|h\|$$

for all $h \in R^n$, we have

$$\|\delta G(x^k, \bullet) - \delta G(x, \bullet)\|_{B(R^n, C^m(\Omega))} \le \max_{\omega \in \Omega} \|g_x(x^k, \omega) - g_x(x, \omega)\| \|h\|.$$

Again, by the continuity of g_x and the compactness of Ω, we obtain $\delta G(x^k, \bullet) \to \delta G(x, \bullet)$ as $x^k \to x$. The operator G is therefore continuously Fréchet differentiable, and its derivative $(dG)_x$ coincides with $\delta G(x, \bullet)$. ❐

Theorem 3.4.1. *The minimum of Problem (3.4.1) is achieved at $x^o \in R^n$ if and only if x^o is feasible and there exists a $p^o \in M^m(\Omega)$ such that*

$$(d\mathcal{L}(\bullet, p^o))_{x^o} = (df)_{x^o} + p^{oT} \bullet (dG)_{x^o} = 0,$$
$$p^{oT} \bullet G(x^o) = 0,$$
$$p^o \in P^{\oplus}_{M^m}.$$

Proof. From the convexity and differentiability, we have

$$G(x) + (dG)_x(\hat{x} - x) \preceq G(\hat{x})$$

for any $\hat{x}, x \in R^n$. Since we assume the existence of $\hat{x}$, satisfying $G(\hat{x}) \prec 0$, Cottle's constraint qualification holds at any $x \in R^n$. The existence of p^o satisfying (3.4.8) then follows from the Kuhn-Tucker theorem (Theorem B.5.1, Appendix B.5). The sufficiency is obvious from the convexity of the problem. ❐

Theorem 3.4.2. *Suppose that x^o is feasible, and that the set of Lagrange multipliers satisfying the KKT conditions is not empty. Then, there exists a measure $p^o = [p_1^o, \ldots, p_m^o]^T \in M^m(\Omega)$ which satisfies the KKT conditions, and each p_i is given by*

$$p_i^o \bullet z = \sum_{j=1}^{k} \lambda_{i,j}^o z(\omega_{i,j}^o),$$

for any $z \in C(\Omega)$, where $\lambda_{i,j}^o \ge 0$, and $\omega_{i,j}^o \in \Omega_c$, $j = 1, \ldots, k$, $k \le n$.

Proof. Suppose that there exists a $p^o \in M^m(\Omega)$ satisfying the KKT conditions. By the continuity of g, the set $\bar{\Omega}(x^o)$ of active points at x^o defined by

$$\bar{\Omega}(x^o) = \{\omega \in \Omega_c : g(x^o, \omega) = 0\}$$

is closed in Ω_c and hence compact. The non-negativity of p^o (i.e. $p^o = 0$ on $\Omega \backslash \Omega_c$) and the complementarity condition $p^{oT} \bullet G(x^o) = 0$ imply that $p^o = 0$ on $\Omega \backslash \bar{\Omega}(x^o)$. This and the stationarity condition $(df)_{x^o} + p^{oT} \bullet (dG)_{x^o} = 0$ imply that

$$-(df)_{x^o} = \sum_{i=1}^{m} \int_{\Omega} g_{i,x}(x^o, \omega) dp_i^o(\omega) = \sum_{i=1}^{m} \int_{\bar{\Omega}(x^o)} g_{i,x}(x^o, \omega) dp_i^o(\omega).$$

Note that, for each $i = 1, \ldots, m$,

$$\int_{\bar{\Omega}(x^o)} g_{i,x}(x^o, \omega) dp_i^o(\omega) \in \text{cone}\{g_{i,x}(x^o, \omega) : \omega \in \bar{\Omega}(x^o)\}.$$

To see this, define

$$s_i \equiv \int_{\bar{\Omega}(x^o)} g_{i,x}(x^o, \omega) dp_i^o(\omega)$$

and

$$S_i \equiv \{g_{i,x}(x^o, \omega) : \omega \in \bar{\Omega}(x^o)\}.$$

Since S_i is compact in $\boldsymbol{R}^n$, by the continuity of g_x and the compactness of $\bar{\Omega}(x^o)$, coneS_i is also compact and hence closed. Now, suppose that s_i is not in coneS_i. Then, by the separation axiom (see Definition A.1.15), there exists a neighborhood of s_i, denoted by $N(s_i)$, which lies outside coneS_i, and from the definition of the Lebesque integral, there exists, for each $i = 1, \ldots, m$,

$$y_i = \sum_{j=1}^{l} g_{i,x}(x^o, \omega_{i,j}) \lambda_{i,j}$$

in the neighbourhood $N(s_i)$ with $\omega_{i,j} \in \bar{\Omega}(x^o)$ and $\lambda_{i,j} > 0$. This is a contradiction since y_i is in coneS_i, because it is a conic combination (see Appendix B.3) of elements of S_i.

Carathéodory's Dimensionality Theorem (Theorem B.3.6, Appendix B.3) then ensures the existence of non-negative $\lambda_{i,j}^o$ and $g_{i,x}(x^o, \omega_{i,j}^o) \in S_i$ (hence $\omega_{i,j}^o \in \overline{\Omega}(x^o)$) for $j = 1, \ldots, k$, $k \leq n$ such that

$$\int_{\overline{\Omega}(x^o)} g_{i,x}(x^o, \omega) dp_i^o(\omega) = \sum_{j=1}^{k} \lambda_{i,j} g_{i,x}(x^o, \omega_{i,j}).$$

The corresponding measure p_i^o can be given by

$$p_i^o \bullet z = \sum_{j=1}^{k} \lambda_{i,j}^o z(\omega_{i,j}^o),$$

and the corresponding vector measure p^o given by

$$p^{oT} \bullet z = \sum_{i=1}^{m} p_i^o \bullet z_i$$

still satisfies the KKT conditions, since

$$p^{oT} \bullet G(x) = \sum_{i=1}^{m} \sum_{j=1}^{k} \lambda_{i,j}^o g_i(x^o, \omega_{i,j}^o) = 0$$

as $\omega_{i,j}^o \in \overline{\Omega}(x^o)$. ❐

***Lemma** 3.4.2. Suppose the minimum of the primal Problem (3.4.1) is achieved at $x^o \in R^n$. Then, the set of multipliers satisfying the KKT conditions coincides with the set of solutions to the dual Problem (3.4.4).*

Proof. Suppose that p^o satisfies the KKT conditions. By convexity and differentiability, we have, for all $x \in R^n$,

$$f(x) \geq f(x^o) + (df)_{x^o}(x - x^o), \tag{3.6.5}$$

$$G(x) \geq G(x^o) + (dG)_{x^o}(x - x^o). \tag{3.6.6}$$

Since $p^o \geq 0$ and

$$[(df)_{x^o} + p^{oT} \bullet (dG)_{x^o}](x - x^o) = 0,$$

for all $x \in R^n$, it follows from (3.6.5) and (3.6.6) that

$$\begin{aligned} \mathcal{L}(x, p^o) &= f(x) + p^o \bullet G(x) \\ &\geq f(x^o) + p^o \bullet G(x^o) + [(df)_{x^o} + p^o \bullet (dG)_{x^o}](x - x^o) \\ &= \mathcal{L}(x^o, p^o). \end{aligned} \tag{3.6.7}$$

Also, since $G(x^o) \leq 0$ and $p^o \bullet G(x^o) = 0$, we have, for all $p \geq 0$,

$$\begin{aligned} \mathcal{L}(x^o, p) &= f(x^o) + p \bullet G(x^o) \\ &\leq f(x^o) + p^o \bullet G(x^o) \\ &= \mathcal{L}(x^o, p^o). \end{aligned} \tag{3.6.8}$$

By (3.6.7) and (3.6.8), it follows that (x^o, p^o) satisfies the saddle point condition:

$$\mathcal{L}(x, p^o) \geq \mathcal{L}(x^o, p^o) \geq \mathcal{L}(x^o, p).$$

This, in turn, implies that

$$\min_x \mathcal{L}(x, p^o) = \mathcal{L}(x^o, p^o) \geq \mathcal{L}(x^o, p) \geq \min_x \mathcal{L}(x, p).$$

Hence, p^o solves the dual Problem (3.4.4). The converse follows directly from the Duality Theorem (Theorem B.5.2, Appendix B.5), since the last relation in (3.4.5) is equivalent to the stationarity condition $(d\mathcal{L}(\bullet, p^o))_{x^o} = 0$. ❐

CHAPTER 4 DISCRETE-TIME EC FILTERING ALGORITHMS

With the mathematical frame work established, we can now focus on the construction and characterization of algorithms for computing solutions. Convex programming is a broad class of problems and there is no general efficient method for finding solutions. Different classes of (convex programming) problems would require different algorithms depending on the specific nature imposed by engineering considerations.

In this chapter, we shall start with the finite structure, discrete-time EC filtering problem, which is a quadratic programming (QP) problem. This is a good starting point since the algorithms are relatively easy to follow, furthermore, it provides a good basis for studying more complex problems.

Section 4.1 presents a problem statement for the discrete-time EC filtering problem and, in particular, the finite dimensional version which can be solved by the methods to be developed in the subsequent sections. In Section 4.2, we summarize one of the most efficient methods for solving QP problems, known as the active set strategy [65]. The active set strategy of updating the filter parameter is, however, not well suited to on-line implementation. Sections 4.3 and 4.4 are devoted to detailed study of two classes of iterative algorithms that are more suitable to on-line automatic filter design; one is based on the primal-dual method and the other based on penalty functions. Section 4.5 examines the discrete-time EC filtering problem with FIR filter, while Section 4.6 looks at discrete-time Laguerre network realization.

4.1 Discrete-time EC Filtering Problem

In Section 2.2, we have presented a problem statement for the analog EC filtering problem. The same can be done in a discrete time setting, which we now recap. In EC filtering, the filter response to a given input s at certain intances, say $\tilde{\Omega}_c \subseteq \tilde{\Omega}_\psi$, are constrained to lie within an envelope described by the upper and lower boundaries $\varepsilon^+, \varepsilon^- \in l_\infty(\tilde{\Omega}_\psi)$. This is a special case of the convexly constrained responses (CCR) filtering problem, where the set $\mathcal{S}$ of allowable inputs is $\{s \in l_2(\tilde{\Omega}_s)\}$, and the set of permissible responses is

$$\Psi = \{\psi \in l_\infty(\tilde{\Omega}_\psi) : \varepsilon^- \preceq \psi \preceq \varepsilon^+\}.$$

Note that the partial ordering $\preceq$ (see Appendix B.3 for more details on partial ordering) on the response space $l_\infty(\tilde{\Omega}_\psi)$ is defined by

$$x, y \in l_\infty(\tilde{\Omega}_\psi), x \preceq y \Leftrightarrow \forall i \in \tilde{\Omega}_c, x(i) \leq y(i).$$

It is straightforward to show that, under this partial ordering, Ψ is closed and convex. Since s has finite energy, the feasible set

$$\mathcal{F} = \{u \in l_2(\tilde{\Omega}_u) : \varepsilon^- \preceq s * u \preceq \varepsilon^+\},$$

where the convolution $s * u$ is defined in (2.5.1), is closed and convex (see Sections 2.1 and 2.5). The cost functional for this problem is a weighted filter norm, because we wish to minimize the effect of input noise (see Section 2.5). The discrete-time EC filtering problem can be stated as

$$\begin{aligned} &\min \ \langle u, Lu \rangle \\ &\text{subject to } \ \varepsilon^- \preceq s * u \preceq \varepsilon^+, \end{aligned}$$

where the cost $\langle u, Lu\rangle$ is given by (2.5.3). Assumptions A-2.2.1 and A-2.2.2 in Section 2.2 are used throughout.

For a finite dimensional filter characterized by an impulse response of the form

$$u = Ka = \boldsymbol{v}^T\boldsymbol{a}, \boldsymbol{v} = [v_0, \ldots, v_{n-1}]^T,$$

the filter response to the excitation s can be expressed as

$$s * Ka = \boldsymbol{y}_s^T\boldsymbol{a},$$

where $\boldsymbol{y}_s = [s * v_0, \ldots, s * v_{n-1}]^T$ is the vector of state variables. Let Q denote the Gram matrix $K^{\dagger}LK$ (see Section 2.5), the finite dimensional discrete-time EC filtering problem can be stated as the following QP problem:

$$\begin{aligned} &\min \boldsymbol{a}^T Q\boldsymbol{a}, \boldsymbol{a} \in \boldsymbol{R}^n \\ &\text{subject to } \varepsilon^-(k) \le \boldsymbol{y}_s^T(k)\boldsymbol{a} \le \varepsilon^+(k), k \in \tilde{\Omega}_c. \end{aligned} \tag{4.1.1}$$

In this chapter, we consider the case where $\tilde{\Omega}_c = \{0, 1, \ldots, N-1\}$, i.e. the constraints are only imposed at a finite number of instances. It is easy to see that only the first N elements of the impulse response sequence are needed as the rest of the sequence $(u(i), i \ge N)$ has no effect on the constraint. Since the filter norm is being minimized, $u(i)$ can be set to zero for all $i \ge N$. The optimum filter in this case is an Nth order FIR filter. However, this type of filter often requires very high order compared to other finite structure filters.

For notational convenience, the constraints of the EC filtering Problem (4.1.1) can be rewritten in matrix-vector form as

$$\varepsilon^- \le Y_s\boldsymbol{a} \le \varepsilon^+, \tag{4.1.2}$$

where

$$\varepsilon^- = \begin{bmatrix} \varepsilon^-(0) \\ \vdots \\ \varepsilon^-(N-1) \end{bmatrix}, \varepsilon^+ = \begin{bmatrix} \varepsilon^+(0) \\ \vdots \\ \varepsilon^+(N-1) \end{bmatrix} \text{ and } Y_s = \begin{bmatrix} \boldsymbol{y}_s^T(0) \\ \vdots \\ \boldsymbol{y}_s^T(N-1) \end{bmatrix}$$

in which $y_s(k)$ is a state vector of the filter at the instances $k = 0, \ldots, N-1$ as shown in Figure 4.3.1. Note that since we are only interested in the instances $k = 0, \ldots, N-1$, the convolution operator $\Xi_s : l_2(\tilde{\Omega}_u) \to l_\infty(\tilde{\Omega}_\psi)$ can be regarded as $\Xi_s : l_2(\tilde{\Omega}_u) \to \boldsymbol{R}^N$ and so $\Xi_s K : \boldsymbol{R}^n \to \boldsymbol{R}^N$ can be represented by the matrix Y_s, i.e.

$$Y_s = \Xi_s K. \tag{4.1.3}$$

4.2 QP via Active Set Strategy

The discrete-time EC filtering problem with finite filter structure has been posed in Section 4.1 as a finite dimensional quadratic programming (QP) problem with affine inequality constraints. Several techniques (see, for example, [27], [37], [65], [110]) are available for solving such problems. Other techniques include the Frank-Wolfe algorithm [6], [9], [31], the reduced gradient, and the gradient projection algorithm [1], [41], [65], [84], [137]. Other more general techniques are also applicable. For example, interior point methods [45], penalty functions [65], [84], [137], the method of centers [84], [137], feasible directions [65], [84], [137], cutting plane algorithms [65], [84]. The books [41], [65] give numerical properties of these algorithms, particularly, the reduced gradient, and gradient projection methods. The QP via active set strategy, see [65], appears to be one of the most successful methods. In this section, we present a summary of this algorithm.

QP with linear inequality constraints

We first consider QP problems with linear equality constraints

$$\begin{aligned} \min\ f(x) &= \frac{1}{2}\boldsymbol{x}^T Q\boldsymbol{x} + \boldsymbol{c}^T\boldsymbol{x} \\ \text{subject to } & A\boldsymbol{x} - \boldsymbol{b} = 0, \end{aligned} \tag{4.2.1}$$

where $\boldsymbol{x}, \boldsymbol{c} \in \boldsymbol{R}^n$, $Q = Q^T \in \boldsymbol{R}^{n\times n}$, $A \in \boldsymbol{R}^{m\times n}$, $\boldsymbol{b} \in \boldsymbol{R}^m$, $m < n$ and $\text{rank}(A) = m$. This problem is referred to as the QPE problem for convenience. The solution method for QPE is needed to solve QP with inequality constraints.

Remark 4.2.1: Without the linear constraints, the unconstrained quadratic programming problem can be solved directly if the Hessian Q is positive definite. The unique minimizing solution in this case is

$$x^o = -Q^{-1}c. \tag{4.2.2}$$

There exist many ways of solving QPE Problem (4.2.1). We shall summarize three well known methods, namely, *direct elimination, generalized elimination* and *Lagrange multipliers*.

Direct elimination: In the direct elimination method, the linear constraints are eliminated by expressing some m independent variables in terms of the remaining $n-m$ variables. This subsequently allows the QPE problem to be reduced to an unconstrained one solvable directly as shown in (4.2.2).

Since A is of full rank, we can rearrange the order of the variables such that the first m columns of A are linearly independent, i.e.

$$A = [A_1|A_2],$$

where $A_1 \in \boldsymbol{R}^{m\times m}$ is non-singular so that if $\boldsymbol{x} = [\boldsymbol{x}_1, \boldsymbol{x}_2]^T$, $\boldsymbol{x}_1 \in \boldsymbol{R}^m$, $\boldsymbol{x}_2 \in \boldsymbol{R}^{n-m}$, we have

$$A\boldsymbol{x} - \boldsymbol{b} = A_1\boldsymbol{x}_1 - \boldsymbol{b} + A_2\boldsymbol{x}_2 = 0,$$

and hence

$$\boldsymbol{x}_1 = A_1^{-1}(\boldsymbol{b} - A_2\boldsymbol{x}_2). \tag{4.2.3}$$

We can then partition the matrix Q and the vector $\boldsymbol{c}$ accordingly as

$$Q = \begin{bmatrix} Q_{11} & Q_{12} \\ Q_{12}^T & Q_{22} \end{bmatrix} \text{ and } \boldsymbol{c} = \begin{bmatrix} \boldsymbol{c}_1 \\ \boldsymbol{c}_2 \end{bmatrix},$$

where $Q_{11} \in \boldsymbol{R}^{m\times m}$, $Q_{12} \in \boldsymbol{R}^{m\times(n-m)}$, $Q_{22} \in \boldsymbol{R}^{(n-m)\times(n-m)}$, $\boldsymbol{c}_1 \in \boldsymbol{R}^m$ and $\boldsymbol{c}_2 \in \boldsymbol{R}^{n-m}$. Now

$$f(\boldsymbol{x}) = \frac{1}{2}\begin{bmatrix}\boldsymbol{x}_1^T & \boldsymbol{x}_2^T\end{bmatrix}\begin{bmatrix}Q_{11} & Q_{12}\\ Q_{12}^T & Q_{22}\end{bmatrix}\begin{bmatrix}\boldsymbol{x}_1\\ \boldsymbol{x}_2\end{bmatrix} + \begin{bmatrix}\boldsymbol{c}_1^T & \boldsymbol{c}_2^T\end{bmatrix}\begin{bmatrix}\boldsymbol{x}_1\\ \boldsymbol{x}_2\end{bmatrix}$$
$$= \frac{1}{2}\boldsymbol{x}_1^T Q_{11}\boldsymbol{x}_1 + \boldsymbol{x}_1^T Q_{12}\boldsymbol{x}_2 + \frac{1}{2}\boldsymbol{x}_2^T Q_{22}\boldsymbol{x}_2 + \boldsymbol{c}_1^T\boldsymbol{x}_1 + \boldsymbol{c}_2^T\boldsymbol{x}_2. \tag{4.2.4}$$

Substituting (4.2.3) into (4.2.4) gives

$$f(\boldsymbol{x}) = \frac{1}{2}(\boldsymbol{b}^T - \boldsymbol{x}_2^T A_2^T)A_1^{-T}Q_{11}A_1^{-1}(\boldsymbol{b} - A_2\boldsymbol{x}_2) + (\boldsymbol{b}^T - \boldsymbol{x}_2^T A_2^T)A_1^{-T}Q_{12}\boldsymbol{x}_2$$
$$+ \frac{1}{2}\boldsymbol{x}_2^T Q_{22}\boldsymbol{x}_2 + \boldsymbol{c}_1^T[A_1^{-1}(\boldsymbol{b} - A_2\boldsymbol{x}_2)] + \boldsymbol{c}_2^T\boldsymbol{x}_2$$
$$= \frac{1}{2}\boldsymbol{x}_2^T(Q_{22} + A_2^T A_1^{-T}Q_{11}A_1^{-1}A_2 - 2A_2^T A_1^{-T}Q_{12})\boldsymbol{x}_2 \tag{4.2.5}$$
$$+ (\boldsymbol{b}^T A_1^{-T}Q_{12} - \boldsymbol{b}^T A_1^{-T}Q_{11}A_1^{-1}A_2 + \boldsymbol{c}_2^T - \boldsymbol{c}_1^T A_1^{-1}A_2)\boldsymbol{x}_2$$
$$+ \frac{1}{2}\boldsymbol{b}^T A_1^{-T}Q_{11}A_1^{-1}\boldsymbol{b} + \boldsymbol{c}_1^T A_1^{-1}\boldsymbol{b}.$$

Clearly, (4.2.4) is now reduced to (4.2.5), which is a function of $\boldsymbol{x}_2$. Thus, we define

$$\tilde{f}(\boldsymbol{x}_2) \equiv \frac{1}{2}\boldsymbol{x}_2^T\tilde{Q}\boldsymbol{x}_2 + \tilde{\boldsymbol{c}}^T\boldsymbol{x}_2 + d, \tag{4.2.6}$$

where

$$\tilde{Q} = Q_{22} + A_2^T A_1^{-T}Q_{11}A_1^{-1}A_2 - 2A_2^T A_1^{-T}Q_{12},$$
$$\tilde{\boldsymbol{c}} = Q_{12}^T A_1^{-1}\boldsymbol{b} - A_2^T A_1^{-T}Q_{11}A_1^{-1}\boldsymbol{b} + \boldsymbol{c}_2 - A_2^T A_1^{-T}\boldsymbol{c}_1,$$
$$d = \frac{1}{2}\boldsymbol{b}^T A_1^{-T}Q_{11}A_1^{-1}\boldsymbol{b} + \boldsymbol{c}_1^T A_1^{-1}\boldsymbol{b}.$$

Differentiate $\tilde{f}$ with respect to $\boldsymbol{x}_2$, and equating the result to zero, we obtain

$$\boldsymbol{x}_2^o = -\tilde{Q}^{-1}\tilde{\boldsymbol{c}}. \tag{4.2.7}$$

If $\tilde{Q}$ is positive definite, then, $\boldsymbol{x}_2^o$ will be a minimizing (both locally and globally) solution of $\tilde{f}$. However, the assumption on the positive definiteness of $\tilde{Q}$ is, in general, not satisfied. The remaining part of solution is constructed from

$$x_1^o = A_1^{-1}(b + A_2\tilde{Q}^{-1}\tilde{c}). \quad (4.2.8)$$

The direct elimination method is easy to understand but nevertheless incurs a large amount of algebraic computation. The method of generalized elimination seeks to reduce this effort.

Generalized elimination: Let E and F be, respectively, real matrices of dimension $n \times m$ and $n \times (n-m)$ such that

(i) $[E|F]$ *is non singular,*

(ii) $AE = I_m$,

(iii) $AF = 0$.

Note that E and F are not necessarily unique. In practice, E and F may be obtained by selecting any matrix $G \in \boldsymbol{R}^{n \times (n-m)}$ such that $[A^T|G]$ is non-singular, and it is easy to verify that

$$[A^T|G]^{-1} = \begin{bmatrix} E^T \\ F^T \end{bmatrix}.$$

Condition (iii) implies that the columns of F are basis vectors for the null space of A, (the null space of A is $\{x : Ax = 0\}$). From conditions (ii) and (iii), we have

$$A(Eb + Fy) - b = 0.$$

Hence, the general solution of the constraint equation can be written as

$$x = Eb + Fy, \quad (4.2.9)$$

where $y \in \boldsymbol{R}^{n-m}$ is arbitrary. Substitution of (4.2.9) into the cost functional yields

$$f(x) = \hat{f}(y) = \frac{1}{2}y^T F^T Q F y + (c + QEb)^T F y + \left(c + \frac{1}{2}QEb\right)^T Eb. \quad (4.2.10)$$

Problem (4.2.1) has now been transformed into the unconstrained problem of minimizing (4.2.10). If $F^T QF$ is positive definite, then the unique minimizer y^o is

$$y^o = -(F^TQF)^{-1}F^T(c + QEb). \tag{4.2.11}$$

Lagrange multipliers: The use of Lagrange multipliers in this method also made it easier to understand. The Lagrangian function for the constrained problem is

$$\mathcal{L}(x, p) = f(x) + p^T(Ax - b). \tag{4.2.12}$$

The necessary conditions for optimality (see Appendix B.5) require that the gradient of the Lagrangian (4.2.12) with respect to both x and p be vanishing, that is,

$$\begin{aligned} x^TQ + c^T + p^TA &= 0, \\ (Ax - b)^T &= 0. \end{aligned}$$

These equations constitute a linear set of simultaneous equations in x and p which can be expressed as

$$\begin{bmatrix} Q & A^T \\ A & 0 \end{bmatrix} \begin{bmatrix} x^o \\ p^o \end{bmatrix} = \begin{bmatrix} -c \\ b \end{bmatrix}. \tag{4.2.13}$$

If Q is positive define and A has full rank, then,

$$\begin{bmatrix} Q & A^T \\ A & 0 \end{bmatrix}$$

is non singular and its inverse is given by

$$\begin{bmatrix} Q & A^T \\ A & 0 \end{bmatrix}^{-1} = \begin{bmatrix} H & P^T \\ P & S \end{bmatrix}, \tag{4.2.14}$$

where

$$\begin{aligned} S &= -(AQ^{-1}A^T)^{-1}, \\ P &= -SAQ^{-1}, \\ H &= Q^{-1} + Q^{-1}A^TSAQ^{-1}. \end{aligned}$$

By (4.2.13) and (4.2.14), the solution to the QPE Problem (4.2.1) can be expressed as

$$x^o = -Hc + P^T b,$$
$$p^o = -Pc + Sb.$$

Substituting for H, P and S gives

$$x^o = Q^{-1}c - Q^{-1}A^T(AQ^{-1}A^T)^{-1}(AQ^{-1}c + b), \tag{4.2.15}$$

which is a much more elegant form than the direct elimination method. The Lagrange multiplier p^o can be obtained similarly.

QP with inequality constraints

In addition to the equality constraints considered previously in the QPE Problem (4.2.1), the general QP problem, which covers the EC filtering Problem (4.1.1), is also subject to inequality constraints. The QP problem is

$$\begin{aligned} \min f(x) &= \frac{1}{2}x^T Q x + c^T x \\ \text{subject to } g_i(x) &= a_i^T x - b_i = 0, \; i \in \mathcal{E}, \\ g_i(x) &= a_i^T x - b_i \le 0, \; i \in I. \end{aligned} \tag{4.2.16}$$

Here, the constraints have been partitioned into two sets, with $\mathcal{E}$ and I denoting the sets of indices corresponding to the equality and inequality constraints, respectively. If Q is positive definite, the problem is convex. In this case, the first order Kuhn-Tucker necessary conditions (Theorem B.5.1, Appendix B.5) are also sufficient. Thus, there exists a vector p^o of Lagrange multipliers such that

$$\begin{aligned} Qx^o + c &= \sum_{i \in \mathcal{E} \cup I} p_i^o a_i, \\ a_i^T x^o - b_i &= 0, \; i \in \mathcal{E}, \\ a_i^T x^o - b_i &\le 0, \; i \in I, \\ p_i^o(a_i^T x^o - b_i) &= 0, \; i \in I, \\ p_i^o &\ge 0, \; i \in I. \end{aligned}$$

Due to the inequality constraints in QP Problem (4.2.16), elimination methods are not applicable. Instead, it is solved by a series of QPE's via the *active set strategy*. Since the constraints are affine, it is easy to compute a feasible solution using Phase I of the simplex algorithm for linear programming.

Suppose that we partition the constraints into those that are to be treated as active and those that are to be treated as inactive. The constraints treated as inactive are essentially ignored. The underlying idea in active set methods is that if the active set were known, the original problem could be replaced by corresponding equality constraints only. Alternatively, suppose that an active set was identified and the corresponding equality constrained problem was solved. If all other constraints were satisfied and the Lagrange multipliers were non-negative, then the resulting solution is optimal.

At each iteration of the algorithm, a set of constraints, to be treated as the active set, is determined. This set is referred to as the working set. At the kth iteration, the working set $\mathcal{W}_k$ is chosen to be a subset of

$$\mathcal{A}_k = \{i \in \mathcal{E} \cup I : g_i(\boldsymbol{x}_k) = 0\},$$

the active set of constraints at the current point $\boldsymbol{x}_k$. Hence, the current point $\boldsymbol{x}_k$ is feasible for the working set. The algorithm then proceeds to search on the surface defined by the working set of constraints for an improved point.

Suppose that the problem with equality constraints

$$\begin{aligned} &\min\ f(\boldsymbol{x}) \\ &\text{subject to } g_i(\boldsymbol{x}) = 0,\ i \in \mathcal{W}_k, \end{aligned} \tag{4.2.17}$$

is solved yielding an $\boldsymbol{x}_w$, which satisfies $g_i(\boldsymbol{x}_w) < 0,\ i \notin \mathcal{W}_k$. The point $\boldsymbol{x}_w$ satisfies the necessary condition

$$\nabla f(\boldsymbol{x}_w) + \sum_{i \in \mathcal{W}_k} \lambda_i \nabla g_i(\boldsymbol{x}_w) = 0.$$

If $\lambda_i \geq 0$ for all $i \in \mathcal{W}_k$, then, $\boldsymbol{x}_w$ is a local solution to the original Problem (4.2.16). On the other hand, if there is an $i \in \mathcal{W}_k$ such that $\lambda_i < 0$, then the objective can be decreased by relaxing constraint i. This follows from the sensitivity interpretation of

Lagrange multipliers: a small decrease in the constraint value from 0 to $-\delta$ would lead to a change in the objective function of $\lambda_i\delta$, which is negative. Thus, by dropping constraint i from the working set, an improved solution can be obtained. The Lagrange multiplier thereby serves as an indication of which constraints should be dropped from the working set.

Since all points defined by the algorithm must be feasible, during the course of minimizing over a given working surface it is necessary to monitor the values of the other constraints to be sure that they are not violated. While moving on a working surface, a new constraint boundary may be encountered, in such a case, this constraint is added to the working set, reducing the dimension of the working surface by one.

Note that, for any $\boldsymbol{x}$,

$$f(\boldsymbol{x}+\boldsymbol{d}) = f(\boldsymbol{x})+\boldsymbol{d}^T(Q\boldsymbol{x}+\boldsymbol{c})+\frac{1}{2}\boldsymbol{d}^TQ\boldsymbol{d},$$

$$g_i(\boldsymbol{x}+\boldsymbol{d}) = \boldsymbol{a}_i^T(\boldsymbol{x}+\boldsymbol{d})-b_i = g_i(\boldsymbol{x})+\boldsymbol{a}_i^T\boldsymbol{d},$$

and so by translating to the kth iterate $\boldsymbol{x}_k$, Problem (4.2.17) takes the form:

$$\begin{aligned} &\min\ \frac{1}{2}\boldsymbol{d}^TQ\boldsymbol{d}+\boldsymbol{d}^T(Q\boldsymbol{x}_k+\boldsymbol{c}) \\ &\text{subject to } \boldsymbol{a}_i^T\boldsymbol{d} = 0, i\in \mathcal{W}_k. \end{aligned} \tag{4.2.18}$$

Suppose that $\boldsymbol{d}_k \neq 0$ is a solution to Problem (4.2.18). If the point $\boldsymbol{x}_k+\boldsymbol{d}_k$ is not feasible, a search of the form $\boldsymbol{x}_k+\alpha_k\boldsymbol{d}_k$ is made and the stepsize α_k is chosen so that no constraints are violated. Since

$$\boldsymbol{a}_i^T\boldsymbol{d}_k = 0,\quad \forall i\in \mathcal{W}_k,$$

we have

$$g_i(\boldsymbol{x}_k+\alpha_k\boldsymbol{d}_k) = 0,\quad \forall i\in \mathcal{W}_k,$$

and the constraints, which may be affected by the choice of α_k, are from the set $I\backslash\mathcal{W}_k$. For each $i\in I\backslash\mathcal{W}_k$, $b_i-\boldsymbol{a}_i^T\boldsymbol{x}_k>0$, if i is such that $\boldsymbol{a}_i^T\boldsymbol{d}_k\le 0$, then

$g_i(\boldsymbol{x}_k + \alpha \boldsymbol{d}_k) \le 0$ for any non-negative α_k and so this constraint is also not affected by the stepsize. On the other hand, if $\boldsymbol{a}_i^T \boldsymbol{d}_k > 0$, then to ensure that

$$g_i(\boldsymbol{x}_k + \alpha_k \boldsymbol{d}_k) \le 0,$$

we require

$$\alpha_k \le (b_i - \boldsymbol{a}_i^T \boldsymbol{x}_k)/\boldsymbol{a}_i^T \boldsymbol{d}_k.$$

By choosing the smallest $(b_i - \boldsymbol{a}_i^T \boldsymbol{x}_k)/\boldsymbol{a}_i^T \boldsymbol{d}_k$ over those $i \in I \backslash \mathcal{W}_k$ such that $\boldsymbol{a}_i^T \boldsymbol{d}_k > 0$, as an upper bound for α_k, we ensure that

$$\alpha_k \boldsymbol{a}_i^T \boldsymbol{d}_k \le b_i - \boldsymbol{a}_i^T \boldsymbol{x}_k,$$

and hence

$$g_i(\boldsymbol{x}_k + \alpha_k \boldsymbol{d}_k) \le 0, \ \ \forall i \in I \cup \mathcal{E}.$$

To have the largest decrease in cost, we choose α_k to be the largest α such that feasibility is maintained, i.e.

$$g_i(\boldsymbol{x}_k + \alpha \boldsymbol{d}_k) \le 0, \ \ \forall i \in I \backslash \mathcal{W}_k.$$

Such α_k is thus given by

$$\alpha_k = \min \{(b_i - \boldsymbol{a}_i^T \boldsymbol{x}_k)/\boldsymbol{a}_i^T \boldsymbol{d}_k : i \in I \backslash \mathcal{W}_k, \boldsymbol{a}_i^T \boldsymbol{d}_k > 0\}. \tag{4.2.19}$$

If $\alpha_k \ge 1$, we set it to 1, otherwise $\alpha_k < 1$. The constraint corresponding to the minimizing index becomes active, and is appended to the new working set.

The active set strategy can be summarized in the following steps:

1. Set $k = 0$, select an initial feasible solution $\boldsymbol{x}_0$ and identify the corresponding working set $\mathcal{W}_0$.

2. Compute the search direction $\boldsymbol{d}_k$ by solving the QPE Problem (4.2.18). If $\boldsymbol{d}_k \ne 0$ solves this problem, then go to step 3, otherwise go to step 4.

3. Compute the stepsize $\alpha_k \le 1$ from (4.2.19) and set $x_{k+1} = x_k + \alpha_k d_k$ such that at least one of the inactive constraints becomes active while not violating the rest of the constraints. If $\alpha_k < 1$, set

$$\mathcal{W}_{k+1} = \mathcal{W}_k \cup \{r\},$$

where $r \in I \backslash \mathcal{W}_k$ achieves the minimum in (4.2.19). Otherwise, ($\alpha_k = 1$) set $\mathcal{W}_{k+1} = \mathcal{W}_k$. Finally, set k to $k+1$ and return to step 2.

4. Since $d = 0$ solves the problem in step 2, x_k solves Problem (4.2.17). Compute the corresponding Lagrange multiplier vector $p_k = [p_{k,i}, i \in \mathcal{W}_k]$ by

$$p_k = -(A_k Q^{-1} A_k^T)^{-1}(b_k + A_k Q^{-1} c),$$

where $A_k = [a_i, i \in \mathcal{W}_k]^T$ and $b_k = [b_i, i \in \mathcal{W}_k]^T$. Select j such that

$$p_{k,j} = \min_{i \in \mathcal{W}_k \cap I} p_{k,i}.$$

If $p_{k,j} \ge 0$, terminate with $x^o = x_k$. Otherwise, set $\mathcal{W}_k = \mathcal{W}_k \backslash \{j\}$ and go to step 2.

It should be noted that if Q is not positive semi-definite, the stationary point may not be a local minima. Special techniques are required for solving this class of indefinite QP problems.

Starting with a given feasible point and its corresponding working set, we minimize the objective function over the corresponding surface by solving a QPE problem for the current working set. If new constraint boundaries are encountered, they are added to the working set but no constraints are dropped. Steps 2 and 3 pertain to the detail of minimizing the cost over the surface associated with the working set. Finally, when a point is obtained which minimizes the f over the current working set of constraints, the corresponding Lagrange multipliers are calculated. If they are all nonnegative, then in accordance with the Kuhn-Tucker conditions (Appendix B.5, Theorem B.5.1), the solution is optimal. Otherwise, if one or more multipliers are negative, then those constraints corresponding to the most negative multipliers are dropped from the working set. This procedure is carried out in Step 4 of the algorithm. The procedure is iterated with the new working set and f will strictly decrease on the next step. To be more specific, such a method is guaranteed to converge to the optimal solution, since after the solution corresponding to one working set is found, a decrease in the objective is made, hence it is not possible to return to that working set. Moreover, as the number of working sets is finite, the process must terminate.

Although numerically efficient, QP via active set method as well as all the algorithms mentioned in this section do not appear to be as well suited to on line-implementation as the iterative algorithms to be presented in the next two sections.

4.3 Iterative Algorithm via The Primal-Dual Method

In iterative algorithms, the actual output and state of the signals are used to update the filter coefficients. This "closed-loop" feature, which non-iterative algorithms such as those mentioned in Section 4.1 lack, allows for implementation errors and quantization in filter realization. Furthermore, the iterative algorithms developed here can also serve as a basis for the development of adaptive algorithms for EC filtering even though adaptive algorithm is not in the scope of this monograph.

In signal processing applications, where the parameters of the underlying signal model are either not known or varying with time, it may be necessary to employ an adaptive filter with parameters that can be adjusted to their optimum values [40], [127]. Figure 4.3.1 shows a general configuration of an adaptive filter structure. Although the filter is a parallel type, the transversal structure assumes an analogous form. In the most common procedure, a known test or training signal s, corrupted by noise, is used as input to the filter and the output ψ is checked against the boundaries ε^+ and ε^- (whereas in the least square approach, the difference between ψ and the desired pulse shape is formed). The result of the comparison together with the state

$$y_s(k) = [y_{s,0}(k), \dots, y_{s,n-1}(k)]^T$$

of the filter at the instances $k = 0, \ldots, N-1$ are then processed in some way and fed back to adjust the filter coefficients. Another (noise corrupted) test pulse arrives and the process is repeated until, for practical purposes, convergence has occurred.

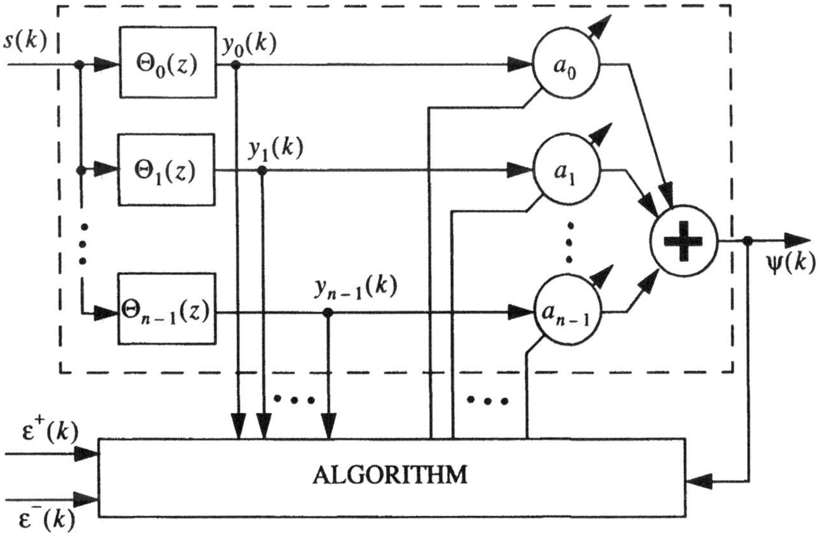

Figure 4.3.1. Configuration for an adaptive EC filter.

Iterative algorithm corresponds to the case where the test pulse s used for updating the filter parameter is not corrupted by noise. Since actual filter outputs and states are used, implementation errors, quantization and other system non-linearities are permitted.

In practice, the signals at the input and output of the filter in Figure 4.3.1 may be contaminated with noise. One approach for dealing with the noise is to lower the noise level by averaging a number of signal sequences in between the application of successive iterations of the update algorithm thereby reducing the system to the noise-free case. However, in general, one is interested in algorithms that work directly with the noisy signals (see [21], [22], [113]). The study of the behaviour of these "adaptive algorithms" is beyond the scope of this monograph. However, the iterative algorithms presented here serve as a platform for the development of adaptive algorithms for EC filtering.

This section presents an iterative algorithm to compute the optimal filter coefficients for the EC filtering problem. In an iterative context, an early approach used to solve the EC filtering problem was the primal-dual algorithm [21] [22], [23]. Essentially this method uses generalized versions of the Kuhn-Tucker Theorem and Duality Theorem (Appendix B.5) to reduce the optimization problem in $\boldsymbol{R}^n$ subject to affine inequality constraints into an unconstrained one in its dual space. The unconstrained dual problem, which has a non-differentiable cost, is then solved using a steepest ascent algorithm based on directional differentials. A novel feature of this algorithm is its surprisingly simple implementation despite cumbersome theoretical manipulations.

Non-smooth dual problem

Results concerning duality addressed in Chapter 3 in a continuous-time scenario carries over to discrete-time. We now summarize those results for discrete-time by outlining the necessary steps. Recall that $\varepsilon^+ = d + \varepsilon$ and $\varepsilon^- = d - \varepsilon$, the Lagrangian for the discrete-time EC filtering Problem (4.1.1) is defined as

$$\begin{aligned}\mathcal{L}(\boldsymbol{a}, \boldsymbol{p}_1, \boldsymbol{p}_2) &= \boldsymbol{a}^T Q \boldsymbol{a} + [Y_s \boldsymbol{a} - \boldsymbol{d} - \varepsilon]^T \boldsymbol{p}_1 - [Y_s \boldsymbol{a} - \boldsymbol{d} + \varepsilon]^T \boldsymbol{p}_2 \\ &= \boldsymbol{a}^T Q \boldsymbol{a} + \boldsymbol{a}^T Y_s^T (\boldsymbol{p}_1 - \boldsymbol{p}_2) - \boldsymbol{d}^T (\boldsymbol{p}_1 - \boldsymbol{p}_2) - \varepsilon^T (\boldsymbol{p}_1 + \boldsymbol{p}_2),\end{aligned}$$

where $\boldsymbol{p}_1, \boldsymbol{p}_2 \geq 0$ are Lagrange multiplier vectors in $\boldsymbol{R}^N$ corresponding to the upper and lower boundaries of the mask, respectively. Because the cost functional is convex and the constraints are affine, the Lagrangian is also convex in $\boldsymbol{a}$. Differentiating the Lagrangian with respect to $\boldsymbol{a}$ and equating the derivative to zero yeild

$$\boldsymbol{a}^o(\boldsymbol{p}_1, \boldsymbol{p}_2) = -\frac{1}{2} Q^{-1} Y_s^T (\boldsymbol{p}_1 - \boldsymbol{p}_2).$$

The dual functional of Problem (4.1.1) is then (see also (3.2.12))

$$\phi(\boldsymbol{p}_1, \boldsymbol{p}_2) = -\frac{1}{4} (\boldsymbol{p}_1 - \boldsymbol{p}_2)^T Y_s Q^{-1} Y_s^T (\boldsymbol{p}_1 - \boldsymbol{p}_2) - \boldsymbol{d}^T (\boldsymbol{p}_1 - \boldsymbol{p}_2) - \varepsilon^T (\boldsymbol{p}_1 + \boldsymbol{p}_2). \quad (4.3.1)$$

Note that $Y_s Q^{-1} Y_s^T$ is a symmetric positive semi-definite matrix. Combining the two positive measures into a single signed measure (see also (3.2.14)) yields an equivalent dual function

$$\eta(p) \equiv \phi(p_1, p_2) = -\frac{1}{4} p^T Y_s Q^{-1} Y_s^T p - d^T p - \varepsilon^T |p|. \tag{4.3.2}$$

Theorem 3.2.1 states that maximizing $\phi(p_1, p_2)$ over $p_1, p_2 \geq 0$ is equivalent to maximizing $\eta(p)$ over all p in the sense that if p^o maximizes η, then (p^{o+}, p^{o-}) maximizes ϕ over $p_1, p_2 \geq 0$, where p^{o+} and p^{o-} are, respectively, the positive and negative variations of p^o. Conversely, if (p_1^o, p_2^o) maximizes ϕ over $p_1, p_2 \geq 0$, then $p_1^o - p_2^o$ maximizes η. This is true even for more general cases where multipliers p_1 and p_2 are not mutually singular (see the proof of Theorem 3.2.1). Moreover,

$$a^T Q a \geq a^{oT} Q a^o = \eta(p^o) \geq \eta(p)$$

holds for all p and for any a which satisfies the constraints. The existence of p^o is implied by the assumption of a feasible solution and the Duality Theorem (Theorem B.5.2, Appendix B.5). Results concerning the uniqueness of p^o are not available. Examples of non-unique p^o can be constructed if the assumption of a strictly feasible solution is relaxed to just a feasible one.

By Theorem 3.1.1, the optimal filter parameter vector a^o can be expressed in terms of p^o, the maximizer of the alternative dual function η, as follows:

$$a^o = -\frac{1}{2} Q^{-1} Y_s^T p^o. \tag{4.3.3}$$

Steepest ascent with directional differentials

The constrained primal Problem (4.1.1) has been transformed into the unconstrained dual problem of maximizing the dual cost (4.3.2). This task is complicated by the presence of the term $\varepsilon^T |p|$, which is not differentiable in the ordinary sense. However, this term is directionally differentiable, and an ascent type algorithm based on directional differentials is developed in this section.

Maximization of the concave dual function η, given in (4.3.2), can be carried out using the directional differential (see Appendix B.4) of η at the point $\boldsymbol{p}$ in the direction $\boldsymbol{h}$ defined by

$$\delta^{+}\eta(\boldsymbol{p};\boldsymbol{h}) \equiv \lim_{\alpha \to 0^{+}} \frac{\eta(\boldsymbol{p}+\alpha\boldsymbol{h}) - \eta(\boldsymbol{p})}{\alpha}.$$

A necessary and sufficient condition for $\boldsymbol{p}^{o}$ to maximize the dual cost η is that

$$\delta^{+}\eta(\boldsymbol{p}^{o};\boldsymbol{h}) \le 0, \; \forall \boldsymbol{h} \in \boldsymbol{R}^{N}. \tag{4.3.4}$$

For convenience, let us denote the mapping of $\boldsymbol{p}$ to $\varepsilon^{T}|\boldsymbol{p}|$ by m, i.e.

$$m(\boldsymbol{p}) = \varepsilon^{T}|\boldsymbol{p}| = \sum_{i=0}^{N-1} \varepsilon_i |p_i|.$$

By inspecting the graph of the real mapping $\mu : x \to |x|$, its directional differential is easily seen to be:

$$\delta^{+}\mu(x, h) \equiv \lim_{\alpha \to 0^{+}} \frac{|x+\alpha h| - |x|}{\alpha} = \begin{cases} h, & \text{if } x > 0 \\ |h|, & \text{if } x = 0 \\ -h, & \text{if } x < 0 \end{cases}.$$

Hence, the directional derivative of the function m is

$$\begin{aligned} \delta^{+}m(\boldsymbol{p};\boldsymbol{h}) \equiv \lim_{\alpha \to 0^{+}} \frac{m(\boldsymbol{p}+\alpha\boldsymbol{h}) - m(\boldsymbol{p})}{\alpha} &= \sum_{i=0}^{N-1} \varepsilon_i \lim_{\alpha \to 0^{+}} \frac{|p_i + \alpha h_i| - |p_i|}{\alpha} \\ &= \sum_{i=0}^{N-1} \varepsilon_i \delta^{+}\mu(p_i, h_i). \end{aligned} \tag{4.3.5}$$

Since the directional differential reduces to an ordinary differential for differentiable functions, the directional differential of the dual function (4.3.2) is

$$\delta^{+}\eta(\boldsymbol{p};\boldsymbol{h}) = -\left[\tfrac{1}{2}\boldsymbol{p}^{T} Y_s Q^{-1} Y_s^{T} + \boldsymbol{d}^{T}\right]\boldsymbol{h} - \delta^{+}m(\boldsymbol{p};\boldsymbol{h}). \tag{4.3.6}$$

By partitioning the set $\tilde{\Omega}_c = \{0, 1, \ldots, N-1\}$ into disjoint sets:

$$\Lambda^+(p) \equiv \{j : p_j > 0\},$$
$$\Lambda^-(p) \equiv \{j : p_j < 0\},$$
$$\Lambda^0(p) \equiv \{j : p_j = 0\},$$

we can express (4.3.5) as

$$\delta^+ m(p;h) = \sum_{i \in \Lambda^+(p)} \varepsilon_i h_i + \sum_{i \in \Lambda^-(p)} -\varepsilon_i h_i + \sum_{i \in \Lambda^0(p)} \varepsilon_i |h_i| . \tag{4.3.7}$$

Using (4.3.7), we can express the directional derivative (4.3.6) as

$$\begin{aligned} \delta^+ \eta(p;h) = & - \sum_{i \in \Lambda^+(p)} [\Delta_i(p) + \varepsilon_i] h_i - \sum_{i \in \Lambda^-(p)} [\Delta_i(p) - \varepsilon_i] h_i \\ & - \sum_{i \in \Lambda^0(p)} [\Delta_i(p) h_i + \varepsilon_i |h_i|], \end{aligned} \tag{4.3.8}$$

where

$$\Delta_i(p) = \frac{1}{2}(Y_s Q^{-1} Y_s^T p)_i + d_i . \tag{4.3.9}$$

The set $\Lambda^0(p)$ can be further partitioned into the disjoint sets:

$$\begin{aligned} \chi^+(p) &\equiv \{ i \in \Lambda^0(p) : -[\Delta_i(p) + \varepsilon_i] > 0 \}, \\ \chi^-(p) &\equiv \{ i \in \Lambda^0(p) : -[\Delta_i(p) - \varepsilon_i] < 0 \}, \\ \chi^0(p) &\equiv \Lambda^0(p) \setminus \chi^+(p) \setminus \chi^-(p). \end{aligned} \tag{4.3.10}$$

Clearly, $-[\Delta_i(p) - \varepsilon_i] < 0$ implies that

$$-[\Delta_i(p) + \varepsilon_i] = -[\Delta_i(p) - \varepsilon_i] - 2\varepsilon_i < 0.$$

Thus, $\chi^-(p)$ and $\chi^+(p)$ are disjoint. The terms in the sum over $\Lambda^0(p)$ may further be expressed as

$$\Delta_i(p) h_i + \varepsilon_i |h_i| = \begin{cases} [\Delta_i(p) + \varepsilon_i] h_i, & \text{if } h_i \geq 0 \\ [\Delta_i(p) - \varepsilon_i] h_i, & \text{if } h_i \leq 0. \end{cases} \tag{4.3.11}$$

Observe from (4.3.10) and (4.3.11) that

$$\Delta_i(\boldsymbol{p})h_i + \varepsilon_i|h_i| \begin{cases} \le 0, & \text{if } i \in \chi^+(\boldsymbol{p}) \text{ and } h_i \ge 0 \\ \ge 0, & \text{if } i \in \chi^+(\boldsymbol{p}) \text{ and } h_i \le 0 \\ \ge 0, & \text{if } i \in \chi^-(\boldsymbol{p}) \text{ and } h_i \ge 0 \\ \le 0, & \text{if } i \in \chi^-(\boldsymbol{p}) \text{ and } h_i \le 0 \\ \ge 0, & \text{if } i \in \chi^0(\boldsymbol{p}) \end{cases}.$$

The objective is to increase η as much as possible, and any $\hat{\boldsymbol{h}}$ which achieves

$$\max_{\|\boldsymbol{h}\|_q \le 1} \delta^+\eta(\boldsymbol{p};\boldsymbol{h})$$

will be called a steepest ascent direction from the point $\boldsymbol{p}$, with respect to the q-norm, $\|\boldsymbol{x}\|_q = \sum_i |x_i|^{1/q}$ (see Lemma B.4.1, Appendix B.4). It is clear that substitution of such an $\hat{\boldsymbol{h}}$ into (4.3.8) must yield no negative terms as shown in the following. If the ith term were negative, then $\hat{h}_i$ could be set to zero, thus increasing $\delta^+\eta(\boldsymbol{p};\boldsymbol{h})$ without increasing $\|\hat{\boldsymbol{h}}\|_q$. If the ith term is zero, then the same reasoning indicates that $\hat{h}_i$ may as well be set to zero. Hence, it can be concluded from (4.3.8) that

$$\operatorname{sgn}(\hat{h}_i) = \begin{cases} -\operatorname{sgn}[\Delta_i(\boldsymbol{p}) + \varepsilon_i], & i \in \Lambda^+(\boldsymbol{p}) \\ -\operatorname{sgn}[\Delta_i(\boldsymbol{p}) - \varepsilon_i], & i \in \Lambda^-(\boldsymbol{p}) \end{cases}. \tag{4.3.12}$$

Furthermore, for $i \in \Lambda^0(\boldsymbol{p})$, taking (4.3.10) and (4.3.11) into consideration leads to

$$\operatorname{sgn}(\hat{h}_i) = \begin{cases} -\operatorname{sgn}[\Delta_i(\boldsymbol{p}) + \varepsilon_i] = +1, & i \in \chi^+(\boldsymbol{p}) \\ -\operatorname{sgn}[\Delta_i(\boldsymbol{p}) - \varepsilon_i] = -1, & i \in \chi^-(\boldsymbol{p}) \\ 0, & i \in \chi^0(\boldsymbol{p}) \end{cases}, \tag{4.3.13}$$

where the sgn function takes value from $\{-1, 0, 1\}$.

Restricting consideration to direction vectors satisfying (4.3.12) and (4.3.13), one may rewrite (4.3.8) as

$$\delta^{+}\eta(p;\hat{h}) = -\sum_{i\in\Lambda^{+}(p)\cup\chi^{+}(p)}[\Delta_i(p)+\varepsilon_i]\hat{h}_i - \sum_{i\in\Lambda^{-}(p)\cup\chi^{-}(p)}[\Delta_i(p)-\varepsilon_i]\hat{h}_i$$
$$= \sum_{i=0}^{N-1} l_i(p)\hat{h}_i = l^T(p)\hat{h}, \tag{4.3.14}$$

where

$$l_i(p) = \begin{cases} -[\Delta_i(p)+\varepsilon_i],\, i\in\Lambda^{+}(p)\cup\chi^{+}(p) \\ -[\Delta_i(p)-\varepsilon_i],\, i\in\Lambda^{-}(p)\cup\chi^{-}(p)\,. \\ 0, \qquad\qquad i\in\chi^{0}(p) \end{cases} \tag{4.3.15}$$

Note that $\mathrm{sgn}(h_i) = \mathrm{sgn}(l_i)$ for all $i\in\tilde{\Omega}_c$. It follows from (4.3.4) that

$$l(p) = 0 \iff p = p^{o}. \tag{4.3.16}$$

Lemma 4.3.1. *If* p *is any non optimal point and* p^{o} *is any optimal point, then the ascent direction* $l(p)$ *has the following properties,*

(i) $-\langle l(p), p-p^{o}\rangle \geq \eta(p^{o})-\eta(p) > 0,$

(ii) $\|l(p)\|^2 \leq 8\|\varepsilon\|^2 + \frac{1}{2}\|Y_s\|^2\|Q^{-1}Y_s^T(p-p^{o})\|^2 \leq 8\|\varepsilon\|^2 + \frac{1}{2}\|Y_s\|^4\|Q^{-1}\|^2\|(p-p^{o})\|^2,$

(iii) $-2\langle l(p), p-p^{o}\rangle \geq \left\|\sqrt{Q^{-1}}\,Y_s^T(p-p^{o})\right\|^2.$

By using (4.3.14) for the directional differential, it is not hard to determine a steepest ascent direction. In fact, any $\hat{h}$ which is aligned with $l(p)$ will do the job. The normalized steepest ascent direction is unique and given by

$$\hat{h} = \frac{l(p)}{\|l(p)\|_2}. \tag{4.3.17}$$

This is a standard result that follows from the Cauchy-Schwartz inequality.

Consider now an ascent algorithm, which uses the un-normalized version of (4.3.17):

$$p_{k+1} = p_k + \alpha_k l(p_k),\; k = 0, 1, 2, \ldots$$

where $l(\bullet)$ is defined as before in (4.3.15), α_k is the stepsize, and p_0 is any convenient starting point (any starting point such as $p_0 = 0$ will do, but convergence can be hastened by making the components of p_0 non-zero at those points where the constraints are likely to be active). Various stepsize selection algorithms, such as any one-dimensional search [83] or the Armijo method [84], are available. Unfortunately, when these algorithms are used with a non-differentiable cost, convergence cannot always be guaranteed [43]. Convergence can be established, however, if the stepsize are chosen so as to satisfy the following conditions.

Theorem 4.3.1. *If the step size α_k satisfies*

(i) $\alpha_k > 0, k = 1, 2, \ldots,$

(ii) $\sum_{k=1}^{\infty} \alpha_k^2 < \infty,$

(iii) $\sum_{k=1}^{\infty} \alpha_k = \infty,$

then the sequence $\{p_k\}_{k=1}^{\infty}$ of iterates converges to an optimal p^o, i.e.

$$\lim_{k \to \infty} \|p_k - p^o\| = 0.$$

Moreover, the corresponding sequence $\{a_k\}_{k=1}^{\infty}$ of filters, where $a_k \equiv -0.5 Y_s^T p_k$, converges to the optimal filter a^o, i.e.

$$\lim_{k \to \infty} \|a_k - a^o\| = 0.$$

The next result establishes the convergence properties of the algorithm when the stepsize is fixed at $\alpha_k = \alpha$. This case turns out to be very important in the adaptive version of the algorithm because fixed step size is most easily implemented in practice.

Theorem 4.3.2. *Suppose that the step size is fixed at*

$$\alpha_k = \alpha < \frac{2}{\|Y_s\|^2},$$

where $\|Y_s\|$ *is the spectral norm of the matrix* Y_s. *Then, the resulting sequence* $\{a_k\}_{k=1}^{\infty}$ *of filters enters a neighbourhood of the optimal filter* a^o *after a finite number of iterations. More precisely,*

$$\limsup_{k \to \infty} \|a_k - a^o\|^2 \le V(\alpha) < \infty,$$

and $\lim_{\alpha \to 0} V(\alpha) = 0$.

The objective here is to form an iterative envelope constrained filter by imbedding the steepest ascent algorithm in a structure of the type shown in Figure 4.3.1. Since the optimal filter is obtained from (4.3.3), it makes sense to define the filter at the kth iteration of the algorithm as

$$a_k = -\frac{1}{2}Q^{-1}Y_s^T p_k.$$

The filter parameter update equation is

$$a_{k+1} = a_k - \frac{1}{2}\alpha_k Q^{-1} Y_s^T l(p_k),$$

and the noiseless filter output is

$$\psi_k = Y_s a_k = -\frac{1}{2}Y_s Q^{-1} Y_s^T p_k. \tag{4.3.18}$$

The steepest ascent direction $l(p)$ defined in (4.3.18) can then be written in terms of the filter output as follows:

$$l_i(p) = \begin{cases} \psi_i - \varepsilon_i^+, i \in \Lambda^+(p) \cup \chi^+(p) \\ \psi_i - \varepsilon_i^-, i \in \Lambda^+(p) \cup \chi^-(p) \\ 0, \quad\quad i \in \chi^0(p) \end{cases}$$

$$= \begin{cases} \psi_i - \varepsilon_i^+, \text{ if } p_i > 0 \text{ or if } p_i = 0 \text{ and } \psi_i > \varepsilon_i^+ \\ \psi_i - \varepsilon_i^-, \text{ if } p_i < 0 \text{ or if } p_i = 0 \text{ and } \psi_i < \varepsilon_i^- \\ 0, \quad\quad \text{if } p_i = 0, \text{ and } \varepsilon_i^- < \psi_i < \varepsilon_i^+ \end{cases}$$

This means that $l(p_k)$ can be determined by passing the signal s through the filter and then comparing the resulting output ψ_k with the constraint boundaries ε^+ and ε^-. Note that $l(p_k)$ is some sort of generalized error vector. It is given by the difference between the output and the upper envelope boundary whenever the Lagrange multiplier is positive; the difference between the output and the lower envelope boundary whenever the Lagrange multiplier is negative; and the constraint violation (upper, lower or none) whenever the Lagrange multiplier is zero.

4.4 Iterative Algorithm using Augmented Cost

Although, the primal-dual algorithm is simple to implement, performance-wise, convergence is slow. In this section, an augmented cost is constructed by composing g_υ (referred to as a penalty allocator) with each constraint function and then added to the original cost. The sub-optimum solution is found by minimizing the augmented cost with descent algorithms, such as steepest descent (SD) and Newton-Raphson (NR) (see [116], [117], [132], [133], [138]). The criteria for choosing g_υ to ensure global convergence are presented. For the NR method with an appropriate line search, a quadratic convergence rate can be achieved.

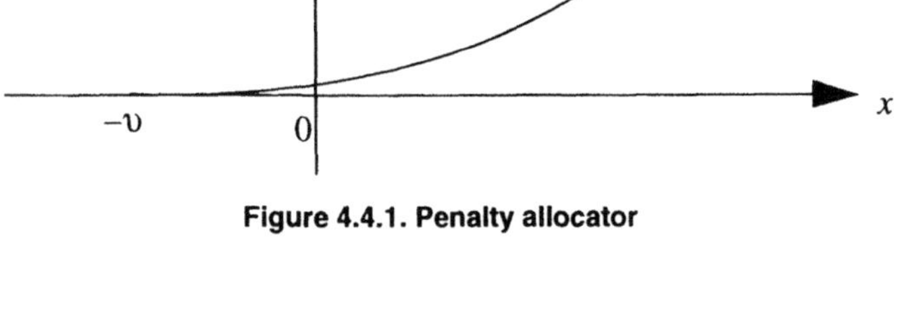

Figure 4.4.1. Penalty allocator

For each instant $j \in \tilde{\Omega}_c = \{0, \ldots, N-1\}$, the upper and lower constraint violations ϕ_j^+, ϕ_j^- are defined by comparing the filter output with the mask values at that instant,

$$\phi_j^+(\boldsymbol{a}) \equiv \boldsymbol{y}_s^T(j)\boldsymbol{a} - \varepsilon^+(j), \quad (4.4.1)$$

$$\phi_j^-(\boldsymbol{a}) \equiv \varepsilon^-(j) - \boldsymbol{y}_s^T(j)\boldsymbol{a}. \quad (4.4.2)$$

It is clear from these definitions that if either ϕ_j^+ or ϕ_j^- is positive, then the filter violates the output constraint at the jth instant. For any $\upsilon > 0$, the penalty function is defined as

$$G_\upsilon(\boldsymbol{a}) \equiv \sum_{j=0}^{N-1} [g_\upsilon(\phi_j^+(\boldsymbol{a})) + g_\upsilon(\phi_j^-(\boldsymbol{a}))], \quad (4.4.3)$$

where the penalty allocator g_υ is any continuous and convex function identically zero on $(-\infty, -\upsilon]$, but increasing otherwise (see Figure 4.4.1).

Contrary to conventional penalty methods in which all feasible points are assigned zero penalty (see e.g. [27], [37], [65], [66]), this penalty scheme does not. In fact, we can show that:

Lemma 4.4.1. *Given any* υ, *if* $G_\upsilon(\boldsymbol{a}) \le g_\upsilon(0)$, *then* $\varepsilon^-(j) \le \boldsymbol{y}_s^T(j)\boldsymbol{a} \le \varepsilon^+(j)$, $\forall j \in \tilde{\Omega}_c$.

For a given $\upsilon > 0$, the set with zero penalty

$$\mathcal{F}_\upsilon \equiv \{\boldsymbol{a} \in \boldsymbol{R}^n\colon G_\upsilon(\boldsymbol{a}) = 0\}$$

can be written as

$$\mathcal{F}_\upsilon = \{\boldsymbol{a} \in \boldsymbol{R}^n\colon \varepsilon^-(j) + \upsilon \le \boldsymbol{y}_s^T(j)\boldsymbol{a} \le \varepsilon^+(j) - \upsilon, j \in \tilde{\Omega}_c\}.$$

Clearly, given any $\upsilon > 0$, $\mathcal{F}_\upsilon$ is not equal to

$$\mathcal{F} \equiv \{\boldsymbol{a} \in \boldsymbol{R}^n\colon \varepsilon^-(j) \le \boldsymbol{y}_s^T(j)\boldsymbol{a} \le \varepsilon^+(j), j \in \tilde{\Omega}_c\},$$

the set of all feasible filters. However, intuitively, as υ tends to zero $\mathcal{F}_\upsilon$ tends towards $\mathcal{F}$.

For each $\upsilon, \gamma > 0$, an augmented cost function $f_{\upsilon,\gamma}$ is defined as

$$f_{\upsilon,\gamma}(\boldsymbol{a}) \equiv \boldsymbol{a}^T Q \boldsymbol{a} + \gamma G_{\upsilon}(\boldsymbol{a}). \tag{4.4.4}$$

The scalars υ and γ are called the accuracy and penalty parameters, respectively. The analytical properties of the augmented cost function are given below.

Lemma 4.4.2. *The augmented cost $f_{\upsilon,\gamma}$ is continuous and strictly convex, and hence has a unique minimum. Moreover;*

(i) *If g_{υ} is once continuously differentiable, then $f_{\upsilon,\gamma}$ is once continuously differentiable and its gradient is given by*

$$\nabla f_{\upsilon,\gamma}(\boldsymbol{a}) = 2Q\boldsymbol{a} + \gamma \sum_{j=0}^{N-1} [g_{\upsilon}'(\phi_j^+(\boldsymbol{a})) - g_{\upsilon}'(\phi_j^-(\boldsymbol{a}))] \boldsymbol{y}_s(j).$$

(ii) *If g_{υ} is twice continuously differentiable, then $f_{\upsilon,\gamma}$ is twice continuously differentiable with positive definite Hessian, given by*

$$\nabla^2 f_{\upsilon,\gamma}(\boldsymbol{a}) = 2Q + \gamma \sum_{j=0}^{N-1} [g_{\upsilon}''(\phi_j^+(\boldsymbol{a})) + g_{\upsilon}''(\phi_j^-(\boldsymbol{a}))] \boldsymbol{y}_s(j) \boldsymbol{y}_s^T(j).$$

(iii) *If g_{υ} has bounded second derivative, i.e. $|g_{\upsilon}''(x)| \le h_{\upsilon}$, $\forall x \in \boldsymbol{R}$, then $f_{\upsilon,\gamma}$ also has bounded Hessian, i.e.*

$$2\|Q\| \le \|\nabla^2 f_{\upsilon,\gamma}(\boldsymbol{a})\| \le 2\|Q\| + 2\gamma h_{\upsilon} \sum_{j=0}^{N-1} \|\boldsymbol{y}_s(j)\|^2.$$

(iv) *If g_{υ} has Lipschitz continuous second derivative, i.e. there exists some $l_{\upsilon} > 0$, such that $|g_{\upsilon}''(x) - g_{\upsilon}''(y)| \le l_{\upsilon}|x - y|$, $\forall x, y \in \boldsymbol{R}$, then $\nabla^2 f_{\upsilon,\gamma}(\boldsymbol{a})$ is also Lipschitz continuous:*

$$\|\nabla^2 f_{\upsilon,\gamma}(\boldsymbol{u}) - \nabla^2 f_{\upsilon,\gamma}(\boldsymbol{v})\| \le 2\gamma l_{\upsilon} \|\boldsymbol{u} - \boldsymbol{v}\| \sum_{j=0}^{N-1} \|\boldsymbol{y}_s(j)\|^3.$$

These properties allow the augmented cost to be efficiently minimized by descent direction based algorithms, such as SD or NR, by choosing an appropriate penalty allocator g_{υ}. But first, let us demonstrate how the augmented cost (4.4.4) can be used to solve the EC filtering Problem (4.1.1).

Approximation results

Suppose that $\boldsymbol{a}^o$ is the solution to Problem (4.1.1). Then, $\boldsymbol{a}^o$ can be approximated by $\boldsymbol{a}^o_{\upsilon,\gamma}$ defined for $\upsilon, \gamma > 0$ as the minimizer of $f_{\upsilon,\gamma}$, i.e.

$$\boldsymbol{a}^o_{\upsilon,\gamma} = \arg\min f_{\upsilon,\gamma}.$$

To justify the approximation, we have the following results.

Theorem 4.4.1. *Let $\boldsymbol{a}^o$ denote the optimum filter coefficient vector, and $\boldsymbol{a}^o_{\upsilon,\gamma}$ denote the minimum of $f_{\upsilon,\gamma}$. Suppose υ is such that $\mathcal{F}_\upsilon \neq \varnothing$. If $\gamma > \boldsymbol{a}_\upsilon^T Q \boldsymbol{a}_\upsilon / g_\upsilon(0)$ for some $\boldsymbol{a}_\upsilon \in \mathcal{F}_\upsilon$, then $\boldsymbol{a}^o_{\upsilon,\gamma}$ satisfies the envelope constraints, i.e.*

$$\varepsilon^-(j) \le \boldsymbol{y}_s^T(j)\boldsymbol{a}^o_{\upsilon,\gamma} \le \varepsilon^+(j), \ \forall j \in \tilde{\Omega}_c.$$

This result implies that for a fixed but small υ, a sufficiently large γ forces the solution of the approximate problem into the feasible region of the EC filtering Problem (4.1.1). This is the principle advantage over other penalty methods, where feasibility is only guaranteed as γ tends to infinity.

Theorem 4.4.2. *Let $\boldsymbol{a}^o$ denote the optimum filter coefficients, $\boldsymbol{a}^o_{\upsilon,\gamma}$ denote the minimum of $f_{\upsilon,\gamma}$, and $\|\boldsymbol{a}\|_Q$ denote the Q-norm $\sqrt{\boldsymbol{a}^T Q \boldsymbol{a}}$. For any $\bar{\boldsymbol{a}} \in \mathcal{F}^o$, define*

$$\delta(\bar{\boldsymbol{a}}) \equiv \min_{j \in \Omega_c} [\min\{-\phi_j^+(\bar{\boldsymbol{a}}), -\phi_j^-(\bar{\boldsymbol{a}})\}]/2\|\bar{\boldsymbol{a}}\|_Q^2.$$

(i) If $0 < \upsilon \le \delta(\bar{\boldsymbol{a}})\min\{\varepsilon, 2\|\bar{\boldsymbol{a}}\|_Q^2\}$ and γ is chosen as in Theorem 4.4.1, then

$$0 \le \|\boldsymbol{a}^o_{\upsilon,\gamma}\|_Q^2 - \|\boldsymbol{a}^o\|_Q^2 \le \varepsilon.$$

(ii) If for each υ, γ is chosen as in part (i), then $\lim_{\upsilon \to 0} \boldsymbol{a}^o_{\upsilon,\gamma} = \boldsymbol{a}^o$.

Proof. The proof is similar to that for the continuous-time case in Chapter 5.

Provided that for each υ, γ is chosen according to Theorem 4.4.1 part (ii) shows that the approximate solution $\boldsymbol{a}^o_{\upsilon,\gamma}$ converges to the true solution $\boldsymbol{a}^o$ as υ tends to zero. Part (i) asserts that if a feasible point $\boldsymbol{a}$ is known, then for any given error bound ε, the accuracy parameter υ can be calculated (without using any information on the solutions $\boldsymbol{a}^o_{\upsilon,\gamma}$ and $\boldsymbol{a}^o$) so that the noise gain deviation

$$\mathrm{NGD} \equiv \|\boldsymbol{a}^o_{\upsilon,\gamma}\|^2_Q - \|\boldsymbol{a}^o\|^2_Q, \tag{4.4.5}$$

is less than the error bound ε.

Update equations

By choosing g_υ appropriately, globally convergent algorithms for minimizing the augmented cost (4.4.4) can be developed. To obtain sub-optimum solutions, the following iterative process is used:

$$\boldsymbol{x}_{k+1} = \boldsymbol{x}_k + \mu_k \boldsymbol{d}_k$$

where $\boldsymbol{x}_k$ is the kth iterate, μ_k is the stepsize, and $\boldsymbol{d}_k$ satisfying $\nabla f^T_{\upsilon,\gamma}(\boldsymbol{x}_k)\boldsymbol{d}_k < 0$ is called the descent direction.

Assuming that g_υ is at least once continuously differentiable, the simplest of the descent directions is the steepest descent (SD) direction,

$$\boldsymbol{d}_k = -\nabla f_{\upsilon,\gamma}(\boldsymbol{x}_k). \tag{4.4.6}$$

The popularity of this method has been favoured due to the comparatively simple implementation and applicability to a wide range of functions.

Theorem 4.4.3. *Suppose that g_υ is twice continuously differentiable for each $\upsilon > 0$ with $|g''_\upsilon(x)| \le h_\upsilon$, $\forall x \in \boldsymbol{R}$. Then, starting from any initial point $\boldsymbol{x}_0$, the sequence $\{\boldsymbol{x}_k\}_{k=0}^{\infty}$ generated by the SD algorithm with constant stepsize μ converges to $\boldsymbol{x}^o_{\upsilon,\gamma}$ if*

$$0 < \mu < \frac{1}{\|Q\| + \gamma h_\upsilon \sum_{j=0}^{N-1} \|\boldsymbol{y}_s(j)\|^2}.$$

Proof. Using Lemma 4.4.2(iii), the bounds on the Hessian of the augmented cost functional (4.4.4) are obtained. Then the conclusion follows readily from Theorem 4.8.1 (see Section 4.8). ❒

The SD algorithm is attractive to a large extent due to its computational simplicity. However, the method has several serious practical drawbacks. Most importantly, convergence is usually slow (see [17], [18], [25], [27], [37], [90]). Another popular method is the Newton-Raphson (NR) algorithm, which uses the second derivative to calculate the descent direction,

$$\boldsymbol{d}_k = -(\nabla^2 f_{\upsilon,\gamma}(\boldsymbol{x}_k))^{-1}\nabla f_{\upsilon,\gamma}(\boldsymbol{x}_k). \tag{4.4.7}$$

It is assumed that g_υ is chosen so that the augmented cost functional (4.4.4) is at least twice continuously differentiable. Since twice continuously differentiable strictly convex functions have positive definite Hessian (see [90]), the Hessian of (4.4.4) is positive definite. A larger stepsize than that for the SD algorithm is allowed as illustrated in the following theorem.

Theorem 4.4.4. *Suppose that g_υ is twice continuously differentiable for each $\upsilon > 0$ with $|g_\upsilon''(x)| \le h_\upsilon$, $\forall x \in \boldsymbol{R}$. Then, starting from any initial point $\boldsymbol{x}_0$, the sequence $\{\boldsymbol{x}_k\}_{k=0}^{\infty}$ generated by the NR algorithm with constant stepsize μ converges to $\boldsymbol{x}_{\upsilon,\gamma}^{o}$ if*

$$0 < \mu < \frac{2}{\|Q\| + \gamma h_\upsilon \sum_{j=0}^{N-1} \|\boldsymbol{y}_s(j)\|^2}.$$

Proof. Again, using Lemma 4.4.2 (iv), the bounds on the Hessian of the augmented cost function are obtained. Then, the conclusion follows readily from Theorem 4.8.2 (see Section 4.8). ❒

Considerably faster convergence can be achieved with the NR method by performing some line search at each iteration to determine the stepsize (an iteration in this context refers to a search direction evaluation). For algorithms with constant stepsize, each search direction evaluation is followed by a single filter update, while algorithms with line search require at least one update to determine the stepsize before the next search direction is calculated. A popular form of line search involves the Goldstein condition (see [18], [27], [37], [90]),

$$f_{\upsilon,\gamma}(\boldsymbol{x}_{k+1}) - f_{\upsilon,\gamma}(\boldsymbol{x}_k) < \alpha\mu_k \nabla f^T_{\upsilon,\gamma}(\boldsymbol{x}_k)\boldsymbol{d}_k,\ \alpha \in (0, 0.5), \tag{4.4.8}$$

and the Wolfe-Powell condition (see [18], [27]),

$$\nabla f^T_{\upsilon,\gamma}(\boldsymbol{x}_k + t_k\boldsymbol{d}_k)\boldsymbol{d}_k \geq \beta \nabla f^T_{\upsilon,\gamma}(\boldsymbol{x}_k)\boldsymbol{d}_k,\ \beta \in (\alpha, 1). \tag{4.4.9}$$

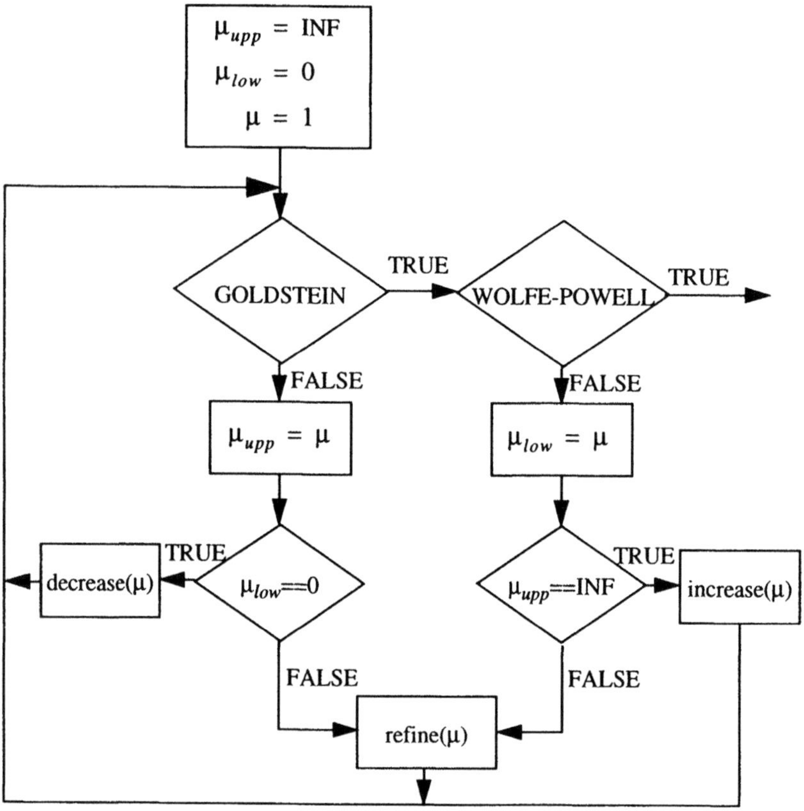

Figure 4.4.2. Flow chart for line search

There are four possible scenarios (see Figure 4.4.2). If $\mu = 1$ satisfies both the Goldstein (4.4.8) and Wolfe-Powell (4.4.9) conditions, then $x_{k+1} = x_k + d_k$, and no further line search calculations are performed.

If $\mu = 1$ is not admissible because it fails the Goldstein condition, then μ will be decreased. This is most often and effectively done by setting the next stepsize to $\max(\mu_m, c\mu)$, where c is typically 0.1 and μ_m is the one-dimensional minimizer of the one-dimensional quadratic approximation $q(\mu)$ to $f_{\upsilon,\gamma}(x_k + \mu d_k)$ that interpolates $f_{\upsilon,\gamma}(x_k)$, $\nabla f^T_{\upsilon,\gamma}(x_k)d_k$ and $f_{\upsilon,\gamma}(x_k + \mu d_k)$, i.e.

$$\mu_m = \frac{-\mu^2 \nabla f^T_{\upsilon,\gamma}(x_k)d_k}{2(f_{\upsilon,\gamma}(x_k + \mu d_k) - f_{\upsilon,\gamma}(x_k) - \mu \nabla f^T_{\upsilon,\gamma}(x_k)d_k)}.$$

This `Decrease(μ)` routine can be repeated if the new stepsize continues to fail the Goldstein condition.

Alternatively, if the stepsize μ satisfies the Goldstein condition but not the Wolfe-Powell condition, it will be increased. Generally, a simple routine such as `Increase(μ)` $= \mu/\rho$, where $\rho \in (0, 1)$ is used although more sophisticated strategies are possible.

After some iterations of `Increase(μ)` or `Decrease(μ)`, either an admissible stepsize is found or the last two values μ_{low} and μ_{upp} of μ, which have been tried bracketed an acceptable stepsize. That is, μ_{low} satisfies Goldstein condition but not the Wolfe-Powell, while μ_{upp} fails the Goldstein condition. In this case, an admissible stepsize must be in the open interval (μ_{low}, μ_{upp}) and is selected by the refine phase. A simple routine such as `Refine`$(\mu, \mu_{low}, \mu_{upp}) = (\mu_{low} + \mu_{upp})/2$ is used here, although there are others.

If g_υ is continuously differentiable, then $f_{\upsilon,\gamma}$ is continuously differentiable and bounded below. Hence, there are $0 < \mu_1 < \mu_2$ such that for any $\mu_k \in [\mu_1, \mu_2]$, $x_{k+1} = x_k + \mu_k d_k$ satisfies both the Goldstein and the Wolfe-Powell conditions (see [18]).

Theorem 4.4.5*. Suppose that g_υ is twice continuously differentiable for each $\upsilon > 0$ with $|g_\upsilon''(x)| \le h_\upsilon$. If $|g_\upsilon''(x) - g_\upsilon''(y)| \le l_\upsilon|x-y|$, $\forall x, y \in \mathbf{R}$, i.e. g_υ'' is Lipschitz continuous, then from any initial point $\mathbf{x}_0$, the sequence $\{\mathbf{x}_k\}_{k=0}^{\infty}$, generated by the NR algorithm with stepsize μ_k chosen by the above line search, converges to $\mathbf{x}_{\upsilon,\gamma}^{o}$ at a quadratic rate.*

Proof. From Lemma 4.4.2 (iii), (iv), the Hessian of the augmented cost function is bounded and satisfies the Lipschitz condition. Thus, applying Theorem 4.8.3 (see Section 4.8) completes the proof. ❒

To find the minimizer $\mathbf{a}_{\upsilon,\gamma}^{o}$ of $f_{\upsilon,\gamma}$ by iterative methods such as SD or NR, the gradient $\nabla f_{\upsilon,\gamma}$ and the Hessian $\nabla^2 f_{\upsilon,\gamma}$ (for NR) are required, the existence of these depends on the choice of the penalty allocator g_υ. Below are some examples of penalty allocators.

$$g_\upsilon(x) = \begin{cases} 0, & x \le -\upsilon \\ (x+\upsilon)^2/4\upsilon, & -\upsilon \le x \le \upsilon \,. \\ x, & x \ge \upsilon \end{cases} \tag{4.4.10}$$

Penalty allocator (4.4.10) generates the approximate problem used in [50], [111], [138]. Note that (4.4.10) is continuously differentiable but has two discontinuities in the second derivative leading to difficulties in convergence analysis.

One of the simplest smooth penalty allocators is given by (4.4.11). This penalty allocator is also continuously differentiable but has one discontinuity in the second derivative. Thus, it suffers from the same analytical drawback as (4.4.10), although, in practice, it displays a surprising improvement in performance over (4.4.10).

$$g_\upsilon(x) = \begin{cases} 0, & x \le -\upsilon \\ (x+\upsilon)^2/4\upsilon, & -\upsilon \le x \end{cases} \,. \tag{4.4.11}$$

Penalty allocator (4.4.12) is constructed to emulate (4.4.11), while eliminating the discontinuity in the second derivative. This penalty allocator also has Lipschitz continuous and bounded second derivative.

$$g_\upsilon(x) = \begin{cases} 0, & x \le -\upsilon \\ -\frac{\upsilon}{\pi}\cos\left(\frac{\pi x}{2\upsilon}\right) + \frac{(x+\upsilon)}{2}, & -\upsilon \le x \le 0 \\ \frac{\pi x^2}{8\upsilon} + \frac{x}{2} + \upsilon\left(\frac{1}{2} - \frac{1}{\pi}\right) & x \ge 0 \end{cases} \tag{4.4.12}$$

Lemma 4.4.2 shows how the gradient and the Hessian of (4.4.4) and hence, the search directions (4.4.6) or (4.4.7) are obtained for each iterate $\boldsymbol{a}_k$ from the corresponding filter output and state variables. Thus, the iterative filter updating procedure:

$$\boldsymbol{a}_{k+1} = \boldsymbol{a}_k + \mu \boldsymbol{d}_k$$

can be easily performed.

The computational load for the NR direction can be reduced by using the matrix inversion lemma in a similar fashion to that presented in [138] to calculate the inverse of the Hessian. Let

$$H_{-1}^{-1}(\boldsymbol{a}) = \frac{1}{2}Q^{-1},$$

and for $j = 0, 1, \ldots, N-1$,

$$H_j^{-1}(\boldsymbol{a}) = H_{j-1}^{-1}(\boldsymbol{a}) - \frac{H_{j-1}^{-1}(\boldsymbol{a})\boldsymbol{y}_s(j)\boldsymbol{y}_s^T(j)H_{j-1}^{-1}(\boldsymbol{a})}{\boldsymbol{y}_s^T(j)H_{j-1}^{-1}(\boldsymbol{a})\boldsymbol{y}_s(j) + \dfrac{1}{\gamma[g_\upsilon''(\phi_j^+(\boldsymbol{a})) + g_\upsilon''(\phi_j^-(\boldsymbol{a}))]}}.$$

Then,

$$H_{N-1}^{-1}(\boldsymbol{a}) = [\nabla^2 f_{\upsilon,\gamma}(\boldsymbol{a})]^{-1}.$$

This reduces the computational load of $[\nabla^2 f_{\upsilon,\gamma}(\boldsymbol{a})]^{-1}$ considerably. Alternatively, Quasi-Newton methods (see [17], [18], [25], [27], [37], [90]) can be adopted to minimize the augmented cost (4.4.4). In this case, the convergence rate would be superlinear.

4.5 Tapped Delay Line FIR Filters

This section examines the discrete-time EC filtering problem where the input signal and the filter impulse response are finite sequences. In particular, two practical examples concerning pulse compression and channel equalization are presented to compare the convergence performance of the three penalty allocators (4.4.10), (4.4.11) and (4.4.12). This problem, first studied in [23] and solved in [22] by the primal-dual method, is perhaps the simplest version of the EC filtering problem computationally and practically. Furthermore, it is a sound platform for understanding the more general problem.

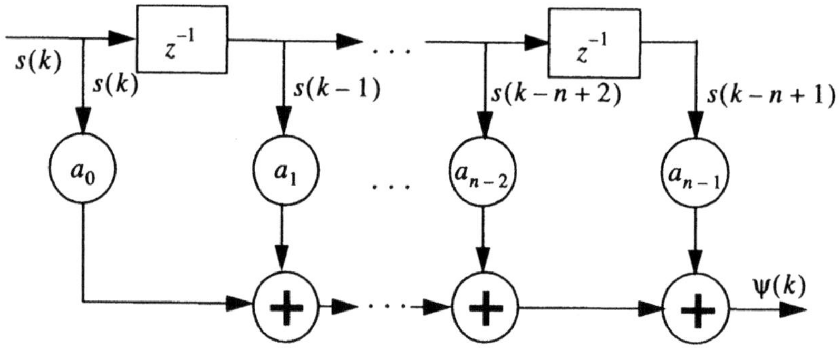

Figure 4.5.1. A tapped delay line FIR filter.

The time invariant finite impulse response (FIR) filter can be realized by the structure shown in Figure 4.5.1. The transfer function for this filter is

$$U(z) = \sum_{k=0}^{n-1} a_k z^{-k}.$$

Hence, the finite impulse response is given by

$$u(k) = a_k,\ k = 0, \ldots, n-1.$$

Suppose that the input signal s has length m, (i.e. $s(k) = 0, \forall k \geq m$). Then, the output has length $m+n-1$ and the linear operator $\Xi_s K: \boldsymbol{R}^n \rightarrow \boldsymbol{R}^{m+n-1}$ can be represented by

$$S \equiv \begin{bmatrix} y_s^T(0) \\ y_s^T(1) \\ \vdots \\ y_s^T(m+n-2) \end{bmatrix} = \begin{bmatrix} s(0) & 0 & \dots & 0 \\ \vdots & s(0) & & \vdots \\ s(m-1) & \vdots & \ddots & 0 \\ 0 & s(m-1) & & s(0) \\ \vdots & & \ddots & \vdots \\ 0 & \dots & 0 & s(m-1) \end{bmatrix}. \tag{4.5.1}$$

The matrix S is normally referred to as the convolution matrix of the signal s. Note that the columns of S are linearly independent if the input s is not identically zero. Hence, for $s \neq 0$, the response space is an n-dimensional subspace of $\boldsymbol{R}^{m+n-1}$. For white input noise, the EC filtering problem for FIR filter can be posed as

$$\min \ \boldsymbol{a}^T\boldsymbol{a}, \boldsymbol{a} \in \boldsymbol{R}^n$$
$$\text{subject to } \ \varepsilon^- \leq S\boldsymbol{a} \leq \varepsilon^+.$$

Lemma 2.5.1 asserts that the mapping $\Xi_s K$, represented by S, has an inverse

$$(\Xi_s K)^{-1} : \boldsymbol{R}^{m+n-1} \to \boldsymbol{R}^n.$$

This inverse is also a linear operator on an n-dimensional space and can be represented by the pseudo-inverse matrix

$$S^+ = S^T(SS^T)^{-1}.$$

Consider the case where $s \neq 0$ and $\tilde{\Omega}_c = \tilde{\Omega}_\psi$, i.e. $N = m + n - 1$. Let

$$\boldsymbol{r} = [r(0), \dots, r(N-1)]^T, \ r(j) \equiv \max\{|\varepsilon^-(j)|, |\varepsilon^+(j)|\}. \tag{4.5.2}$$

Then, for any feasible filter $\boldsymbol{a}$, $|y_s^T(j)\boldsymbol{a}| \leq r(j)$. Consequently, $\|S\boldsymbol{a}\| \leq \|\boldsymbol{r}\|$ and hence,

$$\|\boldsymbol{a}\| \leq \|(\Xi_s K)^{-1}\| \|S\boldsymbol{a}\| \leq \|S^+\| \|\boldsymbol{r}\|,$$

where $\|S^+\|$ is the maximum singular value of S^+, or equivalently, the reciprocal of the minimum singular value of S, denoted as $\sigma(S)$. Using Theorem 4.4.1, a simple criterion for estimating the penalty parameter is obtained.

Corollary 4.5.1. *Suppose* $\upsilon > 0$ *is such that* $\mathcal{F}_\upsilon \neq \emptyset$. *Then,* $a^o_{\upsilon,\gamma}$ *is feasible if*

$$\gamma \geq \frac{\|r\|^2}{g_\upsilon(0)\sigma(S)^2}.$$

Note also that for FIR filters, $\sum_{j=0}^{N-1} \|y_s(j)\|^2$ is simply $n\|s\|^2$, i.e. n times the energy of the input s. Hence, using penalty allocator (4.4.12), convergence is assured for the SD algorithm with constant stepsize μ, if

$$0 < \mu < (\|Q\| + \gamma h_\upsilon n\|s\|^2)^{-1},$$

and for the NR algorithm, μ can be twice as large.

Pulse Compression

The first example involves the compression of a 13-bit Barker-coded signal given by

$$s = [1, 1, 1, 1, 1, -1, -1, 1, 1 - 1, 1, -1, 1]^T \in \boldsymbol{R}^{13}.$$

Using a filter of 27 coefficients and an output mask with sidelobe levels of ± 0.025 and a mainlobe peak of 0.69 ± 0.075, i.e.

$$\varepsilon^+ = [\underbrace{0.025, \ldots, 0.025}_{19}, 0.765, \underbrace{0.025, \ldots, 0.025}_{19}]^T \in \boldsymbol{R}^{39},$$

$$\varepsilon^- = [\underbrace{-0.025, \ldots, -0.025}_{19}, 0.615, \underbrace{-0.025, \ldots, -0.025}_{19}]^T \in \boldsymbol{R}^{39},$$

the optimum EC filter's (obtained by QP via active set strategy) response to the 13-bit Barker-coded input is shown in Figure 4.5.2.

Table 4.5.1 summarizes the results of applying the NR algorithm with line search to minimize the augmented costs generated by various penalty allocators with $\upsilon = 1.5 \times 10^{-5}$ and γ chosen according to Corollary 4.5.1. The sub-optimum solutions for the various penalty allocators are also shown in Figure 4.5.2. Observe that there is no visible difference between the optimal and sub-optimal cases. Note that the relative NGD is defined as the ratio of the NGD (see (4.4.5)) to the noise gain of the optimum EC filter.

From Table 4.5.1 and Figure 4.5.3, observe that only a few iterations are required for the penalty allocators (4.4.11) and (4.4.12), while (4.4.10) takes considerably many more iterations and line search calculations. In all, this is an improvement over the primal-dual algorithm, which takes several thousand iterations to achieve similar precision (see [138]).

Table 4.5.1. Simulation results for 13-bit Barker-coded signal example

Penalty allocator	Penalty parameter	Filter norm	NGD	Relative NGD	Function evaluation	Search direction evaluation
(4.4.10)	$1.5642\text{x}10^4$	0.1719	$2.0767\text{x}10^{-6}$	$7.0281\text{x}10^{-5}$	203	73
(4.4.11)	$1.5642\text{x}10^4$	0.1719	$2.0767\text{x}10^{-6}$	$7.0281\text{x}10^{-5}$	3	3
(4.4.12)	$2.1523\text{x}10^4$	0.1719	$2.0724\text{x}10^{-6}$	$7.0135\text{x}10^{-5}$	15	15

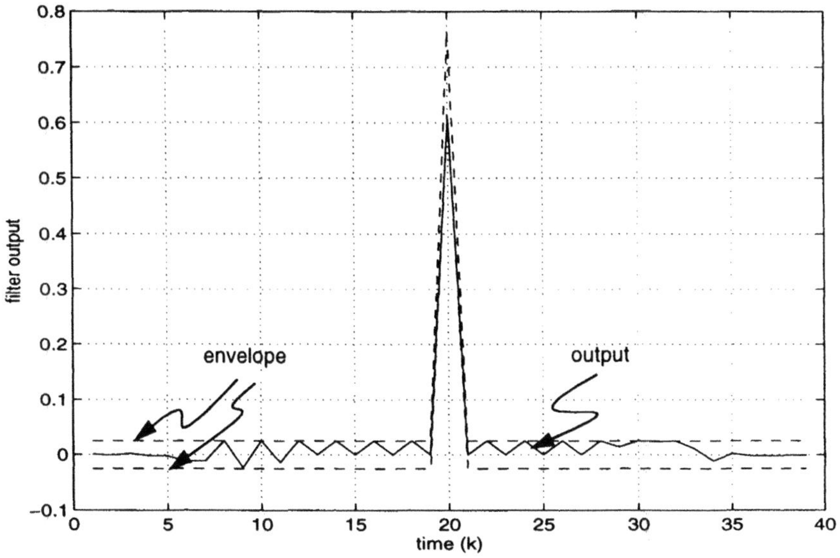

Figure 4.5.2. Optimum EC filter output for a 13-bit Barker-coded input

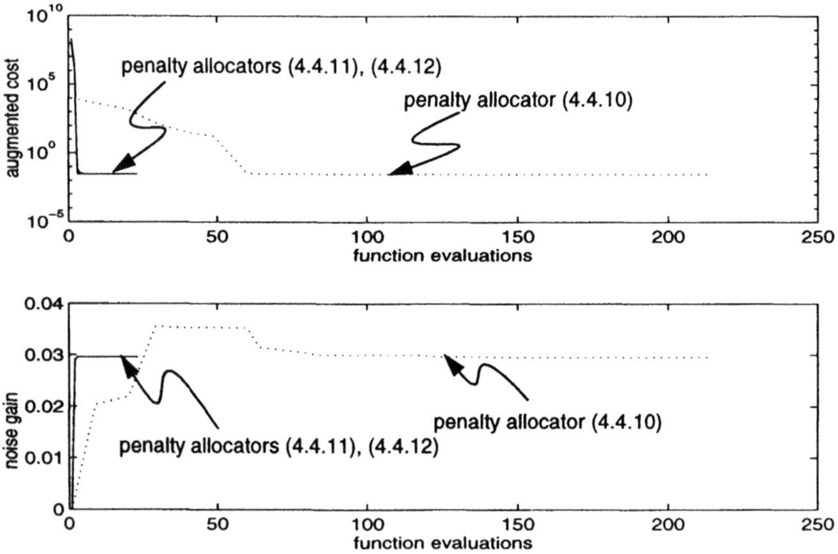

Figure 4.5.3. Augmented cost and noise gain of Barker coded example

The second example considers the compression of a rectangular pulse given by $s = [1, \ldots, 1]^T \in \boldsymbol{R}^{10}$. Using a filter with $n = 21$ coefficients and an output mask with an allowable sidelobe level of ± 0.25 and mainlobe peak of 1 ± 0.15, i.e.

$$\varepsilon^+ = [\underbrace{0.25, \ldots, 0.25}_{14}, 1.15, \underbrace{0.25, \ldots, 0.25}_{15}]^T \in \boldsymbol{R}^{30},$$

$$\varepsilon^- = [\underbrace{-0.25, \ldots, -0.25}_{14}, 0.85, \underbrace{-0.25, \ldots, -0.25}_{15}]^T \in \boldsymbol{R}^{30}.$$

The coefficients of the 'exact' optimal EC filter $\boldsymbol{a}^o$ (obtained by QP from the MATLAB optimization toolbox) and the response of the optimal filter due to the rectangular pulse are shown in Figure 4.5.4. We use $\upsilon = 2 \times 10^{-5}$ and the penalty parameter is calculated from Corollary 4.5.1. From Table 4.5.1 and Figure 4.5.5, observe that only a few iterations are required for the penalty allocators (4.4.11) and (4.4.12), while (4.4.10) takes considerably many more iterations and line search calculations.

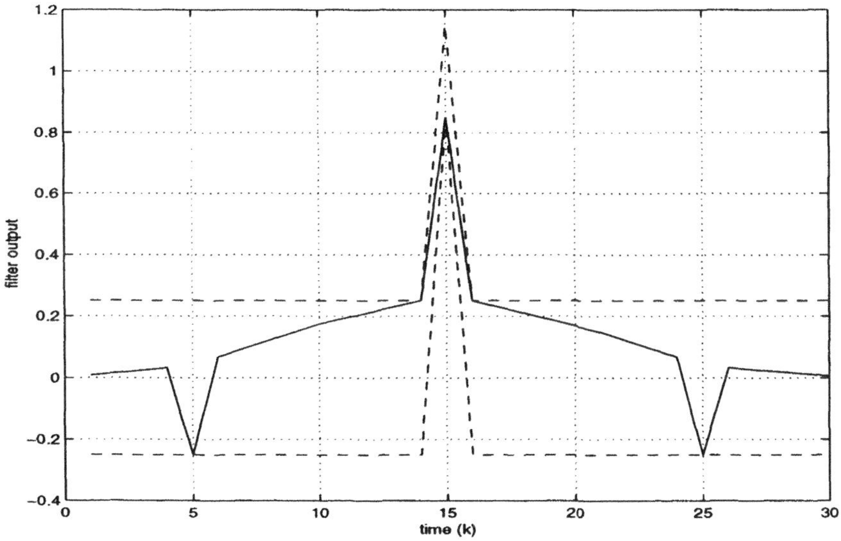

Figure 4.5.4. Optimum EC filter output for a rectangular input

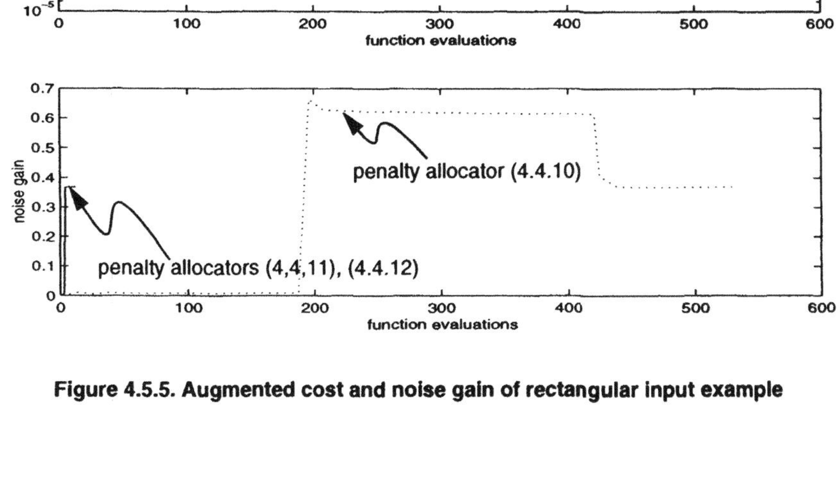

Figure 4.5.5. Augmented cost and noise gain of rectangular input example

Table 4.5.2. results for rectangular pulse example

Penalty allocator	Penalty parameter	Filter norm	NGD	Relative NGD	Function evaluation	Search direction evaluation
(4.4.10)	2.4736×10^6	0.6062	4.8104×10^{-5}	1.3090×10^{-4}	482	65
(4.4.11)	2.4736×10^6	0.6062	4.8104×10^{-5}	1.3090×10^{-4}	6	5
(4.4.12)	3.4037×10^6	0.6062	4.8072×10^{-5}	1.3082×10^{-4}	19	18

Channel Equalization

The third example concerns the equalization of a digital transmission channel consisting of a coaxial cable operating at the DX3 rate (45Mb/s) (see [5], [12]). The coaxial cable has a 30dB attenuation at a normalized frequency of $2\pi/\beta$, where β is the baud period (22.35×10^{-9}s). An equalizing filter is required to shape the impulse response of the cable so that it fits into the envelope given by the DSX3 pulse template (see Figure 4.5.6).

Table 4.5.3. Simulation results for DSX-3 pulse template

Penalty allocator	Penalty parameter	Filter norm	NGD	Relative NGD	Function evaluation	Search direction evaluation
(4.4.10)	4.6645×10^9	3.6812	1.5047×10^{-3}	1.1105×10^{-4}	1443	204
(4.4.11)	4.6645×10^9	3.6812	1.5047×10^{-3}	1.1105×10^{-4}	763	305
(4.4.12)	6.4182×10^9	3.6812	1.5046×10^{-3}	1.1104×10^{-4}	305	148

To have a good discrete-time representation, the analog signal is sampled every $\beta/4$ time unit over $[0, 40\beta]$. The accuracy parameter υ is fixed to 1.4268×10^{-5} (in practice, $\upsilon \approx 10^{-3}$ would suffice) and γ according to Corollary 4.5.1. The starting point of the iteration is the origin. The results of applying the NR algorithm with line search to the unconstrained problems generated by various penalty allocators are summarized in Table 4.5.3. Figure 4.5.6 shows the optimum and three sub-optimum outputs for a 16-tap FIR filter. Note that there is no visible difference between the optimal and sub-optimal cases.

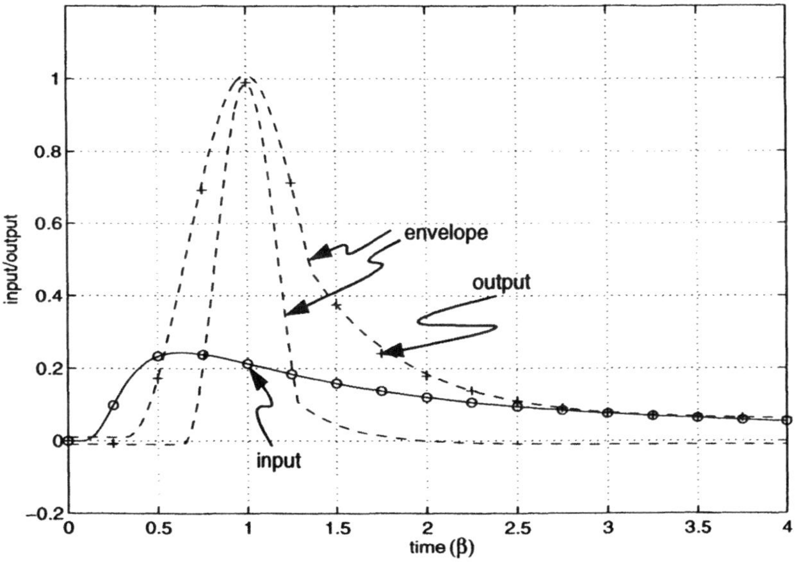

Figure 4.5.6. Optimum and sub-optimum EC filter and their outputs.

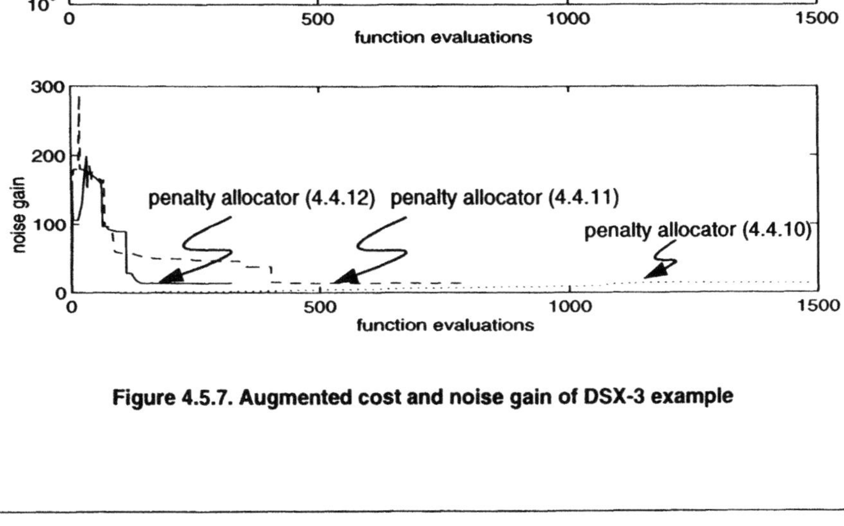

Figure 4.5.7. Augmented cost and noise gain of DSX-3 example

From Table 4.5.3 and Figure 4.5.7, it can be concluded that penalty allocator (4.4.12) displays superior performance compared to the other two. It requires the least number of iterations and line search calculations. This is strongly advocated by the analysis in Section 4.4.

4.6 Discrete-time Laguerre Networks

Although FIR filters are attractive due to their simplicity, they generally require a large number of taps. In this section, an alternative to FIR filters, orthonormal Laguerre networks are used. We compare the performance of these two types of filters via an equalization example. It is demonstrated that for the EC filtering problem, Laguerre networks can offer a more robust, low order alternative to FIR filters.

The use of orthonormal series for signal and filter representation is classical and dates back to the 1930's, see [61] for a summary of this early work. Discrete-time Laguerre functions form a complete orthonormal set in l_2 (see [57], [71], [72], [125]). Any stable transfer function can be expanded using Laguerre functions.

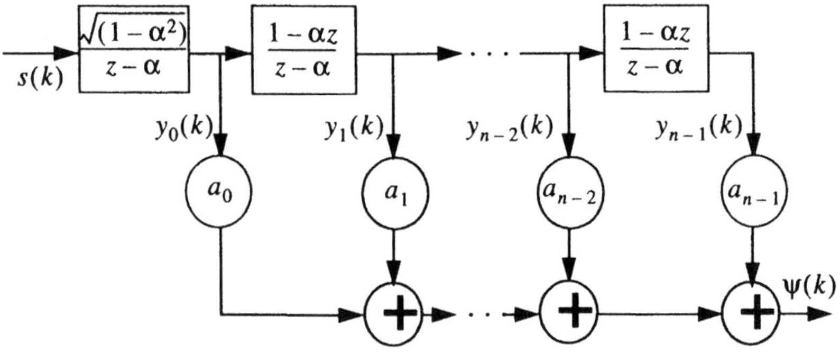

Figure 4.6.1. Discrete-time Laguerre network.

The transfer function of the nth order Laguerre network of Figure 4.6.1 is given by

$$U(z) = \sum_{k=0}^{n-1} a_k \Theta_k^{\alpha}(z),$$

where

$$\Theta_k^{\alpha}(z) = \frac{\sqrt{(1-\alpha^2)}}{z-\alpha}\left(\frac{1-\alpha z}{z-\alpha}\right)^k, \ -1<\alpha<1, \ k = 0, 1, 2, \ldots n-1.$$

For white input noise, the EC filtering problem for finite dimensional Laguerre filter can be posed as

$$\min \boldsymbol{a}^T\boldsymbol{a}, \boldsymbol{a} \in \boldsymbol{R}^n$$

$$\text{subject to } \varepsilon^-(k) \le \boldsymbol{y}_s^T(k)\boldsymbol{a} \le \varepsilon^+(k), k \in \tilde{\Omega}_C.$$

where $\boldsymbol{y}_s(k) = [(s * \theta_0)(k), \ldots, (s * \theta_{n-1})(k)]^T$.

Recall from Section 2.5 that $(\Xi_s K\boldsymbol{a})(k) = \boldsymbol{y}_s^T(k)\boldsymbol{a}$. Assuming the input signal s is non zero, Ξ_s is bounded and invertible by Lemma 2.1.1 and Lemma 2.5.1. Since K is a bounded and invertible linear operator, the operator $\Xi_s K$ is also bounded and invertible. By the Banach inverse theorem (Theorem A.6.2, Appendix A.6)

$$(\Xi_s K)^{-1} : l_\infty(\tilde{\Omega}_\psi) \to \boldsymbol{R}^n$$

is bounded. For each $j \in \tilde{\Omega}_\psi$, let

$$r(j) = \begin{cases} \max\{|\varepsilon^-(j)|, |\varepsilon^+(j)|\}, & j \in \tilde{\Omega}_C \\ w(j), & \text{otherwise} \end{cases},$$

where w is a finite energy sequence which bounds the filter outputs for all time instances not in $\tilde{\Omega}_C$. This is possible if we assume that $\tilde{\Omega}_C$ is sufficiently large. For any feasible filter $\boldsymbol{a}$, $|y_s^T(j)\boldsymbol{a}| \le r(j)$. Consequently, $\|\Xi_s K\boldsymbol{a}\| \le \|r\|$ and hence

$$\|\boldsymbol{a}\| \le \|(\Xi_s K)^{-1}\| \|\Xi_s K\boldsymbol{a}\| \le \|(\Xi_s K)^{-1}\| \|r\|.$$

Using Theorem 4.4.1, a simple criterion for estimating the penalty parameter is given as follows.

Corollary 4.6.1. *Suppose* $\upsilon > 0$ *is such that* $\mathcal{F}_\upsilon \ne \varnothing$. *Then,* $\boldsymbol{a}^o_{\upsilon,\gamma}$ *is feasible if*

$$\gamma \ge \|(\Xi_s K)^{-1}\|^2 \|r\|^2 / g_\upsilon(0).$$

Laguerre networks with $\alpha = 0.53$ and FIR filters are compared for various sampling periods. Penalty allocator (4.4.12) is used here for fast convergence. The accuracy parameter υ is chosen to be 10^{-3}. For FIR filters, the penalty parameters are obtained from Corollary 4.5.1. For the Laguerre filters, assuming that the Y_s matrix is of full rank, the penalty parameters can also be estimated from Corollary 4.6.1.

Table 4.6.1. Laguerre and FIR filters

Filter	Sampling period	Minimum tap number	Penalty parameter	Function evaluation	Filter norm
Laguerre	$\beta/4$	6	1.4270×10^6	84	3.7679
	$\beta/8$	6	6.4112×10^6	126	2.9400
	$\beta/12$	6	2.3450×10^7	102	2.3097
FIR	$\beta/4$	5	1.3023×10^7	73	3.9624
	$\beta/8$	8	1.2665×10^8	180	3.4424
	$\beta/12$	12	3.7331×10^9	271	2.7006

Simulation results are summarized in Table 4.6.1. From this, several remarks are in order:

- The minimum number of taps (or sections) in Laguerre networks is relatively insensitive to the sampling period in comparison with FIR filters.
- Laguerre filters have slightly smaller norms. This can be attributed to better approximation of the filter space by the Laguerre basis.

Figure 4.6.2 plots the input, output and envelope for the case with sampling period $\beta/12$, for the two types of filters.

Another attractive feature of Laguerre filters is that the dominant pole α constitutes an extra design parameter. By varying α, optimum EC filters with considerably lower orders than FIR filters can be obtained as demonstrated in Table 4.6.2. Again, this is due to better approximation of the filter space by the Laguerre basis. Optimization with respect to α is not in the scope of this monograph. Interested readers are referred to [16] for the application of pole placement techniques to Laguerre networks.Note that for $\alpha = 0$, the Laguerre structure reduces to the tapped delay line structure considered in the previous section

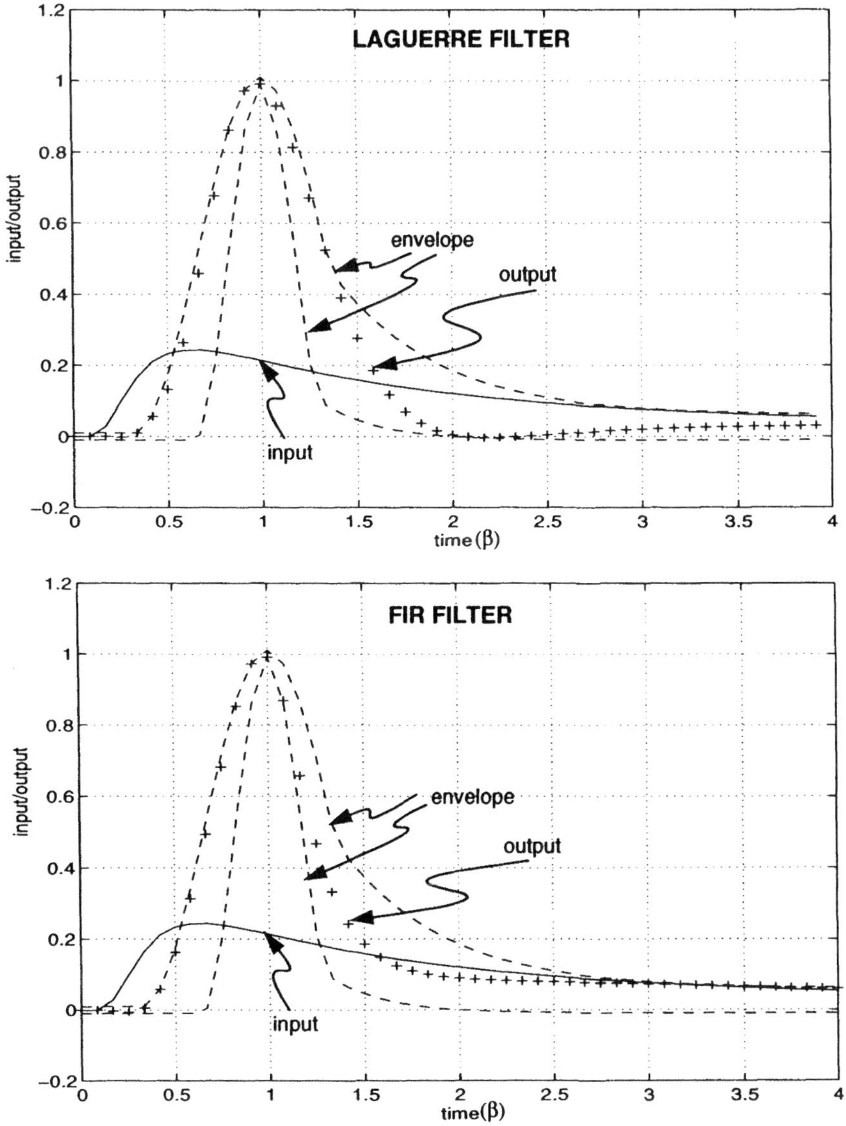

Figure 4.6.2. Signals and output mask for Laguerre and FIR EC filter.

Table 4.6.2. Performance of Laguerre filters for various dominant pole values.

Sampling period	Minimum tap number	Penalty parameter	Function evaluations	Filter norm	α
$\beta/4$	3	1.2111×10^6	61	3.7782	0.23
$\beta/8$	4	5.2560×10^6	18	3.7267	0.38
$\beta/12$	5	2.3915×10^7	108	2.7165	0.45

4.7 Appendix A

Lemma 4.3.1. *If* $\boldsymbol{p}$ *is any non optimal point and* $\boldsymbol{p}^o$ *is any optimal point, then the ascent direction* $\boldsymbol{l}(\boldsymbol{p})$ *has the following properties,*

(i) $-\langle \boldsymbol{l}(\boldsymbol{p}), \boldsymbol{p}-\boldsymbol{p}^o\rangle \geq \eta(\boldsymbol{p}^o)-\eta(\boldsymbol{p}) > 0$

(ii) $\|\boldsymbol{l}(\boldsymbol{p})\|^2 \leq 8\|\varepsilon\|^2+\frac{1}{2}\|Y_s\|^2\|Q^{-1}Y_s^T(\boldsymbol{p}-\boldsymbol{p}^o)\|^2 \leq 8\|\varepsilon\|^2+\frac{1}{2}\|Y_s\|^4\|Q^{-1}\|^2\|(\boldsymbol{p}-\boldsymbol{p}^o)\|^2$

(iii) $-2\langle \boldsymbol{l}(\boldsymbol{p}), \boldsymbol{p}-\boldsymbol{p}^o\rangle \geq \left\|\sqrt{Q^{-1}}\,Y_s^T(\boldsymbol{p}-\boldsymbol{p}^o)\right\|^2$

Proof. It is useful to write

$$\boldsymbol{l}(\boldsymbol{p}) = -\frac{1}{2}Y_sQ^{-1}Y_s^T\boldsymbol{p}-\boldsymbol{d}-\boldsymbol{f}(\boldsymbol{p}),$$

where $\boldsymbol{f}(\boldsymbol{p})$ is defined by

$$f_i(\boldsymbol{p}) \equiv \begin{cases} +\varepsilon_i, & i\in\Lambda^+(\boldsymbol{p})\cup\chi^+(\boldsymbol{p}) \\ -\varepsilon_i, & i\in\Lambda^-(\boldsymbol{p})\cup\chi^-(\boldsymbol{p}) \\ -\frac{1}{2}(Y_sQ^{-1}Y_s^T\boldsymbol{p})_i-d_i, & i\in\chi^0(\boldsymbol{p}) \end{cases}.$$

From the definitions of $\chi^+(\boldsymbol{p}), \chi^-(\boldsymbol{p}), \chi^0(\boldsymbol{p})$ given in (4.3.10), we see that $|f_i(\boldsymbol{p})| \leq \varepsilon_i$, for all i. Now, from the definition of $\boldsymbol{f}$, it is easy to see that

$$\boldsymbol{f}^T(\boldsymbol{p})\boldsymbol{p} = \varepsilon^T|\boldsymbol{p}|,$$

and for $\boldsymbol{p}\neq\boldsymbol{q}$

$$\boldsymbol{f}^T(\boldsymbol{p})\boldsymbol{q} \leq \varepsilon^T|\boldsymbol{q}|.$$

(i) This follows because η is concave and $\boldsymbol{l}(\boldsymbol{p})$ belongs to the subdifferential of η at $\boldsymbol{p}$. Alternatively,

$$
\begin{aligned}
-\langle \boldsymbol{l}(\boldsymbol{p}), \boldsymbol{p}-\boldsymbol{p}^o\rangle &= \langle \tfrac{1}{2}Y_sQ^{-1}Y_s^T\boldsymbol{p}+\boldsymbol{d}+\boldsymbol{f}(\boldsymbol{p}), \boldsymbol{p}-\boldsymbol{p}^o\rangle \\
&= \tfrac{1}{2}\boldsymbol{p}^TY_sQ^{-1}Y_s^T\boldsymbol{p}+\boldsymbol{d}^T\boldsymbol{p}+\varepsilon^T|\boldsymbol{p}|-\tfrac{1}{2}\boldsymbol{p}^{oT}Y_sQ^{-1}Y_s^T\boldsymbol{p}-\boldsymbol{d}^T\boldsymbol{p}^o-\boldsymbol{f}^T(\boldsymbol{p})\boldsymbol{p}^o \\
&= \tfrac{1}{4}\boldsymbol{p}^TY_sQ^{-1}Y_s^T\boldsymbol{p}+\boldsymbol{d}^T\boldsymbol{p}+\varepsilon^T|\boldsymbol{p}| \\
&\quad -\tfrac{1}{4}\boldsymbol{p}^{oT}Y_sQ^{-1}Y_s^T\boldsymbol{p}^o-\boldsymbol{d}^T\boldsymbol{p}^o-\varepsilon^T|\boldsymbol{p}^o| \\
&\quad +\tfrac{1}{4}\boldsymbol{p}^TY_sQ^{-1}Y_s^T\boldsymbol{p}-\tfrac{1}{2}\boldsymbol{p}^{oT}Y_sQ^{-1}Y_s^T\boldsymbol{p}+\tfrac{1}{4}\boldsymbol{p}^{oT}Y_sQ^{-1}Y_s^T\boldsymbol{p}^o \\
&\quad +\varepsilon^T|\boldsymbol{p}^o|-\boldsymbol{f}^T(\boldsymbol{p})\boldsymbol{p}^o \\
&= \eta(\boldsymbol{p}^o)-\eta(\boldsymbol{p})+\tfrac{1}{4}(\boldsymbol{p}-\boldsymbol{p}^o)^TY_sQ^{-1}Y_s^T(\boldsymbol{p}-\boldsymbol{p}^o)+\varepsilon^T|\boldsymbol{p}^o|-\boldsymbol{f}^T(\boldsymbol{p})\boldsymbol{p}^o.
\end{aligned}
$$

The quadratic term is clearly non-negative, $\eta(\boldsymbol{p}^o)-\eta(\boldsymbol{p})$ is, by definition, non-negative, and the remaining term $\varepsilon^T|\boldsymbol{p}^o|-\boldsymbol{f}^T(\boldsymbol{p})\boldsymbol{p}^o$ is also non-negative. Thus, the result follows.

(ii) Since $\boldsymbol{l}(\boldsymbol{p}^o) = 0$ by (4.3.16), we can write:

$$\boldsymbol{l}(\boldsymbol{p}) = \boldsymbol{l}(\boldsymbol{p})-\boldsymbol{l}(\boldsymbol{p}^o) = -\frac{1}{2}Y_sQ^{-1}Y_s^T(\boldsymbol{p}-\boldsymbol{p}^o)-\boldsymbol{f}(\boldsymbol{p})+\boldsymbol{f}(\boldsymbol{p}^o).$$

Using $\|a+b\|^2 = \|a\|^2+\|b\|^2+2\langle a, b\rangle$, we have $\|a\|^2+\|b\|^2 \geq -2\langle a, b\rangle$. Hence,

$$\|a-b\|^2 = \|a\|^2+\|b\|^2-2\langle a, b\rangle \leq 2\|a\|^2+2\|b\|^2.$$

Since $|f_i(\boldsymbol{p})| \leq \varepsilon_i, \forall i,$ we have

$$
\begin{aligned}
\|\boldsymbol{l}(\boldsymbol{p})-\boldsymbol{l}(\boldsymbol{p}^o)\|^2 &\leq 2\left\|\tfrac{1}{2}Y_sQ^{-1}Y_s^T(\boldsymbol{p}-\boldsymbol{p}^o)\right\|^2+2\|\boldsymbol{f}(\boldsymbol{p}^o)-\boldsymbol{f}(\boldsymbol{p})\|^2 \\
&\leq \tfrac{1}{2}\left\|Y_sQ^{-1}Y_s^T(\boldsymbol{p}-\boldsymbol{p}^o)\right\|^2+8\|\varepsilon\|^2.
\end{aligned}
$$

(iii) The inner product of $-2(\boldsymbol{p}-\boldsymbol{p}^o)$ with $\boldsymbol{l}(\boldsymbol{p})$ gives

$$-2\langle p-p^o, l(p)\rangle = (p-p^o)^T Y_s Q^{-1} Y_s^T (p-p^o) + 2\langle f(p)-f(p^o), (p-p^o)\rangle$$
$$= \left\|\sqrt{Q^{-1}}Y_s^T(p-p^o)\right\|^2 + 2f^T(p)(p-p^o) + 2f^T(p^o)(p^o-p)$$
$$= \left\|\sqrt{Q^{-1}}Y_s^T(p-p^o)\right\|^2 + 2[f^T(p)p - f^T(p^o)p + f^T(p^o)p^o - f^T(p)p^o].$$

The two differences on the right has the form $f^T(p)p - f^T(q)p$, which is non-negative because $f^T(p)p = \varepsilon^T|p|$ and $f^T(q)p \le \varepsilon^T|p|$. Thus, the conclusion follows. ❐

Lemma 4.7.1. *Let $0 < A < \infty$, $0 < B < \infty$, $\alpha_k > 0$ be such that*

$$\sum_{k=1}^{\infty} \alpha_k = \infty \textit{ and } \sum_{k=1}^{\infty} \alpha_k^2 < \infty.$$

If the sequence $\{z_k\}_{k=0}^{\infty}$, with $z_0 \ge 0$, satisfies

$$z_k \le z_{k-1}(1 + A\alpha_k^2) + B\alpha_k^2,$$

then $\{z_k\}_{k=0}^{\infty}$ is bounded and converges.

Proof. Clearly,

$$\lim_{k\to\infty} z_k \le z_0 \prod_{k=0}^{\infty}(1 + A\alpha_k^2) + \sum_{k=0}^{\infty} B\alpha_k^2 \prod_{i=k+1}^{\infty}(1 + A\alpha_i^2).$$

Now since $\sum_{k=1}^{\infty}\alpha_k^2 < \infty$, it follows that $\prod_{k=0}^{\infty}(1 + A\alpha_k^2)$ converges and is bounded. Also

$$\sum_{k=0}^{\infty} B\alpha_k^2 \prod_{i=k+1}^{\infty}(1 + A\alpha_i^2) \le \sum_{k=0}^{\infty} B\alpha_k^2 \prod_{i=1}^{\infty}(1 + A\alpha_i^2) < \infty$$

Thus, the sequence $\{z_k\}_{k=0}^{\infty}$ is bounded. Since $\alpha_k \to 0$, $B\alpha_k^2 \to 0$ and $\{1 + A\alpha_k^2\} \to 1$, $\{z_k\}_{k=0}^{\infty}$ converges to a finite limit. ❐

Lemma 4.7.2*. Let f be a closed proper convex function which attains its infimum at a unique point x. If $\{x_k\}_{k=1}^{\infty}$ is any sequence of vectors such that $\{f(x_k)\}_{k=1}^{\infty}$ converges to $\inf f$, then $\{x_k\}_{k=1}^{\infty}$ converges to x (see Corollary 27.2.2 of [96]).*

Theorem 4.3.1*. If the step size α_k satisfies*

(i) $\alpha_k > 0, k = 1, 2, \ldots,$

(ii) $\sum_{k=1}^{\infty} \alpha_k^2 < \infty,$

(iii) $\sum_{k=1}^{\infty} \alpha_k = \infty,$

then the sequence $\{\boldsymbol{p}_k\}_{k=1}^{\infty}$ of iterates converges to an optimal $\boldsymbol{p}^o$, i.e.

$$\lim_{k \to \infty} \|\boldsymbol{p}_k - \boldsymbol{p}^o\| = 0.$$

Moreover, the corresponding sequence $\{\boldsymbol{a}_k\}_{k=1}^{\infty}$ of filters, where $\boldsymbol{a}_k \equiv -0.5 Y_s^T \boldsymbol{p}_k$, converges to the optimal filter $\boldsymbol{a}^o$, i.e.

$$\lim_{k \to \infty} \|\boldsymbol{a}_k - \boldsymbol{a}^o\| = 0.$$

Proof. Assume for the moment that $\boldsymbol{p}^o$ is unique. Define $\boldsymbol{e}_k = \boldsymbol{p}_k - \boldsymbol{p}^o$ so that

$$\boldsymbol{e}_{k+1} = \boldsymbol{e}_k + \alpha_k \boldsymbol{l}(\boldsymbol{p}_k),$$

$$\|\boldsymbol{e}_{k+1}\|^2 = \|\boldsymbol{e}_k\|^2 + 2\alpha_k \langle \boldsymbol{l}(\boldsymbol{p}_k), \boldsymbol{e}_k \rangle + \alpha_k^2 \|\boldsymbol{l}(\boldsymbol{p}_k)\|^2.$$

The center term is negative by Lemma 4.3.1 (i). Now, using Lemma 4.3.1 (ii) yields

$$\|\boldsymbol{e}_{k+1}\|^2 \le \|\boldsymbol{e}_k\|^2 + \alpha_k^2 8 \|\varepsilon\|^2 + \frac{1}{2}\alpha_k^2 \|Y_s\|^4 \|\boldsymbol{e}_k\|^2$$

Using Lemma 4.7.1, $\|\boldsymbol{e}_k\|^2$ converges. Also,

$$\|\boldsymbol{e}_{\infty}\|^2 = \|\boldsymbol{e}_0\|^2 + 2\sum_{k=0}^{\infty} \alpha_k \langle \boldsymbol{l}(\boldsymbol{p}_k), \boldsymbol{e}_k \rangle + \sum_{k=0}^{\infty} \alpha_k^2 \|\boldsymbol{l}(\boldsymbol{p}_k)\|^2. \tag{4.7.1}$$

Since the left side of (4.7.1) is bounded, the right side must be bounded also. Furthermore, the last term on the right is bounded by virtue of $\sum_{k=1}^{\infty}\alpha_k^2 < \infty$ and Lemma 4.3.1 (ii). Thus, $\sum_{k=0}^{\infty}\alpha_k\langle \boldsymbol{l}(\boldsymbol{p}_k), \boldsymbol{e}_k\rangle$ is bounded.

Now, since $\sum_{k=1}^{\infty}\alpha_k = \infty$ and $-\langle \boldsymbol{l}(\boldsymbol{p}_k), \boldsymbol{e}_k\rangle > 0$, it follows that

$$\liminf_{k\to\infty}\{-\langle \boldsymbol{l}(\boldsymbol{p}_k), \boldsymbol{e}_k\rangle\} = 0.$$

Thus, from Lemma 4.3.1 (i), we obtain

$$\liminf_{k\to\infty}|\eta(\boldsymbol{p}^o) - \eta(\boldsymbol{p}_k)| = 0.$$

This, in turn implies, by virtue of Lemma 4.7.2, that

$$\liminf_{k\to\infty}\|\boldsymbol{e}_k\|^2 = 0,$$

and since the sequence converges, we have

$$\lim_{k\to\infty}\|\boldsymbol{e}_k\|^2 = 0.$$

The above argument assumes that $\boldsymbol{p}^o$ is unique. If not, then it can be shown (under the assumption that the feasible set has a non-empty interior) that the set of optimal $\boldsymbol{p}^o$ is bounded. The above arguments then hold if $\|\boldsymbol{e}_k\|^2$ is replaced at appropriate points by the minimum distance from $\boldsymbol{p}_k$ to the optimal set. ❐

Theorem 4.3.2. *Suppose the step size is fixed at*

$$\alpha_k = \alpha < \frac{2}{\|Y_s\|^2},$$

where $\|Y_s\|$ is the spectral norm of the matrix Y_s. Then, the resulting sequence $\{\boldsymbol{a}_k\}_{k=1}^{\infty}$ of filters enters a neighbourhood of the optimal filter $\boldsymbol{a}^o$ after a finite number of iterations. More precisely,

$$\limsup_{k \to \infty} \|a_k - a^o\|^2 \le V(\alpha) < \infty$$

and $\lim_{\alpha \to 0} V(\alpha) = 0.$

Proof. For a fixed stepsize α,

$$\|e_{k+1}\|^2 = \|e_k\|^2 + 2\alpha \langle l(p_k), e_k \rangle + \alpha^2 \|l(p_k)\|^2.$$

The positive sequence $\|e_k\|^2$ will decrease monotonically until

$$-2\langle l(p_k), e_k \rangle \le \alpha \|l(p_k)\|^2.$$

Upon using Lemma 4.3.1 (ii) and (iii), this inequality implies

$$\|Y_s^T e_k\|^2 \le \alpha \left(8\|\varepsilon\|^2 + \frac{1}{2}\|Y_s\|^2 \|Y_s^T e_k\|^2 \right),$$

or equivalently,

$$\|Y_s^T e_k\|^2 (1 - 0.5\alpha \|Y_s\|^2) \le \alpha 8 \|\varepsilon\|^2.$$

Since $\alpha < 2/\|Y_s\|^2$, we have $1 - 0.5\alpha\|Y_s\|^2 > 0$. Thus,

$$\|Y_s^T e_k\|^2 \le \frac{\alpha 8 \|\varepsilon\|^2}{1 - 0.5\alpha \|Y_s\|^2} \equiv \hat{V}(\alpha).$$

Clearly, $\hat{V}(\alpha) < \infty$ and $\lim_{\alpha \to 0} \hat{V}(\alpha) = 0$.

Now, using $\|a\|^2 + \|b\|^2 \ge 2\langle a, b \rangle$ yields

$$\begin{aligned}
\|Y_s^T e_{k+1}\|^2 &= \|Y_s^T e_k\|^2 + 2\alpha \langle Y_s^T l(p_k), Y_s^T e_k \rangle + \alpha^2 \|Y_s^T l(p_k)\|^2 \\
&\le \|Y_s^T e_k\|^2 + \alpha^2 \|Y_s^T l(p_k)\|^2 + \alpha \|Y_s^T l(p_k)\|^2 + \alpha \|Y_s^T e_k\|^2 \\
&\le \|Y_s^T e_k\|^2 (1 + \alpha) + \|Y_s\|^2 (\alpha + \alpha^2) \|l(p_k)\|^2.
\end{aligned}$$

Using Lemma 4.3.1 (ii) for $\|l(p_k)\|^2$, it follows that

$$\left\|Y_s^T e_{k+1}\right\|^2 \leq \left\|Y_s^T e_k\right\|^2\left[1+\alpha+\frac{1}{2}(\alpha+\alpha^2)\left\|Y_s\right\|^4\right]+8\|\varepsilon\|^2\left\|Y_s\right\|^2(\alpha+\alpha^2).$$

Thus, the error vector will decrease in length until

$$\left\|Y_s^T e_k\right\|^2 \leq \hat{V}(\alpha),$$

and then

$$\left\|Y_s^T e_{k+1}\right\|^2 \leq \hat{V}(\alpha)\left[1+\alpha+\frac{1}{2}(\alpha+\alpha^2)\left\|Y_s\right\|^4\right]+8\|\varepsilon\|^2\left\|Y_s\right\|^2(\alpha+\alpha^2) \equiv 4V(\alpha).$$

Since $4V(\alpha) \to 0$ as $\alpha \to 0$ and $a_k - a^o = -0.5 Y_s^T e_k$, this proves that

$$\limsup_{k \to \infty}\left\|a_k - a^o\right\|^2 \leq V(\alpha). \quad \square$$

4.8 Appendix B

Lemma 4.4.1. *Given any* υ, *if* $G_\upsilon(\boldsymbol{a}) \le g_\upsilon(0)$, *then* $\varepsilon^-(j) \le \boldsymbol{y}_s^T(j)\boldsymbol{a} \le \varepsilon^+(j)$, $\forall j \in \tilde{\Omega}_c$.

Proof. Suppose that $\boldsymbol{a}$ is not feasible, i.e. there is an integer $k \in \tilde{\Omega}_c$ such that

$$\varepsilon^-(k) > \boldsymbol{y}_s^T(k)\boldsymbol{a} \text{ or } \boldsymbol{y}_s^T(k)\boldsymbol{a} > \varepsilon^+(k).$$

Then, either $\phi_k^+(\boldsymbol{a}) > 0$ or $\phi_k^-(\boldsymbol{a}) > 0$. Thus, from the increasing property of g_υ,

$$g_\upsilon(\phi_k^+(\boldsymbol{a})) + g_\upsilon(\phi_k^-(\boldsymbol{a})) > g_\upsilon(0),$$

that is

$$\sum_{j \in \tilde{\Omega}_c} [g_\upsilon(\phi_j^+(\boldsymbol{a})) + g_\upsilon(\phi_j^-(\boldsymbol{a}))] > g_\upsilon(0).$$

Clearly, this is a contradiction, and the proof is complete. ❐

Proof of Lemma 4.4.2. (i) Continuity of $f_{\upsilon,\gamma}$ is trivial. Since the ϕ_j^+ 's are affine mappings, we have, for each $j \in \tilde{\Omega}_c$, and, for any $\alpha \in [0, 1]$,

$$\phi_j^+(\alpha\boldsymbol{u} + (1-\alpha)\boldsymbol{v}) = \alpha\phi_j^+(\boldsymbol{u}) + (1-\alpha)\phi_j^+(\boldsymbol{v}).$$

Also, since g_υ is convex, it follows that

$$\begin{aligned} g_\upsilon(\phi_j^+(\alpha\boldsymbol{u} + (1-\alpha)\boldsymbol{v})) &= g_\upsilon(\alpha\phi_j^+(\boldsymbol{u}) + (1-\alpha)\phi_j^+(\boldsymbol{v})) \\ &\le \alpha g_\upsilon(\phi_j^+(\boldsymbol{u})) + (1-\alpha)g_\upsilon(\phi_j^+(\boldsymbol{v})). \end{aligned}$$

Hence, for each $j \in \tilde{\Omega}_c$, $g_\upsilon \circ \phi_j^+$ is convex. Similarly, $g_\upsilon \circ \phi_j^-$ is also convex. Since the sum of convex functions is convex and a positively scaled convex function is convex, it follows that γG_υ is convex. Moreover, strict convexity of $f_{\upsilon,\gamma}$ follows from the fact that the sum of a strictly convex function (f) and a convex function (γG_υ) is strictly convex.

(ii), (iii) The expressions for the gradient and Hessian of $f_{\upsilon,\gamma}$ follow from the chain rule, and the differentiablity of g_υ and each $\phi_j^-, \phi_j^+, j \in \tilde{\Omega}_c$.

(iv) Using (iii) and the triangle inequality yield

$$\begin{aligned}\|\nabla^2 f_{\upsilon,\gamma}(\boldsymbol{a})\| &\le 2\|Q\| + \gamma\sum_{j=0}^{N-1}\left|g_\upsilon''(\phi_j^+(\boldsymbol{a})) + g_\upsilon''(\phi_j^-(\boldsymbol{a}))\right|\left\|\boldsymbol{y}_s(j)\boldsymbol{y}_s^T(j)\right\| \\ &\le 2\|Q\| + 2\gamma h_\upsilon\sum_{j=0}^{N-1}\left\|\boldsymbol{y}_s(j)\boldsymbol{y}_s^T(j)\right\|.\end{aligned}$$

Now,

$$\begin{aligned}\left\|\boldsymbol{y}_s(j)\boldsymbol{y}_s^T(j)\right\| &= \max_{\boldsymbol{x}\ne 0}\frac{[\boldsymbol{x}^T\boldsymbol{y}_s(j)\boldsymbol{y}_s^T(j)\boldsymbol{y}_s(j)\boldsymbol{y}_s^T(j)\boldsymbol{x}]^{1/2}}{\|\boldsymbol{x}\|} \\ &= \max_{\boldsymbol{x}\ne 0}\frac{\|\boldsymbol{y}_s(j)\|\left\|\boldsymbol{y}_s^T(j)\boldsymbol{x}\right\|}{\|\boldsymbol{x}\|}.\end{aligned}$$

By the Cauchy-Schwartz inequality

$$\max_{\boldsymbol{x}\ne 0}\frac{\|\boldsymbol{y}_s(j)\|\left\|\boldsymbol{y}_s^T(j)\boldsymbol{x}\right\|}{\|\boldsymbol{x}\|} = \|\boldsymbol{y}_s(j)\|^2.$$

Hence,

$$\|\nabla^2 f_{\upsilon,\gamma}(\boldsymbol{a})\| \le 2\|Q\| + 2\gamma h_\upsilon\sum_{j=0}^{N-1}\|\boldsymbol{y}_s(j)\|^2.$$

To show that $\|\nabla^2 f_{\upsilon,\gamma}(\boldsymbol{u})\| \ge 2\|Q\|$, let

$$\gamma\sum_{j=0}^{N-1}[g_\upsilon''(\phi_j^+(\boldsymbol{a})) + g_\upsilon''(\phi_j^-(\boldsymbol{a}))]\boldsymbol{y}_s(j)\boldsymbol{y}_s^T(j) = M(\boldsymbol{a}).$$

Then,

$$\|\nabla^2 f_{\upsilon,\gamma}(\boldsymbol{a})\| = \|2Q + M(\boldsymbol{a})\|.$$

Since

$$\max_{\boldsymbol{x}\ne 0}\frac{\|A\boldsymbol{x}\|}{\|\boldsymbol{x}\|} = \max_{\boldsymbol{x}\ne 0}\sqrt{\frac{\boldsymbol{x}^T A^T A\boldsymbol{x}}{\boldsymbol{x}^T\boldsymbol{x}}} = \sqrt{\lambda_{max}(A^T A)},$$

the matrix norm induced by the Euclidean vector norm is, in fact, the spectral norm:

$$\|A\| = \sqrt{\lambda_{max}(A^T A)}.$$

It is clear from the positive definiteness of Q and $M(\boldsymbol{a})$ that

$$\|\nabla^2 f_{\upsilon,\gamma}(\boldsymbol{a})\| = \|2Q + M(\boldsymbol{a})\| \geq 2\|Q\|.$$

Thus, the conclusion follows.

(v) Again, using (iii), we have

$$\nabla^2 f_{\upsilon,\gamma}(\boldsymbol{u}) - \nabla^2 f_{\upsilon,\gamma}(\boldsymbol{v}) = \gamma \sum_{j=0}^{N-1} [\, g_\upsilon'' \circ \phi_j^+(\boldsymbol{u}) - g_\upsilon'' \circ \phi_j^+(\boldsymbol{v}) + g_\upsilon'' \circ \phi_j^-(\boldsymbol{u}) - g_\upsilon'' \circ \phi_j^-(\boldsymbol{v}) \,] \boldsymbol{y}_s(j) \boldsymbol{y}_s^T(j).$$

The triangle inequality and the Lipschitz condition $|g_\upsilon''(x) - g_\upsilon''(y)| \leq l_\upsilon |x - y|$ give:

$$\begin{aligned} \|\nabla^2 f_{\upsilon,\gamma}(\boldsymbol{u}) - \nabla^2 f_{\upsilon,\gamma}(\boldsymbol{v})\| &\leq 2\gamma \sum_{j=0}^{N-1} l_\upsilon |\boldsymbol{y}_s^T(j)(\boldsymbol{u} - \boldsymbol{v})| \|\boldsymbol{y}_s(j)\|^2 \\ &\leq 2\gamma \sum_{j=0}^{N-1} l_\upsilon \|(\boldsymbol{u} - \boldsymbol{v})\| \|\boldsymbol{y}_s^T(j)\| \|\boldsymbol{y}_s(j)\|^2 \\ &= 2\gamma l_\upsilon \|(\boldsymbol{u} - \boldsymbol{v})\| \sum_{j=0}^{N-1} \|\boldsymbol{y}_s(j)\|^3. \; \square \end{aligned}$$

Theorem 4.4.1. *Let $\boldsymbol{a}^o$ denote the optimum filter coefficient vector, and $\boldsymbol{a}^o_{\upsilon,\gamma}$ denote the minimum of $f_{\upsilon,\gamma}$. Suppose υ is such that $\mathcal{F}_\upsilon \neq \emptyset$. If $\gamma > \boldsymbol{a}_\upsilon^T Q \boldsymbol{a}_\upsilon / g_\upsilon(0)$ for some $\boldsymbol{a}_\upsilon \in \mathcal{F}_\upsilon$, then $\boldsymbol{a}^o_{\upsilon,\gamma}$ satisfies the envelope constraints, i.e.*

$$\varepsilon^-(j) \leq \boldsymbol{y}_s^T(j) \boldsymbol{a}^o_{\upsilon,\gamma} \leq \varepsilon^+(j), \; \forall j \in \tilde{\Omega}_c.$$

Proof. From the definition of $\mathcal{F}_\upsilon$, it follows that

$$\forall \boldsymbol{a} \in \mathcal{F}_\upsilon, \; G_\upsilon(\boldsymbol{a}) = 0,$$

consequently,

$$f_{\upsilon,\gamma}(\boldsymbol{a}^o_{\upsilon,\gamma}) \leq f_{\upsilon,\gamma}(\boldsymbol{a}_\upsilon) = \boldsymbol{a}_\upsilon^T Q \boldsymbol{a}_\upsilon.$$

Hence, using the definition of $f_{\upsilon,\gamma}$ and rearranging yield

$$\gamma G_{\upsilon}(a^{o}_{\upsilon,\gamma}) \le a^{T}_{\upsilon}Qa_{\upsilon} - a^{oT}_{\upsilon,\gamma}Qa^{o}_{\upsilon,\gamma} \le a^{T}_{\upsilon}Qa_{\upsilon}{}^{2}$$

$$\Rightarrow G_{\upsilon}(a^{o}_{\upsilon,\gamma}) \le \frac{a^{T}_{\upsilon}Qa_{\upsilon}}{\gamma} < g_{\upsilon}(0).$$

Thus, $a^{o}_{\upsilon,\gamma}$ satisfies the envelope constraints by Lemma 4.4.1. ❐

Theorem 4.8.1. *Suppose that f is twice continuously differentiable with*

$$0 < m\|h\|^2 \le h^T\nabla^2 f(x)h \le M\|h\|^2, \quad \forall x, h \in R^n.$$

Then, starting from any initial point x_0, the sequence $\{x_k\}_{k=0}^{\infty}$ generated by the SD algorithm with constant stepsize μ converges to the minimum x^o if

$$0 < \mu < \frac{2}{M}.$$

Proof. See the proof given for Theorem 1.3 of [90], pp. 50. ❐

Theorem 4.8.2. *Suppose that f is twice continuously differentiable with*

$$0 < m\|h\|^2 \le h^T\nabla^2 f(x)h \le M\|h\|^2, \quad \forall x, h \in R^n.$$

Then, starting from any initial point x_0, the sequence $\{x_k\}_{k=0}^{\infty}$ generated by the NR algorithm with constant stepsize μ converges to the minimum x^o if

$$0 < \mu < \frac{4}{M}.$$

Proof. See the proof given for Theorem 1.6 of [90]. ❐

Theorem 4.8.3. *Suppose that f is twice continuously differentiable with*

$$0 < m\|\boldsymbol{h}\|^2 \le \boldsymbol{h}^T \nabla^2 f(\boldsymbol{x})\boldsymbol{h} \le M\|\boldsymbol{h}\|^2, \ \forall \boldsymbol{x}, \boldsymbol{h} \in \boldsymbol{R}^n,$$

and $\nabla^2 f$ is Lipschitz continuous, i.e.

$$\|\nabla^2 f(\boldsymbol{x}) - \nabla^2 f(\boldsymbol{y})\| \le R\|\boldsymbol{x} - \boldsymbol{y}\|, \ \forall \boldsymbol{x}, \boldsymbol{y} \in \boldsymbol{R}^n.$$

Then from any initial point $\boldsymbol{x}_0$, the sequence $\{\boldsymbol{x}_k\}_{k=0}^{\infty}$, generated by the NR algorithm with variable stepsize μ_k converges to the minimum $\boldsymbol{x}^o$ at a quadratic rate if μ_k satisfies the Goldstein condition:

$$f(\boldsymbol{x}_{k+1}) - f(\boldsymbol{x}_k) < \alpha\mu_k \nabla f^T(\boldsymbol{x}_k)\boldsymbol{d}_k, \ \alpha \in (0, 0.5).$$

Proof. This is a standard result which can be found in [18], [27], [37], [90]. See the proof given for Theorem 2.2 of [90], pp. 61. ❐

CHAPTER 5 NUMERICAL METHODS FOR CONTINUOUS-TIME EC FILTERING

In Chapter 2, the continuous-time EC filtering problem has been formulated for both a purely analog filter structure and a hybrid filter structure that includes analog and digital signal processing components. Both of these formulations resulted in a semi-infinite programming (SIP) problem with quadratic cost. In this chapter, three methods for solving this SIP problem are presented. These methods can be considered as extensions of the methods developed in Chapter 4 for solving the discrete-time EC filtering problem. One approach is to make use of the active set method by approximating the continuum of constraints with a finite number of constraints. A criterion is derived for selecting the granularity of the discretization to ensure feasibility of the approximate solution. This approach is attractive for its simplicity and efficiency. However, the performance tends to deteriorates with finer discretization. When this happens, we can turn to the primal-dual or penalty method. The primal-dual algorithm can be extended to the continuous-time case by discretizing the infinite dimensional dual problem (the continuum of constraints of primal problem means that the dual problem is infinite dimensional). There is no guarantee that primal solution obtained by this approach is feasible. While the active set method and primal-dual involves discretization of the primal and dual problems respectively, the penalty approach introduces approximation that does not require discretization. Moreover, feasibility of the approximate solutions are guaranteed.

The chapter is organized in the same way as Chapter 4. Section 5.1 recaps the problem statement. Section 5.2 shows how the primal problem can be discretized, so that the approximate solution is feasible. Section 5.3 presents the extended primal-dual algorithm for solving the EC filtering problem. In Section 5.4, we focus on the penalty method for solving a range of SIP's including the EC filtering problem. Sections 5.5 and 5.6 present numerical examples with analog Laguerre filters and hybrid filters, respectively. Proofs for the results in this chapter are presented in Section 5.7.

5.1 Continuous-time EC Filtering

In continuous-time EC filtering, the filter response to a given input s on a certain interval say $\Omega_c \subseteq \Omega_\psi$, are constrained to lie within an envelope described by upper and lower boundaries ε^+, $\varepsilon^- \in L_\infty(\Omega_\psi)$. In Section 2.2, we have presented a problem statement for the analog EC filtering problem. However, continuous-time filters can also be realized by hybrid structures involving analog and digital components as shown in Section 2.6. In this section, we summarize the problem statement for both the EC filtering problem with analog and hybrid filters, including finite dimensional filters. The assumptions A-2.2.1 and A-2.2.2 (Section 2.2) are used throughout this chapter.

Analog filters

The EC filtering problem is a special case of the analog CCR filtering problem, where the set $\mathcal{S}$ of allowable inputs is $\{s \in L_2(\Omega_s)\}$ and the set Ψ of permissible responses is

$$\Psi = \{\psi \in L_\infty(\Omega_\psi) : \varepsilon^- \preceq \psi \preceq \varepsilon^+\}.$$

The signal s is assumed to have finite energy. The partial ordering $\preceq$ on the response space $L_\infty(\Omega_\psi)$ is defined by

$$x, y \in L_\infty(\Omega_\psi), x \preceq y \Leftrightarrow \forall t \in \Omega_c, x(t) \le y(t).$$

Under this partial ordering, Ψ is a convex set in $L_\infty(\Omega_\psi)$. Moreover, Ψ is closed since s has finite energy. Hence, the feasible set

$$\mathcal{F} = \{u \in L_2(\Omega_u) : \varepsilon^- \preceq s * u \preceq \varepsilon^+\},$$

where the convolution $s * u$, defined in (2.1.2), is closed and convex (see Section 2.1). The cost functional for this problem is a weighted filter norm, because we wish to minimize the effect of input noise (see Section 2.2). The continuous-time EC filtering problem can be stated as

$$\begin{aligned} &\min \langle u, Lu \rangle \\ &\text{subject to } \varepsilon^- \preceq \Xi_s u \preceq \varepsilon^+, \end{aligned} \tag{5.1.1}$$

where Ξ_s is defined by (2.1.3), and L is defined by (2.1.5).

For a finite dimensional filter characterized by the impulse response of the form

$$u = Ka = v^T a, \; v = [v_0, \ldots, v_{n-1}]^T,$$

the filter response to the excitation s can be expressed as

$$s * Ka = y_s^T a \tag{5.1.2}$$

where $y_s = [s * v_0, \ldots, s * v_{n-1}]^T$ is the vector of state variables. Let Q denote the Gram matrix $K^\dagger L K$ (see (2.4.5)), the finite dimensional analog EC filtering problem can be stated as the following quadratic semi-infinite programming (QSIP) problem.

$$\begin{aligned} &\min a^T Q a, a \in R^n \\ &\text{subject to } \varepsilon^-(t) \le y_s^T(t) a \le \varepsilon^+(t), t \in \Omega_c \end{aligned} \tag{5.1.3}$$

Hybrid filters

For hybrid filters, the EC filtering problem for hybrid systems takes on the same form as Problem (5.1.1), where Ξ_s is defined by (2.6.4) and L is defined by (2.6.12).

Similar to the analog problem, for finite hybrid filter structures, the EC filtering problem can be written as the QSIP Problem (5.1.3) in which $Q = K^{\dagger}LK$ is given by (2.7.4) and

$$\begin{aligned} y^T(t)a &= [(\Xi_s v_0)(t), \ldots, (\Xi_s v_{n-1})(t)]a \\ &= (\Xi_s Ka)(t). \end{aligned} \tag{5.1.4}$$

Note that the hybrid EC filtering problem can be cast in the same form as its analog counterpart. Whereas $y_k(t)$ is a state of the analog system which can be observed, the time function $y_k(t) = (\Xi_s v_k)(t)$ is not a state of the hybrid system which can be observed. However, for both systems, y_k can be interpreted as the response to the excitation s when the filter coefficient a_k is set to unity, while all other coefficients are set to zero. Since the post-filter is assumed to be stable, each y_k is a bounded function on $C(\Omega_\psi)$.

5.2 Non-iterative Method

In this section, the QP problem with affine functional inequality constraint is solved by approximating the continuum of constraints. The resulting problem is a conventional QP problem with a finite number of affine inequality constraints. A criterion is presented for selecting the granularity of the discretization so that the sub-optimum solution satisfies the continuum of constraints.

In this section, without loss of generality, we consider the case where $\Omega_c = [0, T_c]$. The functional constraints can be rewritten in an alternative form as follows:

$$\phi^+(a, t) = y^T(t)a - \varepsilon^+(t) \le 0, \ \forall t \in [0, T_c]$$

$$\phi^-(a, t) = \varepsilon^-(t) - y^T(t)a \le 0, \ \forall t \in [0, T_c]$$

To make use of available QP techniques, the simplest way of solving this problem is to discretize the constraints. Instead of constraining the output on the entire interval $[0, T_c]$, we only constrain it at a finite number of points $t_j \in [0, T_c]$ for $j = 0, 1, \ldots, M$ say. Typically, $t_j = j\Delta$, $j = 0, 1, \ldots, M$, $M = \lfloor T_c/\Delta \rfloor$, where

$\lfloor x \rfloor$ denotes the largest integer not exceeding x. This type of approximation results in a problem that can be efficiently dealt with by off-line methods such as QP via active set strategy (see Section 4.2). However, there is no guarantee that the constraints are satisfied everywhere on $[0, T_c]$ (assuming there is such a solution).

To ensure that the functional constraints are satisfied, the same discretization procedure can be used, but the constraints are modified by introducing a parameter $\upsilon > 0$ as follows:

$$\varepsilon^{-}(t_j) + \upsilon \le \boldsymbol{y}^T(t_j)\boldsymbol{a} \le \varepsilon^{+}(t_j) - \upsilon,\ t_j = j\Delta,\ j = 0, 1, \ldots, M. \tag{5.2.1}$$

This effectively tightens the mask up at the points t_k, $k = 0, \ldots, M$.

Using the band limited property of the output signal, and the fact that any feasible point $\boldsymbol{a}$ would generate a response lying inside the envelope, an upper bound η for the time derivative of the output can be derived.

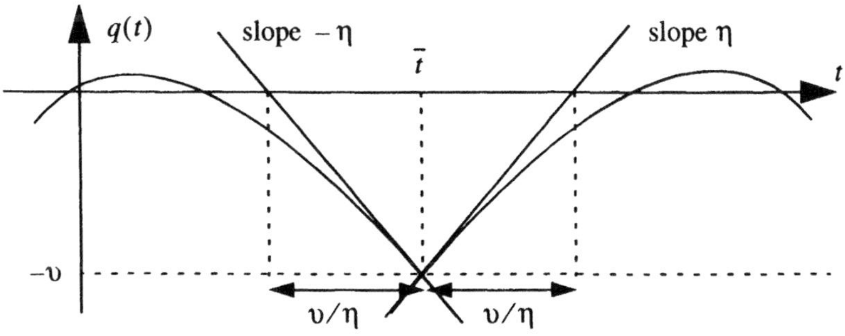

Figure 5.2.1. Bound on slopes

Figure 5.2.1 shows a function with bounded directional derivative at $\bar{t}$. Intuitively, $q(t) \le 0$ for all t on $[\bar{t} - \upsilon/\eta, \bar{t} + \upsilon/\eta]$.

Lemma 5.2.1. *Suppose that q is Lipschitz continuous on a compact interval Ω, i.e.*

$$|q(\omega_2) - q(\omega_1)| \le \eta|\omega_2 - \omega_1|,\ \forall \omega_2, \omega_1 \in \Omega.$$

Then,

$$\exists\bar{\omega} \in \Omega \textit{ such that } q(\bar{\omega}) \le -\upsilon \;\Rightarrow\; q(\omega) \le 0, \forall \omega \in [\bar{\omega} - \upsilon/\eta, \bar{\omega} + \upsilon/\eta].$$

It follows from Lemma 5.2.1 that, if $\boldsymbol{a}$ satisfies the approximate constraints (5.2.1) for $\Delta \le 2\upsilon/\eta$, then

$$\phi^{+}(\boldsymbol{a}, t) \le 0, \phi^{-}(\boldsymbol{a}, t) \le 0, \;\forall t \in [0, \hat{T}_c],$$

where $\hat{T}_c = \min\{T_c, (M + 0.5)\Delta\}$.

Theorem 5.2.1. *Suppose that the output ψ and the envelope boundaries ε^{+}, ε^{-} are band limited to ω_c. If $\boldsymbol{a}$ satisfies (5.2.1), in which Δ is chosen so that*

$$\Delta^2 \sum_{k=0}^{M} [\varepsilon(j\Delta) - \upsilon] \le 2\pi\upsilon/\omega_c^2,$$

then

$$\varepsilon^{-}(t) \le \boldsymbol{y}^T(t)\boldsymbol{a} \le \varepsilon^{+}(t), \;\forall t \in [0, T_c].$$

Theorem 5.2.1 shows how to discretize the constraint interval so that the entire continuum of constraints can be satisfied. Let $\hat{\varepsilon} = \sup_{t \in \Omega}\varepsilon(t)$. Then, $T_c(\hat{\varepsilon} - \upsilon)$ is an upper bound for

$$\Delta \sum_{k=0}^{M} [\varepsilon(j\Delta) - \upsilon].$$

Hence, a loose estimate of the discretization interval Δ can be given by

$$\Delta \le \frac{2\pi\upsilon}{\omega_c^2 T_c(\hat{\varepsilon} - \upsilon)} = \frac{\upsilon}{2\pi f_c^2 T_c(\hat{\varepsilon} - \upsilon)}.$$

Better estimates can be obtained if bounds on the second derivatives are known. This type of approximation results in a problem which can be efficiently dealt with by off-line methods such as QP via active set strategy. This approach is attractive for its simplicity and efficiency. However, for finer discretizations, the number of constraints increases which can cause numerical problems.

5.3 Primal Dual Method

The primal-dual approach proposed by Fortmann and Evans in [29] for analog pulse shaping filters was probably the earliest approach used to tackle the EC filtering problem. The technique involves transforming the constrained optimization problem into an unconstrained dual problem with a non differentiable cost and then solved by using a steepest ascent algorithm based on directional differentials. In Section 3.2, the EC filtering problem in Hilbert space has been converted into an unconstrained one in the dual space of regular Borel measures. By discretizing the dual space, the infinite dimensional dual problem is reduced to a finite dimensional one.

Discretization of dual problem for analog filters

One method of solving the continuous time EC filtering problem is to exploit the primal-dual relation and solve the dual problem as for the discrete-time case. However, the continuum of constraints means that the dual problem is infinite dimensional regardless of the dimension of the primal problem. Discretizing the dual space leads to a finite dimensional problem that can be solved by the steepest ascent method (see Section 4.3). As the finite dimensional problem is only an approximation, its solution is not optimal and may not even be feasible. However, it converges to the solution of the infinite dimensional dual problem as the dimension increases.

We consider here the analog EC filtering problem, where the constraints are specified on an interval, as follows:

$$\begin{aligned} &\min \ \langle u, Lu \rangle, u \in L_2(\Omega_u) \\ &\text{subject to } \left|(\Xi_s u)(t) - d(t)\right| \le \varepsilon(t), \forall t \in [t_0, t_0 + T_c]. \end{aligned}$$

Recall from (3.2.14) that the dual functional is

$$\eta(p) \equiv -\frac{1}{4}\langle L^{-1}\hat{\alpha}(p), \hat{\alpha}(p)\rangle - p \bullet d - \|p\|_{\varepsilon}, \, p \in M[t_0, t_0 + T_c], \tag{5.3.1}$$

where $\hat{\alpha}(p) \in L_2(\Omega_u)$ is given by

$$[\hat{\alpha}(p)](\tau) = \int_{t_0}^{t_0+T_c} s(t-\tau)dp(t). \tag{5.3.2}$$

The optimum filter is given by

$$u^o = -\frac{1}{2}L^{-1}\hat{\alpha}(p^o),$$

where p^o is a minimizer of the dual function (5.3.1).

For computational purpose, the space $M[t_0, t_0 + T_c]$ of measures is discretized by partitioning the time interval $[t_0, t_0 + T_c]$ into n sub-intervals $(t_{j-1}, t_j]$, $j = 1, 2, \ldots, n$, where

$$t_j = t_0 + j\frac{T_c}{n}, j = 0, 1, 2, \ldots, n.$$

Next define the n-dimensional subspace of $M[t_0, t_0 + T_c]$ by

$$M_n[t_0, t_0 + T_c] = \left\{p: p \bullet x = \int_{t_0}^{t_0+T_c} x(t)p'(t)dt = \sum_{j=1}^{n} p_j x_j, x \in C[t_0, t_0 + T_c]\right\}$$

where

$$x_j = \int_{t_{j-1}}^{t_j} x(t)dt,$$

and $p'(t)$ is the Radon-Nikodym derivative (Theorem A.8.8, Appendix A.8) of the measure p,

$$p'(t) = p_j, t \in (t_{j-1}, t_j], \ j = 1, 2, \ldots, n.$$

A measure in $M_n[t_0, t_0 + T_c]$ is represented by an ordinary function that is constant on each of the interval of length T_c/n. Such a measure is completely specified by the parameters $p_j, j = 1, 2, \ldots, n$. Its ε-weighted norm is

$$\|p\|_\varepsilon = \int_{t_0}^{t_0+T_c} \varepsilon(t)|p'(t)|dt = \sum_{j=1}^{n} |p_j|\varepsilon_j.$$

Now (5.3.2) can be written as

$$[\hat{\alpha}(p)](\tau) = \sum_{j=1}^{n} p_j \int_{t_{j-1}}^{t_j} s(t-\tau)dt.$$

Hence, using the notation $s_t(\tau) = s(t-\tau)$, we have

$$L^{-1}\hat{\alpha}(p) = \sum_{j=1}^{n} p_j \int_{t_{j-1}}^{t_j} L^{-1} s_t dt,$$

and the dual cost functional (5.3.1) in the subspace $M_n[t_0, t_0 + T_c]$ becomes:

$$\begin{aligned}\eta(p) &= -\frac{1}{4}\langle L^{-1}\sum_{i=1}^{n} p_i \int_{t_{i-1}}^{t_i} s_t dt, \sum_{j=1}^{n} p_j \int_{t_{j-1}}^{t_j} s_\lambda d\lambda\rangle - \sum_{j=1}^{n} p_j d_j - \sum_{j=1}^{n} \varepsilon_j |p_j| \\ &= -\frac{1}{4}\sum_{i=1}^{n}\sum_{j=1}^{n} p_i p_j \varsigma_{i,j} - \sum_{j=1}^{n} p_j d_j - \sum_{j=1}^{n} \varepsilon_j |p_j|,\end{aligned}$$

where

$$\varsigma_{i,j} = \int_{t_{i-1}}^{t_i}\int_{t_{j-1}}^{t_j} \langle L^{-1}s_t, s_\lambda\rangle dt d\lambda.$$

The subspace $\bigcup_{n=1}^{\infty} M_n[t_0, t_0 + T_c]$ is dense in the weak-star topology of $M[t_0, t_0 + T_c]$ (see Appendix A.7). This means that for sufficiently large n, the approximating subspace always contain near optimum elements. More concisely, we have

***Lemma* 5.3.1.** *To every $\delta > 0$, there exists an $N(\delta)$ such that for $n \geq N(\delta)$, $M_n[t_0, t_0 + T_c]$ contains an element p^n for which*

$$\left|[\hat{\alpha}(p^n)](t) - [\hat{\alpha}(p^o)](t)\right| < \delta,\ t \in [t_0, t_0 + T_c],$$

$$\eta(p^o) - \eta(p^n) < \delta,$$

where p^o is a minimizer of the dual function (5.3.1), and

$$[\hat{\alpha}(p^n)](t) = \int_{t_0}^{t_0 + T_c} s(\tau - t)dp^n(\tau),$$

$$[\hat{\alpha}(p^o)](t) = \int_{t_0}^{t_0 + T_c} s(\tau - t)dp^o(\tau).$$

Discretization of dual problem for hybrid filters

For hybrid filter configuration, $\hat{\alpha}(p) \in l_2(\tilde{\Omega}_u)$ is given by

$$[\hat{\alpha}(p)](k) = \int_{t_0}^{t_0+T_c} \sum_{i=0}^{\infty} s[(i-k)T_s]\Lambda(t-iT_s)dp(t),$$

and the same analysis as in the analog case is applicable. Let

$$\tilde{s}_t(k) = \sum_{i=0}^{\infty} s[(i-k)T_s]\Lambda(t-iT_s)$$

so that

$$(\Xi_s u)(t) = \langle \tilde{s}_t, u \rangle .$$

Restricting to the n -dimensional subspace $M_n[t_0, t_0+T_c]$ of $M[t_0, t_0+T_c]$, $\hat{\alpha}(p)$ is given by

$$\hat{\alpha}(p) = \sum_{j=1}^{n} p_j \int_{t_{j-1}}^{t_j} \tilde{s}_t dt.$$

Hence,

$$L^{-1}\hat{\alpha}(p) = \sum_{j=1}^{n} p_j \int_{t_{j-1}}^{t_j} L^{-1}\tilde{s}_t dt,$$

and the dual cost functional in the subspace $M_n[t_0, t_0+T_c]$ becomes:

$$\begin{aligned}\eta(p) &= -\frac{1}{4}\langle L^{-1}\sum_{i=1}^{n} p_i \int_{t_{i-1}}^{t_i} \tilde{s}_t dt, \sum_{j=1}^{n} p_j \int_{t_{j-1}}^{t_j} \tilde{s}_\lambda d\lambda\rangle - \sum_{j=1}^{n} p_j d_j - \sum_{j=1}^{n} \varepsilon_j |p_j| \\ &= -\frac{1}{4}\sum_{i=1}^{n}\sum_{j=1}^{n} p_i p_j \varsigma_{i,j} - \sum_{j=1}^{n} p_j d_j - \sum_{j=1}^{n} \varepsilon_j |p_j|,\end{aligned}$$

where

$$\varsigma_{i,j} = \int_{t_{i-1}}^{t_i}\int_{t_{j-1}}^{t_j} \langle L^{-1}\tilde{s}_t, \tilde{s}_\lambda\rangle dt d\lambda$$

and

$$L^{-1}\tilde{s}_t = \sum_{i=0}^{\infty} \Lambda(t - iT_s)L^{-1}s_{iT_s}.$$

Finite filter structures

Recall (3.3.9) of Chapter 3:

$$\eta(p) = -\frac{1}{4}\langle p \bullet y_s, (K^{\dagger}LK)^{-1}p \bullet y_s\rangle - p \bullet d - \|p\|_{\varepsilon}, \tag{5.3.3}$$

where $y_s = [\Xi_s v_0, \ldots, \Xi_s v_{n-1}]^T$. Let p^o be a minimizer of the dual functional η given by (5.3.3). Then, the optimal filter parameter vector a^o is given by (3.3.10), i.e.

$$a^o = -\frac{1}{2}(K^{\dagger}LK)^{-1}K^{\dagger}\hat{\alpha}(p^o) = -\frac{1}{2}(K^{\dagger}LK)^{-1}\int_{\Omega_c} y_s dp^o.$$

Discretize the dual space $M[t_0, t_0 + T_c]$ to a subspace $M_n[t_0, t_0 + T_c]$ as before. Then, substituting

$$p \bullet y_s = \int_{t_0}^{t_0+T_c} y_s dp = \sum_{i=1}^{n} p_i \int_{t_{i-1}}^{t_i} y_s(t)dt$$

into the expression for the dual cost (5.3.3) yields

$$\begin{aligned}\eta(p) &= -\frac{1}{4}\langle (K^{\dagger}LK)^{-1}\sum_{i=1}^{n} p_i \int_{t_{i-1}}^{t_i} y_s(t)dt, \sum_{i=1}^{n} p_i \int_{t_{i-1}}^{t_i} y_s(\lambda)d\lambda\rangle - \sum_{j=1}^{n} p_j d_j - \sum_{j=1}^{n} \varepsilon_j |p_j| \\ &= -\frac{1}{4}\sum_{i=1}^{n}\sum_{j=1}^{n} p_i p_j \varsigma_{i,j} - \sum_{j=1}^{n} p_j d_j - \sum_{j=1}^{n} \varepsilon_j |p_j|\end{aligned}$$

where

$$\varsigma_{i,j} = \int_{t_{i-1}}^{t_i}\int_{t_{j-1}}^{t_j} \langle (K^{\dagger}LK)^{-1} y_s(t), y_s(\lambda)\rangle dt d\lambda.$$

Steepest ascent algorithm

A steepest ascent search procedure is used to maximize the dual cost functional $\eta(p)$. The functional to be maximized is not Fréchet (strong sense) differentiable because it contains the ε-weighted norm

$$\|p\|_{\varepsilon} = \int_{t_0}^{t_0+T_c} \varepsilon(t)|p'(t)|dt = \sum_{j=1}^{n} \varepsilon_j|p_j|.$$

This weighted norm does, however, have a directional Gateaux (weak sense) differential, which is convex (but not always linear) in its increment (see Appendix B.4). This differential at p in the direction of h, where $p, h \in M_n[t_0, t_0 + T_c]$, is defined as

$$\delta^+\|p;h\|_{\varepsilon} = \lim_{\alpha \to 0^+} \frac{\|p + \alpha h\|_{\varepsilon} - \|p\|_{\varepsilon}}{\alpha}$$

Since this is a finite dimensional problem, the procedure is nearly the same as the discrete-time case from which it follows that

$$\delta^+\|p;h\|_{\varepsilon} = \sum_{i \in \Lambda^+(p)} \varepsilon_i h_i - \sum_{i \in \Lambda^-(p)} \varepsilon_i h_i + \sum_{i \in \Lambda^0(p)} \varepsilon_i |h_i|$$

where the set of integers $\{1, \ldots, n\}$ is partitioned into disjoint sets:

$$\Lambda^+(p) \equiv \{i : p_i > 0\}$$

$$\Lambda^-(p) \equiv \{i : p_i < 0\}$$

$$\Lambda^0(p) \equiv \{i : p_i = 0\}$$

The directional differential of $\eta(p)$ is

$$\delta^+\eta(p;h) = -\frac{1}{2}\sum_{j=1}^{n}\sum_{i=1}^{n} p_j h_i \varsigma_{i,j} - \sum_{i=1}^{n} h_i d_i - \delta^+\|p;h\|_{\varepsilon}.$$

By defining $\Delta_i(p)$ to be

$$\Delta_i(p) = \frac{1}{2}\sum_{j=1}^{n} p_j \varsigma_{i,j} + d_i,$$

we can also write

$$\begin{aligned}\delta^+\eta(p;h) = &-\sum_{i\in\Lambda^+(p)}(\Delta_i(p)+\varepsilon_i)h_i - \sum_{i\in\Lambda^-(p)}(\Delta_i(p)-\varepsilon_i)h_i \\ &-\sum_{i\in\Lambda^0(p)}\Delta_i(p)h_i + \varepsilon_i|h_i|.\end{aligned} \tag{5.3.4}$$

The set $(\Lambda^0(p))$ can be further partitioned into disjoint sets:

$$\chi^+(p) \equiv \{i \in \Lambda^0(p) : -(\Delta_i(p)+\varepsilon_i) > 0\}$$
$$\chi^-(p) \equiv \{i \in \Lambda^0(p) : -(\Delta_i(p)-\varepsilon_i) < 0\}$$
$$\chi^0(p) \equiv \Lambda^0(p) - \chi^+(p) - \chi^-(p)$$

Since, any $\hat{h}$, which achieves $\max_{\|h\|\le 1}\delta^+\eta(n;h)$, is a direction of steepest ascent from the point p, it follows from the same arguments as for the discrete-time case given in Section 4.3 that

$$\mathrm{sgn}(\hat{h}_i) = \begin{cases} -\mathrm{sgn}[\Delta_i(p)+\varepsilon_i],\ i\in\Lambda^+(p) \\ -\mathrm{sgn}[\Delta_i(p)-\varepsilon_i],\ i\in\Lambda^-(p)\end{cases} \tag{5.3.5}$$

and

$$\mathrm{sgn}(\hat{h}_i) = \begin{cases} -\mathrm{sgn}[\Delta_i(p)+\varepsilon_i] = +1,\ i\in\chi^+(p) \\ -\mathrm{sgn}[\Delta_i(p)-\varepsilon_i] = -1,\ i\in\chi^-(p) \\ 0, \qquad\qquad\qquad\qquad\quad i\in\chi^0(p)\end{cases} \tag{5.3.6}$$

Restricting consideration to direction vectors that satisfy (5.3.5) and (5.3.6), one may rewrite the directional differential as

$$\delta^+\eta(p;h) = \langle l(p), h\rangle$$

where

$$l_i(p) = \begin{cases} -(\Delta_i(p) + \varepsilon_i), & i \in \Lambda^+(p) \cup \chi^+(p) \\ -(\Delta_i(p) - \varepsilon_i), & i \in \Lambda^-(p) \cup \chi^-(p) \\ 0, & i \in \chi^0(p) \end{cases}.$$

Since $u = -0.5L^{-1}\hat{\alpha}(p)$, the output ψ of this filter is

$$\psi = -0.5\Xi_s L^{-1}\hat{\alpha}(p).$$

It can be shown that

$$\frac{1}{2}\sum_{j=1}^{n} p_j \varsigma_{i,j} = -\int_{t_{i-1}}^{t_i} \psi(t)dt, \; i = 1, 2, \ldots, n. \tag{5.3.7}$$

To see this, consider the analog case first. Noting $(\Xi_s L^{-1} s_t)(\lambda) = \langle L^{-1} s_t, s_\lambda \rangle$, it follows that

$$\psi(\lambda) = -\frac{1}{2}[\Xi_s L^{-1}\hat{\alpha}(p)](\lambda) = -\frac{1}{2}\sum_{j=1}^{n} p_j \int_{t_{j-1}}^{t_j} (\Xi_s L^{-1} s_t)(\lambda)dt$$

$$= -\frac{1}{2}\sum_{j=1}^{n} p_j \int_{t_{j-1}}^{t_j} \langle L^{-1} s_t, s_\lambda \rangle dt$$

Similarly, for the hybrid case, we have

$$\psi(\lambda) = -0.5\sum_{j=1}^{n} p_j \int_{t_{j-1}}^{t_j} \langle L^{-1}\tilde{s}_t, \tilde{s}_\lambda \rangle dt.$$

For the case of finite dimensional filters,

$$\boldsymbol{a} = -0.5(K^\dagger L K)^{-1}\int_{\Omega_c} y_s dp.$$

Thus,

$$\psi(\lambda) = -\frac{1}{2} y_s^T(\lambda)(K^\dagger L K)^{-1}\int_{\Omega_c} y_s dp$$

$$= -\frac{1}{2}\sum_{j=1}^{n} p_j \int_{t_{j-1}}^{t_j} y_s^T(\lambda)(K^\dagger L K)^{-1} y_s(t)dt$$

Hence, for all three cases, integrating both sides yields (5.3.7).

By defining $\psi_i = \int_{t_{i-1}}^{t_i} \psi(t)dt$, it is clear that $l_i(p)$ can be written as

$$l_i(p) = \begin{cases} \psi_i - \varepsilon_i^+, \, i \in \Lambda^+(p) \cup \chi^+(p) \\ \psi_i - \varepsilon_i^-, \, i \in \Lambda^+(p) \cup \chi^-(p) \\ 0, \qquad i \in \chi^0(p) \end{cases}$$

$$= \begin{cases} \psi_i - \varepsilon_i^+, \text{ if } p_i > 0 \text{ or if } p_i = 0 \text{ and } \psi_i > \varepsilon_i^+ \\ \psi_i - \varepsilon_i^-, \text{ if } p_i < 0 \text{ or if } p_i = 0 \text{ and } \psi_i < \varepsilon_i^- \\ 0, \qquad \text{if } p_i = 0, \text{ and } \varepsilon_i^- < \psi_i < \varepsilon_i^+ \end{cases}$$

In this form, the steepest ascent direction is the same as that for the discrete-time case, and hence, the convergence results similar to the discrete-time case apply. The solution obtained by discretizing the dual problem is an approximation of the dual solution. Thus, the corresponding primal solution is not guaranteed to satisfy the envelope constraints. Only in the limiting case, i.e. infinitesimal discretization, would the constraints be satisfied. The limitation of the primal-dual method can be alleviated by the penalty approach introduced in the next section.

5.4 Penalty Approach for Semi-Infinite Programming

This section presents a new numerical method that can be used to solve the EC filtering problem with finite dimensional cost functional. For completeness, results concerning infinite dimensional costs are also presented. The techniques developed in this chapter are not only applicable to EC filtering problems but also useful to other optimization problems (see [119]).

First, we provide some backgrounds on a general problem with functional inequality constraints. This problem is then approximated by conventional constrained sub-problems. Although this is not the approach for solving the EC filtering problem, it provides necessary results for the next section in which conventional unconstrained sub-problems are used for the approximation. We then narrow our scope down to those problems with quadratic cost and convex functional constraints for which a closed form expression for the approximation error is derived. Next we narrow the scope further down to the class of quadratic programs with affine constraints, which contains the EC filtering problem. When the decision variable is finite dimension, we have a semi-infinite program and a class of iterative algorithms is developed by applying the steepest descent direction methods to the associated unconstrained sub-problem.

Consider the class of optimization problems where the continuously differentiable cost function f is to be minimized over a compact subset Θ of a Hilbert space H subject to the functional inequality constraints:

$$\begin{aligned} &\min \ f(x) \\ &\text{subject to } \ \phi_j(x) \preceq 0, j = 1, \ldots, m, \end{aligned} \tag{5.4.1}$$

where $x \in H$, $\phi_j : H \to W$ is continuously differentiable on H, and W is the space of $L_2[0, \infty)$ functions that are Lipschitz continuous on Ω_c. As before, the partial ordering $\preceq$ on W is defined by

$$y, z \in W, \ y \preceq z \ \Leftrightarrow \ \forall t \in \Omega_c, \ y(t) \leq z(t),$$

where Ω_c is a compact interval in $\boldsymbol{R}$. For convenience, this problem is referred to as Problem $(\boldsymbol{P})$ and any of its solutions is denoted by x^o.

The functional constraints can also be expressed as

$$g_j(x) \equiv \max_{t \in \Omega_c} [\phi_j(x)](t) \leq 0.$$

Hence, the feasible set can be defined by

$$\mathcal{F} = \{x \in \Theta : g_j(x) \leq 0, j = 1, \ldots, m\}.$$

Observe that $\mathcal{F}$ is closed since it is an intersection of closed sets, which happen to be inverse images of the closed semi-infinite interval $(-\infty, 0]$ under the continuous maps g_j's. It is assumed that the following conditions are satisfied,

Assumption A-5.4.1. $\mathcal{F}^o \equiv \{x \in \Theta^o : \phi_j(x) \prec 0, j = 1, \ldots, m\} \neq \emptyset$,

Assumption A-5.4.2. $\exists \bar{x} \in \mathcal{F}^o$ *such that* $x_\alpha \equiv \alpha\bar{x} + (1-\alpha)x^o \in \mathcal{F}^o$, $\forall \alpha \in (0, 1]$.

Note that convexity of $\mathcal{F}$ is sufficient for Assumption A-5.4.2 to hold.

Approximations by conventional constrained problems

In this section, a class of conventional inequality constrained optimization problems is constructed to approximate Problem (***P***). These approximate problems do not suffer from failure of constraint qualifications as the one considered in [109] and can also be chosen to be smooth. Thus, they can be readily solved by existing methods such as those in [25], [27], [37], [84] and Section 4.4.

Let $G_\upsilon : W \to \boldsymbol{R}$ be a (nonlinear) functional on W defined by

$$G_\upsilon(w) = \int_{\Omega_c} g_\upsilon(w(t))dt \qquad (5.4.2)$$

where $g_\upsilon : \boldsymbol{R} \to \boldsymbol{R}$ is an integrable function with the following properties (see Figure 4.4.1):

(P1) $g_\upsilon(z) = 0, \forall z \leq -\upsilon$,

(P2) $g_\upsilon(z)$ is strictly increasing $\forall z > -\upsilon$.

Clearly, G_υ is continuously differentiable in w if g_υ is continuously differentiable. The Fréchet derivative of G_υ at w, i.e. $G'_\upsilon(w) = (dG_\upsilon)_w : W \to \boldsymbol{R}$ a linear functional on W (see Appendix B.4) defined by

$$G'_\upsilon(w)h = \int_{\Omega_c} g'_\upsilon(w(t))h(t)dt, \text{ for any } h \in W. \qquad (5.4.3)$$

For each $j = 1, \ldots, m$, and $\upsilon > 0$, define a mapping $G_{j,\upsilon}: H \to \boldsymbol{R}$ by

$$G_{j,\upsilon} \equiv G_{\upsilon} \circ \phi_j .$$

Each $G_{j,\upsilon}$ is continuously differentiable in x if g_{υ} is continuously differentiable. For each $\upsilon > 0$ and $\tau \geq 0$ an approximate Problem $(\boldsymbol{P}_{\upsilon,\tau})$ is posed as follows:

$$\begin{gathered} \min f(x) \\ \text{subject to } G_{j,\upsilon}(x) \leq \tau, j = 1, \ldots, m. \end{gathered} \tag{5.4.4}$$

The feasible set for Problem $(\boldsymbol{P}_{\upsilon,\tau})$ is given by

$$\mathcal{F}_{\upsilon,\tau} = \{x \in \Theta : G_{j,\upsilon}(x) \leq \tau,\ j = 1, \ldots, m\}.$$

For each $\upsilon > 0$, note firstly that $\mathcal{F}_{\upsilon,0} \subset \mathcal{F}$, and secondly $\varsigma \leq \tau$ implies $\mathcal{F}_{\upsilon,\varsigma} \subseteq \mathcal{F}_{\upsilon,\tau}$. These suggest that there is some critical $\tau_{\upsilon} > 0$ such that if $\tau < \tau_{\upsilon}$, then, $\mathcal{F}_{\upsilon,\tau}$ is contained in the feasible region $\mathcal{F}$.

Theorem 5.4.1. *Let*

$$\tau_{\upsilon} \equiv \sup_{\xi \in (0,\upsilon)} \{g_{\upsilon}(-\xi)\min(|\Omega_c|, \xi/\eta)\}, \tag{5.4.5}$$

where η *is a number satisfying*

$$\eta|t_2 - t_1| \geq |\phi_j(x)(t_2) - \phi_j(x)(t_1)|$$

for all $t_2, t_1 \in \Omega_c, x \in \Theta, j \in \{1, \ldots, m\}$. *Then,* $0 < \tau < \tau_{\upsilon} \Rightarrow \mathcal{F}_{\upsilon,\tau} \subset \mathcal{F}$

The parameter τ can be chosen by calculating τ_{υ} directly or by finding a positive lower bound for it. Figure 5.4.1 graphically depicts $g_{\upsilon}(-\xi)$ and $\min(|\Omega_c|, \xi/\eta)$ as functions of ξ. This assumes that $|\Omega_c|\eta < \upsilon$. It is intuitive that for any $\rho \in (0, 1)$, $g_{\upsilon}(-\rho|\Omega_c|\eta)\rho|\Omega_c|$ is a positive lower bound for τ_{υ}. In general, the parameter τ can be estimated according to the following result.

Corollary 5.4.1. *Let* $\bar{\upsilon} = \min(|\Omega_c|\eta, \upsilon)$ *where* η *is given in Theorem 5.4.1. Then, for any* $\rho \in (0, 1)$, $\tau = g_{\upsilon}(-\rho\bar{\upsilon})\rho\bar{\upsilon}/\eta \Rightarrow \mathcal{F}_{\upsilon,\tau} \subset \mathcal{F}$.

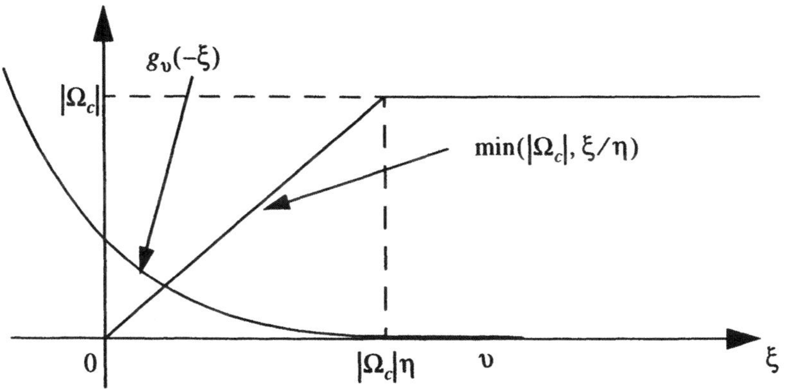

Figure 5.4.1. Lower bound.

Using the notation $x^o_{\upsilon,\tau}$ to denote a solution of Problem $(\boldsymbol{P}_{\upsilon,\tau})$. The following theorem illustrates the relationship between Problems $(\boldsymbol{P}_{\upsilon,\tau})$ and $(\boldsymbol{P})$ by showing that $f(x^o_{\upsilon,\tau})$ approaches $f(x^o)$ as the parameter υ tends to zero (and τ is chosen appropriately).

Theorem 5.4.2. *If $\tau < \tau_\upsilon$, then, $\lim_{\upsilon \to 0} f(x^o_{\upsilon,\tau}) = f(x^o)$, and any accumulation point of $\{x^o_{\upsilon(n),\tau}\}_{n=1}^{\infty}$ is a solution of $(\boldsymbol{P})$, where $\{\upsilon(n)\}_{n=1}^{\infty}$ is any positive sequence that converges to zero.*

The smoothness of Problem $(\boldsymbol{P}_{\upsilon,\tau})$ depends on the choice of g_υ. Smooth problems are generally easier to handle than their non-smooth counterparts. This technique is also useful in optimal control applications, see [111] for details. However, it is not as useful as the unconstrained approximation method to be presented in the next section.

Approximations by unconstrained problems

In this section, a penalty function associated with the constraints of Problem $(\boldsymbol{P})$ (see (5.4.1)) is added to the original cost to form an augmented cost. The resulting class of conventional unconstrained problems can then be used to approximate $(\boldsymbol{P})$. The advantage of this method is that unconstrained problems are much easier to handle than constrained problems. The unconstrained optimization Problem $(\bar{\boldsymbol{P}}_{\upsilon,\gamma})$ is defined for each $\upsilon, \gamma > 0$ as follows:

$$\min f_{\upsilon,\gamma}(x) \equiv f(x) + \gamma \sum_{j=1}^{m} G_{j,\upsilon}(x) \tag{5.4.6}$$

The notation $x^{o}_{\upsilon,\gamma}$ is used to denote a solution of Problem $(\bar{\boldsymbol{P}}_{\upsilon,\gamma})$. The following results show that for a sufficiently large penalty parameter γ, $x^{o}_{\upsilon,\gamma}$ is a feasible point for Problem $(\boldsymbol{P})$. Note that the parameter υ must be sufficiently small so that $\mathcal{F}_{\upsilon} \equiv \mathcal{F}_{\upsilon,0}$ is non-empty.

Theorem 5.4.3. *Let $\tilde{x} \in \Theta$ be such that $f(\tilde{x}) \le f(x)$, $\forall x \in \Theta$, let υ be such that $\mathcal{F}_{\upsilon}$ is non-empty and let $x_{\upsilon} \in \mathcal{F}_{\upsilon}$. If $\gamma > \gamma_{\upsilon} \equiv (f(x_{\upsilon}) - f(\tilde{x}))/\tau_{\upsilon}$, in which τ_{υ} is given by (5.4.5), then $x^{o}_{\upsilon,\gamma} \in \mathcal{F}$.*

Corollary 5.4.2. *Let $\bar{\upsilon} = \min(|\Omega_c|\eta, \upsilon)$, where η is given in Theorem 5.4.1. Then for any $\rho \in (0, 1)$,*

$$\gamma = \frac{(f(x_{\upsilon}) - f(\tilde{x}))\eta}{g_{\upsilon}(-\rho\bar{\upsilon})\rho\bar{\upsilon}} \Rightarrow x^{o}_{\upsilon,\gamma} \in \mathcal{F}.$$

The proof follows from Corollary 5.4.1 and Theorem 5.4.3.

The approximation of Problem $(\boldsymbol{P})$ by Problem $(\bar{\boldsymbol{P}}_{\upsilon,\gamma})$ can be justified by showing that $f(x^{o}_{\upsilon,\gamma})$ approaches $f(x^{o})$ as the parameter υ tends to zero (and γ is chosen appropriately). Note that as υ tends to zero, the parameter γ tends to infinity.

Theorem 5.4.4. *If $\gamma > \gamma_{\upsilon}$, then, $\lim_{\upsilon \to 0} f(x^{o}_{\upsilon,\gamma}) = f(x^{o})$, and any accumulation point of $\{x^{o}_{\upsilon(n),\gamma}\}_{n=1}^{\infty}$ is a solution of $(\boldsymbol{P})$, where $\{\upsilon(n)\}_{n=1}^{\infty}$ is any positive sequence that converges to zero.*

Since the cost function is smooth, the augmented cost function is smooth if g_{υ} is also smooth. Smooth unconstrained problem can be readily solved by existing gradient based algorithms. Examples of smooth g_{υ} are given in (4.4.10)-(4.4.12) (see Chapter 4). For problems where the original cost is twice continuously differentiable, a continuously differentiable g_υ yields a twice continuously differentiable augmented cost.

Approximating convex problems

The functional inequality constrained problems considered in the previous sections are quite general. This section narrows the scope down to those problems for which the cost is strictly convex, while the functional constraints are convex. More specifically, we consider problems of the form:

$$\begin{array}{c} \min f(x) \\ \text{subject to } \phi_j(x) \preceq 0, j = 1, \ldots, m, \end{array} \tag{5.4.7}$$

where f is a strictly convex cost functional and the ϕ_j's are convex.

Since the constraint functions are convex, the feasible set for this problem is also convex. Convexity of the feasible region and strict convexity of the cost imply that Problem (***P***) has a unique solution x^o, assuming that a solution exists.

Lemma 5.4.1. *If g_υ is continuous and convex, then $f_{\upsilon,\gamma}$ is continuous and strictly convex.*

Strict convexity of the augmented cost $f_{\upsilon,\gamma}$ implies that $x^o_{\upsilon,\gamma}$ is unique. This uniqueness also implies that not only $f(x^o_{\upsilon,\gamma}) \to f(x^o)$ as in Theorem 5.4.4 but also $x^o_{\upsilon,\gamma} \to x^o$ as in the following theorem.

Theorem 5.4.5. *For each $\upsilon > 0$ such that $\mathcal{F}_\upsilon \neq \varnothing$, if $\gamma > f(x_\upsilon)/\tau_\upsilon$ for some $x_\upsilon \in \mathcal{F}_\upsilon$, then $x^o_{\upsilon,\gamma} \in \mathcal{F}$. Moreover, $\lim_{\upsilon \to 0} x^o_{\upsilon,\gamma} = x^o$.*

Suppose that the cost functional is quadratic, i.e.

$$f(x) = \langle x, Lx \rangle \equiv \|x\|_L^2,$$

where L is positive definite. By exploiting the convexity of inequality constraints and the quadratic cost function, it can be shown that, for any given error bound ε, the accuracy parameter υ can be chosen (without using any information on the solutions $x^o_{\upsilon,\gamma}$ and x^o) so that the approximation error $\|x^o_{\upsilon,\gamma}\|^2_L - \|x^o\|^2_L$, is less than the error bound ε. This is presented in the following theorem.

Theorem 5.4.6. *For any $\bar{x} \in \mathcal{F}^o$, let*

$$\delta(\bar{x}) \equiv \min_j \min_{t \in \Omega_c} \{-[\phi_j(\bar{x})](t)/2\|\bar{x}\|^2_L\}.$$

If $\upsilon > 0$ is such that $\mathcal{F}_\upsilon$ is non-empty, and $\gamma > \|x_\upsilon\|^2_L/\tau_\upsilon$ for some $x_\upsilon \in \mathcal{F}_\upsilon$, then,

$$0 < \upsilon \le \delta(\bar{x})\min\{\varepsilon, 2\|\bar{x}\|^2_L\} \Rightarrow 0 \le \|x^o_{\upsilon,\gamma}\|^2_L - \|x^o\|^2_L \le \varepsilon.$$

When an interior point is known, given any tolerance on the cost, the accuracy parameter can be easily selected according to Theorem 5.4.6 so that $\|x^o_{\upsilon,\gamma}\|^2_L - \|x^o\|^2_L$ is less than the tolerance. This is useful in implementing interior point algorithms [45] for Problem (***P***).

Affine functional inequality constrained problems

This section presents a class of iterative algorithms that can be used to solve the EC filtering problem. The EC filtering problem with known input can be posed as a QP problem with affine functional inequality constraints (see Section 2.2). The constraints of the EC filtering problem are not only convex but affine. For generality, we consider Problem (3.1.1) of Chapter 3, i.e.

$$\begin{aligned} &\min \langle x, Lx\rangle + \langle b, x\rangle \\ &\text{subject to } \phi_j(x) \equiv A_j x - c_j \preceq 0, j = 1, \ldots, m, \end{aligned} \tag{5.4.8}$$

where $x, b \in H$; $c_i \in W$; L is a positive definite linear operator on H; and each A_i is a bounded linear operator from H to W.

All the results of the previous section apply since the affine constraints are convex. The following lemma summarizes some useful analytical properties of the augmented cost function.

Lemma 5.4.2.

(i) *If g_υ is once continuously differentiable, then $f_{\upsilon,\gamma}$ is once continuously differentiable and its gradient $\nabla f_{\upsilon,\gamma}(x)$ at x is defined by*

$$\langle \nabla f_{\upsilon,\gamma}(x), h\rangle = 2\langle Lx, h\rangle + \langle b, h\rangle + \gamma \sum_{j=1}^{m} \int_{\Omega_c} g'_\upsilon(\phi_j(x)(t))(A_j h)(t)dt.$$

(ii) *If g_υ is twice continuously differentiable, then $f_{\upsilon,\gamma}$ is twice continuously differentiable, and its Hessian $\nabla^2 f_{\upsilon,\gamma}(x)$ at x is defined by*

$$\langle \nabla^2 f_{\upsilon,\gamma}(x)k, h\rangle = 2\langle Lk, h\rangle + \gamma \sum_{j=1}^{m} \int_{\Omega_c} g''_\upsilon(\phi_j(x)(t))(A_j h)(t)(A_j k)(t)dt.$$

(iii) *If g_υ has bounded second derivative, i.e. $|g''_\upsilon(\omega)| \le h_\upsilon, \forall \omega \in \boldsymbol{R}$, then $f_{\upsilon,\gamma}$ also has bounded Hessian, i.e.*

$$\|L\| \le \|\nabla^2 f_{\upsilon,\gamma}(x)\| \le 2\|L\| + \gamma h_\upsilon \sum_{j=1}^{m} \|A_j\|_2^2.$$

By choosing g_υ appropriately, globally convergent algorithms for minimizing the augmented cost can be developed. To obtain sub-optimum solutions, the following iterative process is used:

$$x_{k+1} = x_k + \mu_k d_k$$

where x_k is the kth iterate, μ_k is the stepsize, and d_k satisfying $\langle \nabla f_{\upsilon,\gamma}(x_k), d_k\rangle < 0$ is called the descent direction.

Assuming that g_υ is at least once continuously differentiable, the simplest of the descent directions is the steepest descent (SD) direction, $-\nabla f_{\upsilon,\gamma}(x_k)$. The popularity of this method has been favored due to the comparatively simple implementation and applicability to a wide range of functions.

Theorem 5.4.7*. Suppose that g_υ is twice continuously differentiable for each $\upsilon > 0$ with $|g''_\upsilon(\omega)| \le h_\upsilon$, $\forall \omega \in \boldsymbol{R}$. Then, starting from any initial point x_0, the sequence $\{x_k\}_{k=0}^{\infty}$ generated by the SD algorithm with constant stepsize μ converges to $x^o_{\upsilon,\gamma}$ if*

$$0 < \mu < \frac{2}{2\|L\| + \gamma h_\upsilon \sum_{j=1}^{m} \|A_j\|^2}.$$

Proof. Using Lemma 5.4.2(iii), the bounds on the Hessian of the augmented cost function are obtained. Then the conclusion follows readily from Theorem 4.8.1. ❐

Making the identifications:

$$\phi^+(u) = \Xi_s u - \varepsilon^+,$$
$$\phi^-(u) = -\Xi_s u + \varepsilon^-,$$

the augmented cost for the EC filtering problem is

$$f_{\upsilon,\gamma}(u) = \langle u, Lu\rangle + \gamma\int_{\Omega_c} [g_\upsilon(\phi^+(u)(t)) + g_\upsilon(\phi^-(u)(t))]dt.$$

It is clear that the SD algorithm converges to a feasible solution when the step size satisfies

$$0 < \mu < \frac{1}{\|L\| + \gamma h_\upsilon \|\Xi_s\|^2}.$$

SIP with quadratic cost and affine constraints

As discussed in Section 2.4, the infinite dimensional problem is only of theoretical interest. For practical circuits with finite structures, the resulting cost is finite dimensional. In this section, we confine ourselves to finite dimensional decision variables, i.e., Problem (3.5.1) of Chapter 3.

$$\min \frac{1}{2}\boldsymbol{x}^T Q\boldsymbol{x} + \boldsymbol{b}^T\boldsymbol{x}$$
$$\text{subject to } \boldsymbol{a}_j^T\boldsymbol{x} - c_j \preceq 0, j = 1, \ldots, m, \tag{5.4.9}$$

where $Q \in \boldsymbol{R}^{n\times n}$ is positive definite, $\boldsymbol{b} \in \boldsymbol{R}^n$, $c_j \in W$, and $\boldsymbol{a}_j \in W^n$ is given by $\boldsymbol{a}_j^T(\omega) = [a_{j,1}(\omega), \ldots, a_{j,n}(\omega)], j = 1, \ldots, m$.

The following summarizes the properties of the augmented cost for the finite dimensional case:

(i) *If g_υ is once continuously differentiable, then $f_{\upsilon,\gamma}$ is once continuously differentiable and its gradient is given by*

$$\nabla f_{\upsilon,\gamma}(\boldsymbol{x}) = 2Q\boldsymbol{x} + \boldsymbol{b} + \gamma \sum_{j=1}^{m} \int_{\Omega_c} g'_\upsilon(\phi_j(\boldsymbol{x}, t))\boldsymbol{a}_j(t)dt .$$

(ii) *If g_υ is twice continuously differentiable, then $f_{\upsilon,\gamma}$ is twice continuously differentiable, and its Hessian is given by*

$$\nabla^2 f_{\upsilon,\gamma}(\boldsymbol{x}) = 2Q + \gamma \sum_{j=1}^{m} \int_{\Omega_c} g''_\upsilon(\phi_j(\boldsymbol{x}, t))\boldsymbol{a}_j(t)\boldsymbol{a}_j^T(t)dt .$$

(iii) *If g_υ has bounded second derivative i.e. $|g''_\upsilon(\omega)| \le h_\upsilon, \forall \omega \in \boldsymbol{R}$, then $f_{\upsilon,\gamma}$ also has bounded Hessian, i.e.*

$$2\|Q\| \le \|\nabla^2 f_{\upsilon,\gamma}(\boldsymbol{x})\| \le 2\|Q\| + \gamma h_\upsilon \sum_{j=1}^{m} \int_{\Omega_c} \|\boldsymbol{a}_j(t)\|^2 dt .$$

(iv) *If g_υ has Lipschitz continuous second derivative, i.e. there exists some l_υ, such that $|g''_\upsilon(x) - g''_\upsilon(y)| \le l_\upsilon |x - y|$, $\forall x, y \in \boldsymbol{R}$, then $\nabla^2 f_{\upsilon,\gamma}$ is also Lipschitz continuous. More precisely,*

$$\|\nabla^2 f_{\upsilon,\gamma}(\boldsymbol{u}) - \nabla^2 f_{\upsilon,\gamma}(\boldsymbol{v})\| \le \gamma l_\upsilon \|\boldsymbol{u} - \boldsymbol{v}\| \sum_{j=1}^{m} \int_{\Omega_c} \|\boldsymbol{a}_j(t)\|^3 dt .$$

The SD algorithm is attractive to a large extent due to its computational simplicity. However, the method has several serious practical drawbacks. Most importantly, convergence is usually slow (see [17], [18], [25], [27], [37], [90]). Another popular method is the Newton-Raphson (NR) algorithm, which uses the second derivative to calculate the descent direction,

$$\boldsymbol{d}_k = -(\nabla^2 f_{\upsilon,\gamma}(\boldsymbol{x}_k))^{-1} \nabla f_{\upsilon,\gamma}(\boldsymbol{x}_k) .$$

It is assumed that g_υ is chosen so that the augmented cost functional is twice continuously differentiable, and hence its Hessian is positive definite (twice continuously differentiable strictly convex functions have positive definite Hessian [90]). A larger stepsize than the SD algorithm is allowed as illustrated in the following theorem.

Theorem 5.4.8. *Suppose that g_υ is twice continuously differentiable for each $\upsilon > 0$ with $|g''_\upsilon(\omega)| \le h_\upsilon$, $\forall \omega \in \boldsymbol{R}$. Then, starting from any initial point $\boldsymbol{x}_0$, the sequence $\{x_k\}_{k=0}^{\infty}$ generated by the NR algorithm with constant stepsize μ converges to $\boldsymbol{x}^o_{\upsilon,\gamma}$ if*

$$0 < \mu < \frac{4}{2\|Q\| + \gamma h_\upsilon \sum_{j=1}^{m} \int_{\Omega_c} \|\boldsymbol{a}_j(t)\|^2 dt}.$$

Proof. Again, using Lemma 5.4.2 (iv), the bounds on the second derivative of the augmented cost function are obtained. Then, the conclusion follows readily from Theorem 4.8.2. ❒

If $g''_\upsilon(x)$ is bounded and Lipschitz continuous, and the line search procedure outlined in Section 4.4 is used to calculate the stepsize, then the NR algorithm has a quadratic convergence rate as asserted by Theorem 4.4.5.

Application to finite dimensional EC filters

Recall from Section 2.4 that EC filtering problem for finite dimensional filters can be written as

$$\begin{gathered} \min \boldsymbol{a}^T Q \boldsymbol{a}, \boldsymbol{a} \in \boldsymbol{R}^n \\ \text{subject to } \varepsilon^-(t) \le \boldsymbol{y}_s^T(t)\boldsymbol{a} \le \varepsilon^+(t), \forall t \in [0, T_c], \end{gathered} \tag{5.4.10}$$

where $Q = K^\dagger L K$ and $\boldsymbol{y}_s^T(t)$ is given by (3.3.4) in Chapter 3. Assume that the feasible set of Problem (5.4.10) has a non-empty interior, i.e. $\mathcal{F}^o \ne \emptyset$. For each $\upsilon, \gamma > 0$, the augmented cost function $f_{\upsilon,\gamma}$ is defined as

$$f_{\upsilon,\gamma}(\boldsymbol{a}) = \boldsymbol{a}^T Q \boldsymbol{a} + \gamma \int_{\Omega_c} [g_\upsilon(\phi^+(\boldsymbol{a}, t)) + g_\upsilon(\phi^-(\boldsymbol{a}, t))] dt.$$

Remark 5.4.1. From Lemma 5.4.1, the augmented cost $f_{\upsilon,\gamma}$ is continuous and strictly convex, and hence attains a unique minimum.

Let $\boldsymbol{a}^o$ be the optimum filter parameter vector of Problem (5.4.10), and let $\boldsymbol{a}^o_{\upsilon,\gamma}$ be the minimizer of $f_{\upsilon,\gamma}$. For any $\bar{\boldsymbol{a}} \in \mathcal{F}^o$, let

$$\delta(\bar{a}) = \min_{t \in \Omega_c} \min\{-\phi^+(\bar{a}, t), -\phi^-(\bar{a}, t)\}/2\|\bar{a}\|_Q^2,$$

and, for any υ such that $\mathcal{F}_\upsilon \neq \emptyset$, let

$$\tau_\upsilon = \sup_{\xi > 0}\{g_\upsilon(-\xi)\min(T_c, \xi/\eta)\},$$

where η is a number satisfying

$$\eta|t_2 - t_1| \geq \max\{|\phi^+(a, t_2) - \phi^+(a, t_1)|, |\phi^-(a, t_2) - \phi^-(a, t_1)|\}, \forall t_2, t_1 \in \Omega_c.$$

Applying Theorems 5.4.5 and 5.4.6 yield:

(i) *If* $\gamma > \|a_\upsilon\|_Q^2/\tau_\upsilon$ *for some* $a_\upsilon \in \mathcal{F}_\upsilon$ *and* τ_υ *is given by (5.4.5),, then* $a_{\upsilon,\gamma}^o$ *satisfies the envelope constraints.*

(ii) *If for each* υ, γ *is chosen as in (i), then* $\lim_{\upsilon \to 0} a_{\upsilon,\gamma}^o = a^o$,

(iii) *For any* $\bar{a} \in \mathcal{F}^o$, *if* $0 < \upsilon \leq \delta(\bar{a})\min\{\varepsilon, 2\|\bar{a}\|_Q^2\}$ *and* γ *is chosen as in (i), then* $0 \leq \|a_{\upsilon,\gamma}^o\|_Q^2 - \|a^o\|_Q^2 \leq \varepsilon$.

These results share the same interpretations as for the discrete-time case. Part (i) asserts that for a fixed but small positive υ, a sufficiently large but finite γ yields a feasible solution for the EC filtering Problem (5.4.10). Provided that for each υ, the γ is chosen according to part (i), part (ii) shows that the approximate solution $a_{\upsilon,\gamma}^o$ converges to the true solution a^o as υ tends to zero (and γ tends to infinity in the process). For a given positive υ, part (iii) provides an estimate for the noise gain deviation $\|a_{\upsilon,\gamma}^o\|_Q^2 - \|a^o\|_Q^2$, assuming a feasible point $\bar{a}$ is known.

If g_υ is once continuously differentiable, then $f_{\upsilon,\gamma}$ is once continuously differentiable and the gradient is given by

$$\nabla f_{\upsilon,\gamma}(a) = 2Qa + \gamma\int_{\Omega_c}[g_\upsilon'(\phi^+(a, t)) - g_\upsilon'(\phi^-(a, t))]y(t)dt. \tag{5.4.11}$$

If g_υ is chosen to be twice continuously differentiable, then $f_{\upsilon,\gamma}$ is also twice continuously differentiable with positive definite Hessian given by

$$\nabla^2 f_{\upsilon,\gamma}(\boldsymbol{a}) = 2Q + \gamma\int_{\Omega_c}[g_\upsilon''(\phi^+(\boldsymbol{a},t)) + g_\upsilon''(\phi^-(\boldsymbol{a},t))]\boldsymbol{y}(t)\boldsymbol{y}^T(t)dt. \qquad (5.4.12)$$

Thus, the search directions are easily obtained for each iterate $\boldsymbol{a}_k$ from the corresponding filter output and state variables. This enables the filter updating procedure

$$\boldsymbol{a}_{k+1} = \boldsymbol{a}_k + \mu\boldsymbol{d}_k$$

to be performed in an on-line or adaptive fashion.

Suppose that g_υ is twice continuously differentiable for each $\upsilon > 0$ with $|g_\upsilon''(x)| \le h_\upsilon$, $\forall x \in \boldsymbol{R}$, (e.g (4.4.12)). Then, it can be shown (see Theorem 5.4.7) that starting from any initial point $\boldsymbol{a}_0$, the sequence $\{\boldsymbol{a}_k\}_{k=0}^{\infty}$ generated by the SD algorithm with constant step size μ converges to the sub-optimum solution $\boldsymbol{a}_{\upsilon,\gamma}^{o}$ if

$$0 < \mu < \left(\|Q\| + \gamma h_\upsilon\int_{\Omega_c}\|y(t)\|^2 dt\right)^{-1}.$$

With the NR algorithm, the stepsize can be twice as large (see Theorem 5.4.8).

If g_υ'' is bounded and Lipschitz continuous, then the sequence $\{\boldsymbol{a}_k\}_{k=0}^{\infty}$ generated by the NR algorithm with the line search procedure outlined in Section 4.4 converges to $\boldsymbol{a}_{\upsilon,\gamma}^{o}$ at a quadratic rate. Alternatively, Quasi-Newton methods (see [17], [18], [25], [27], [37], [90]) can be adopted. In this case, the convergence rate would be super-linear.

5.5 Laguerre Networks in Continuous-time EC Filtering

The use of Laguerre methods in prediction theory [61] and specifically, the use of Laguerre networks in filter realization problems has been well documented in [55] and [101]. The Laguerre methods provide concrete and novel techniques for approximating infinite dimensional systems by finite dimensional systems (see [67], [68]). In this section, the orthonormal Laguerre basis is introduced and applied to a practical example in data channel equalization.

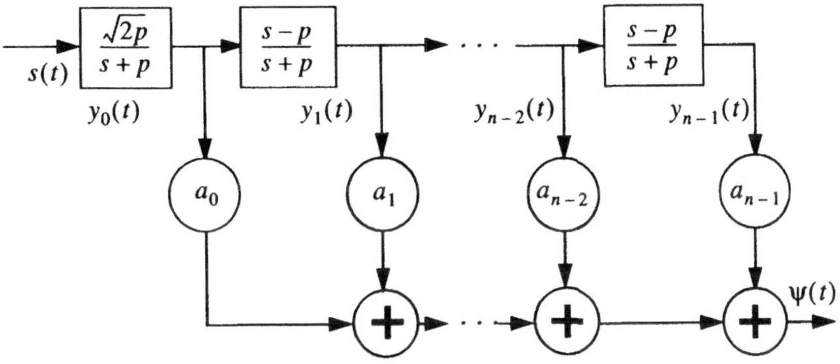

Figure 5.5.1. Block diagram of a Laguerre network

The Laguerre polynomials $L_k(t)$ are defined (see [67], [68], [106]) for each integer $k \geq 0$ as

$$L_k(t) = \frac{e^t}{k!}\frac{d^k}{dt^k}(t^k e^{-t}).$$

Using Leibnitz's formula, $L_k(t)$ can be expressed explicitly as

$$L_k(t) = \sum_{i=0}^{k} \binom{k}{k-i} \frac{(-t)^i}{i!}.$$

For a given $p > 0$ and any non-negative integer k, let

$$\theta_k^p(t) = \sqrt{2p}e^{-pt}L_k(2pt).$$

Then the sequence $\{\theta_k^p\}_{k=0}^{\infty}$ forms an orthonormal basis, called the Laguerre basis, for $L_2[0, \infty)$ (see [67], [68], [106]). In other words, the span of $\{\theta_k^p\}_{k=0}^{\infty}$ is dense in $L_2[0, \infty)$.

By taking the Laplace transform of θ_k^p, the frequency domain expression is

$$\Theta_k^p(s) = \frac{\sqrt{2p}}{s+p}\left(\frac{s-p}{s+p}\right)^k, \; k = 0, 1, 2, \ldots \tag{5.5.1}$$

The sequence $\{\Theta_k^p\}_{k=0}^{\infty}$ also forms an orthonormal basis for the Hardy space $\mathcal{H}_2$ (see [68], [106]) since the Laplace transform preserves inner products. A Laguerre network with transfer function

$$U(s) = \frac{\sqrt{2p}}{s+p}\sum_{k=0}^{n-1} a_k\left(\frac{s-p}{s+p}\right)^k$$

can be realized as shown in Figure 5.5.1.

The filter's response to an input signal s can be expressed as

$$\boldsymbol{y}^T(t)\boldsymbol{a},$$

where

$$\begin{aligned}\boldsymbol{y}^T(t) &= [y_0(t), y_1(t), \ldots, y_{n-1}(t)]\\ &= [(\theta_0^p * s)(t), (\theta_1^p * s)(t), \ldots, (\theta_{n-1}^p * s)(t)].\end{aligned}$$

Note that orthonormality of the θ_k^p's implies that the Gram matrix $K^{\dagger}K$ is an $n \times n$ identity matrix. Thus, the noise gain of the filter (for white noise), and hence the cost functional for the optimization problem is simply $\boldsymbol{a}^T\boldsymbol{a}$.

Application to channel equalization

Consider a digital transmission channel consisting of a coaxial cable operating at the DX3 rate, i.e. 45Mb/s (see [5], [12]). The cable has a 30dB attenuation at the baud frequency $2\pi/\beta$, where β denotes the baud interval (22.35×10^{-9} s). An equalizing filter is required to shape the impulse response of the cable so that it fits in the envelope given by the DSX-3 pulse template (see also Figure 5.5.4). The duration of the input signal and the output mask for this example is taken to be 20 baud intervals, i.e. $T_c = 20\beta$.

Using the Laguerre network of Figure 5.5.1 with 8 coefficients, the equalizers can be sought by applying the Newton-Raphson algorithm to the augmented cost. For computational purposes, the integrals resulting from the penalty term are approximated by finite sums, i.e.

$$f_{\upsilon,\gamma}(a) \cong a^T Q a + \gamma \frac{T_c}{N} \sum_{j=0}^{N-1} \left(g_{\upsilon}\left(\phi^+\left(a, \frac{jT_c}{N}\right)\right) + g_{\upsilon}\left(\phi^-\left(a, \frac{jT_c}{N}\right)\right)\right),$$

$$\nabla f_{\upsilon,\gamma}(a) \cong 2Qa + \gamma \frac{T_c}{N} \sum_{j=0}^{N-1} \left(g'_{\upsilon}\left(\phi^+\left(a, \frac{jT_c}{N}\right)\right) - g'_{\upsilon}\left(\phi^-\left(a, \frac{jT_c}{N}\right)\right)\right) y\left(\frac{jT_c}{N}\right),$$

$$\nabla^2 f_{\upsilon,\gamma}(a) \cong 2Q + \gamma \frac{T_c}{N} \sum_{j=0}^{N-1} \left(g''_{\upsilon}\left(\phi^+\left(a, \frac{jT_c}{N}\right)\right) + g''_{\upsilon}\left(\phi^-\left(a, \frac{jT_c}{N}\right)\right)\right) y\left(\frac{jT_c}{N}\right) y^T\left(\frac{jT_c}{N}\right).$$

These approximations lead to a solution that would at least satisfy the constraints at times $t = jT_c/N$. Coarse partitioning in the integral approximations will lead to erroneous solution. Using $N = 640$ would give a reasonably fine partitioning.

Figure 5.5.2 plots the value of the augmented cost function at each search direction evaluation. After 376 function evaluations and 211 search direction evaluations, the NR algorithm with line search yields a sub-optimum filter with a noise gain of 61.32. Figure 5.5.4 shows the DSX3 mask, the filter output and input from 0 to 4β (the time axis are marked in units of β). The output of the sub-optimum EC filter fits inside the DSX3 template at all points on $[0, 20\beta]$. Note that the constraints could not be met by simply scaling the input signal. Figure 5.5.3 shows the magnitude response of the cable, the equalizer and the equalized cable. Note that the equalized response is flatter and has sharper roll-off.

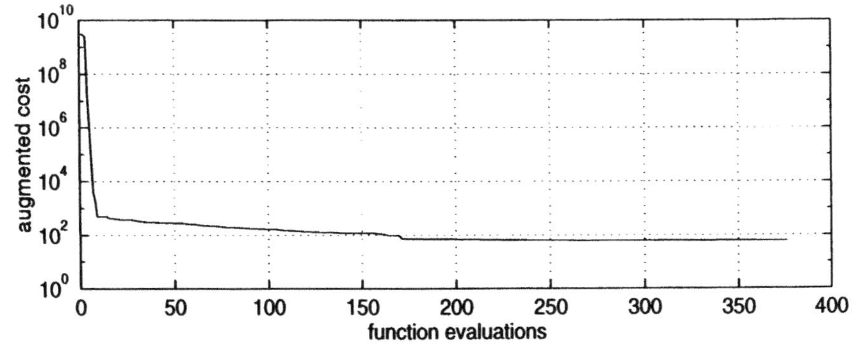

Figure 5.5.2. Augmented cost function for coaxial cable example

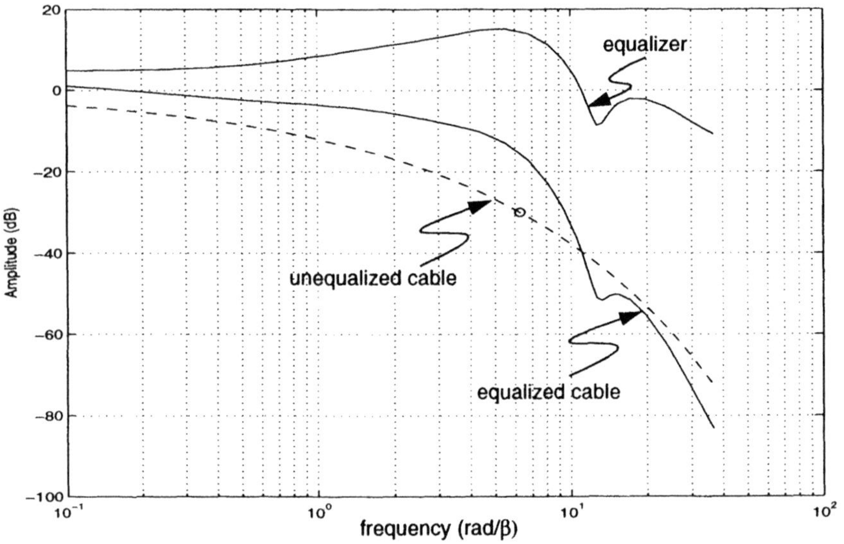

Figure 5.5.3. Magnitude responses of equalizer, unequalized and equalized cable

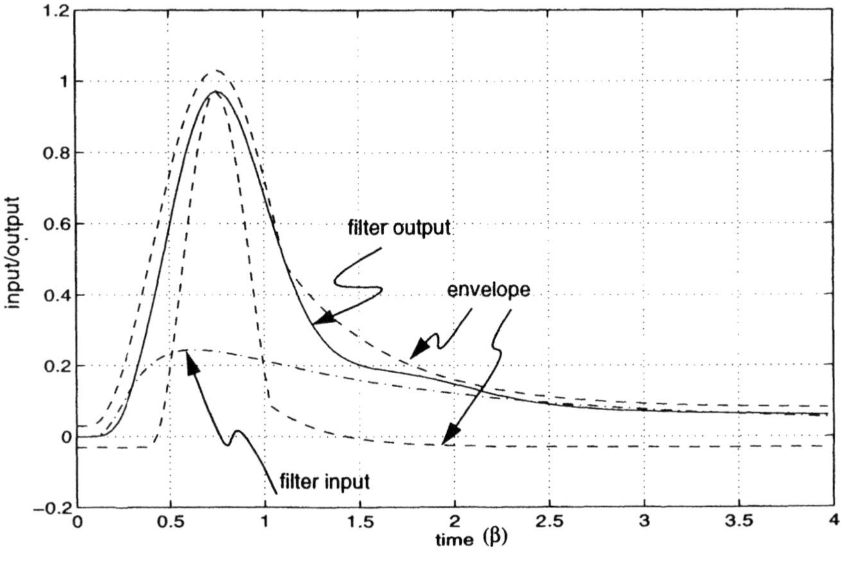

Figure 5.5.4. DSX3 pulse template, coaxial cable response and filter output

As another example with Laguerre filter, the coaxial cable is replaced by a fourth order circuit approximating a raised cosine filter with a roll-off factor of 0.7. The algorithm yields a sub-optimum filter with a noise gain of 178.7 after 258 function evaluations and 155 search direction evaluations. Figure 5.5.5 shows the behavior of the augmented cost. Figure 5.5.6 shows the sub-optimal EC filter output on $[0,4\beta]$ and Figure 5.5.7 shows the magnitude responses of the cable and the equalizer.

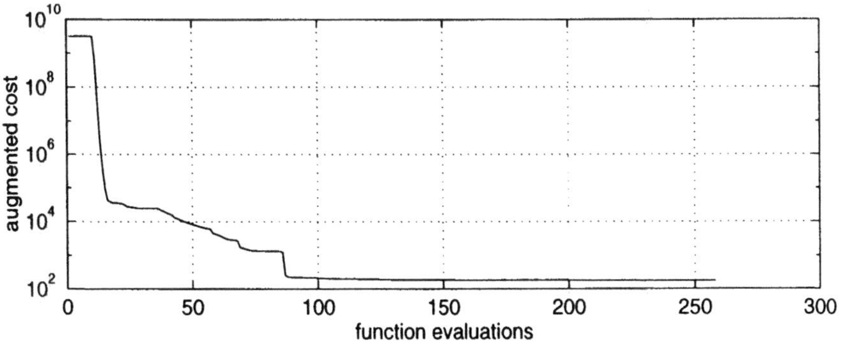

Figure 5.5.5. Augmented cost function for fourth order circuit example

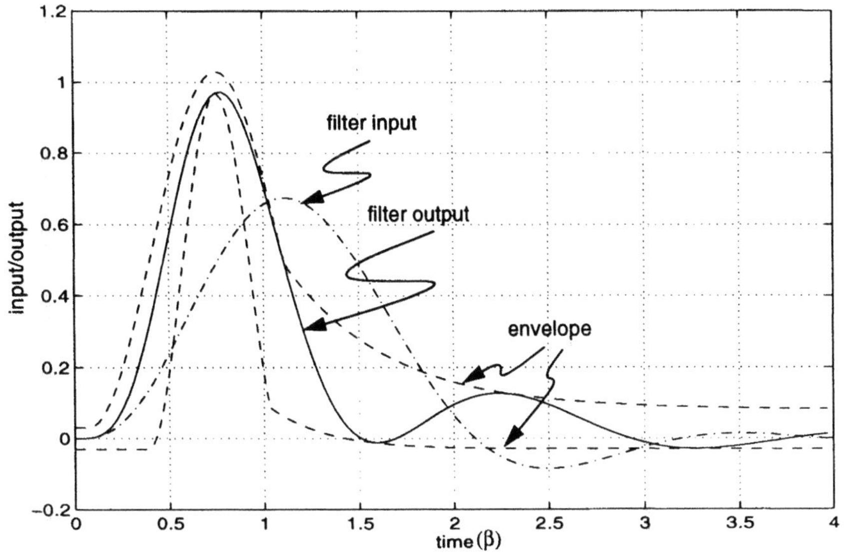

Figure 5.5.6. DSX3 pulse template, filter input and output

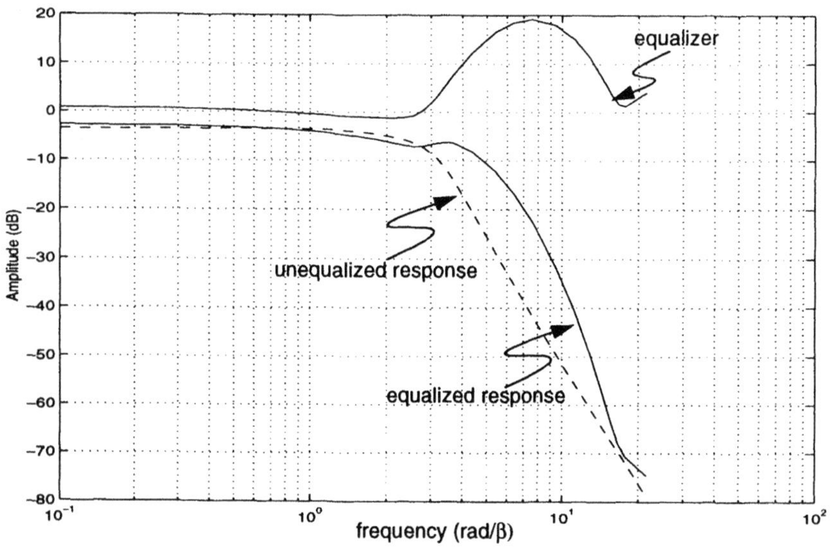

Figure 5.5.7. Magnitude responses of equalizer, unequalized and equalized circuit

5.6 Hybrid filter with FIR Digital Components

In this section, hybrid EC filters are applied to the coaxial cable equalization example of Section 5.5. In particular, we confine ourselves to hybrid filters with digital components being FIR filters and examine their performances for various post-filters.

To have a good representation of the signal, the input is sampled at every $\tau = \beta/4$. In other words, the sampling frequency is $8\pi/\beta$ (four times the baud frequency). The solution is sought by applying the NR algorithm with line search on the associated augmented cost with $\upsilon = 0.001$ and $\gamma = 3.2 \times 10^7$. Again, for computational purposes, the integrals resulting from the penalty term are approximated by finite sums with $N = 640$.

Linear interpolator

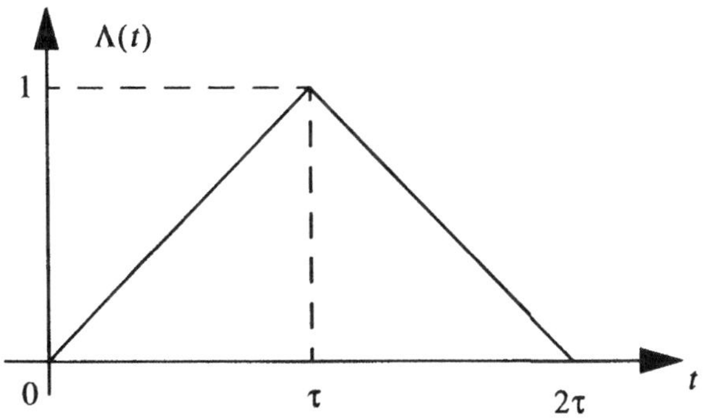

Figure 5.6.1. Rectangular pulse response of post-filter.

From a theoretical perspective, the combined function of the D/A converter and the post-filter is to interpolate the output of the discrete-time filter. Linear interpolation essentially fits a line between two adjacent points. The corresponding post-filter has impulse response $h(t) = (1/\tau)\Pi(t)$ (see [118]) and rectangular pulse response $\Lambda(t)$ shown in Figure 5.6.1.

Using an 8-tap FIR filter, after 134 function evaluations and 74 search direction evaluations, the algorithm yields a sub-optimum EC filter with a noise enhancement of 6.30. Figure 5.6.2 shows the DSX3 mask, the input and the optimum EC output. The output fits inside the envelope on $[0, 20\beta]$, but it is not a smooth waveform.

There are many more interpolation techniques. However, the important issue is whether they can be implemented with existing circuitry. In this context, realizable lowpass filters such as Butterworth, Bessel, Chebyshev and elliptic filters offer valuable alternatives.

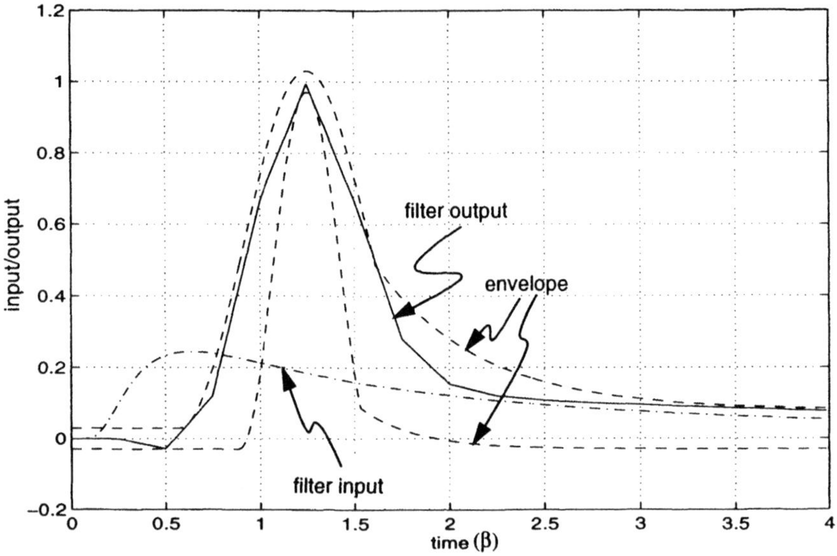

Figure 5.6.2. FIR-linear-interpolator filter output.

Butterworth and Bessel post-filtering

Still using an 8-tap FIR filter, we compare the performance of linear interpolation with more practical post-filters such as Butterworth and Bessel. With these post-filters the cut-off frequency is set to $4\pi/\beta$, i.e. half the sampling frequency.

Figure 5.6.3 shows the sub-optimum hybrid filter outputs on the critical interval $[0, 4\beta]$ for a 5th order Butterworth and a Bessel post-filter. Table 5.6.1 shows the noise enhancement and convergence behavior for the three post-filters considered. Although the linear interpolator exhibits faster convergence, it also yields the largest noise enhancement.

Figure 5.6.4 shows the magnitude response of the post-filters and Figure 5.6.5 shows the magnitude responses of the cable for various equalizers. Observe that these post filters have roughly the same response for low frequencies. However, the Butterworth and Bessel filters are much better than the linear interpolator at rejecting high frequencies (above $4/\beta$). This largely accounts for the lower noise enhancement and smoother output. Note that the notches in the magnitude response of the linear interpolator occur at multiples of the sampling frequency $8\pi/\beta$.

Table 5.6.1. Simulation data for linear interpolator, Butterworth and Bessel post-filters

Post-filter	Function evaluations	Search direction evaluations	Noise power gain
Linear interpolator	134	74	6.3014
5th-order Butterworth	432	131	0.0148
5th-order Bessel	785	256	0.0154

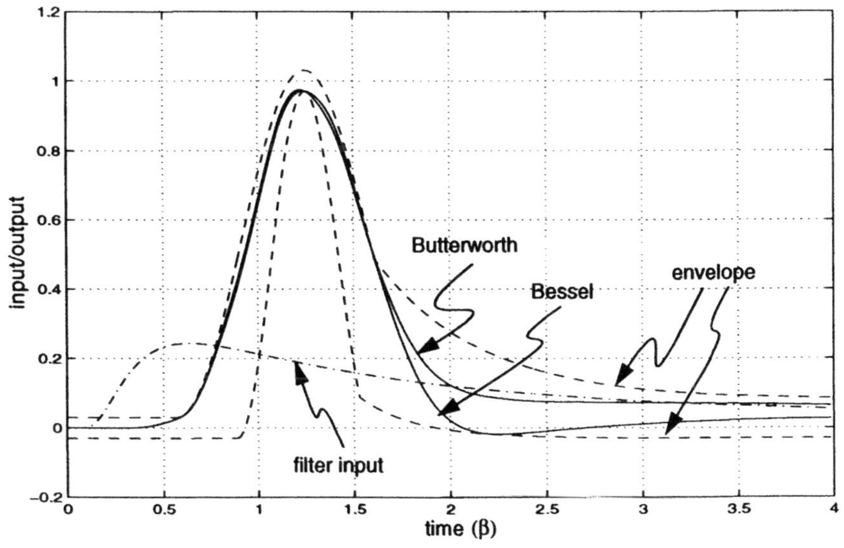

Figure 5.6.3. FIR-Butterworth and FIR-Bessel filter outputs.

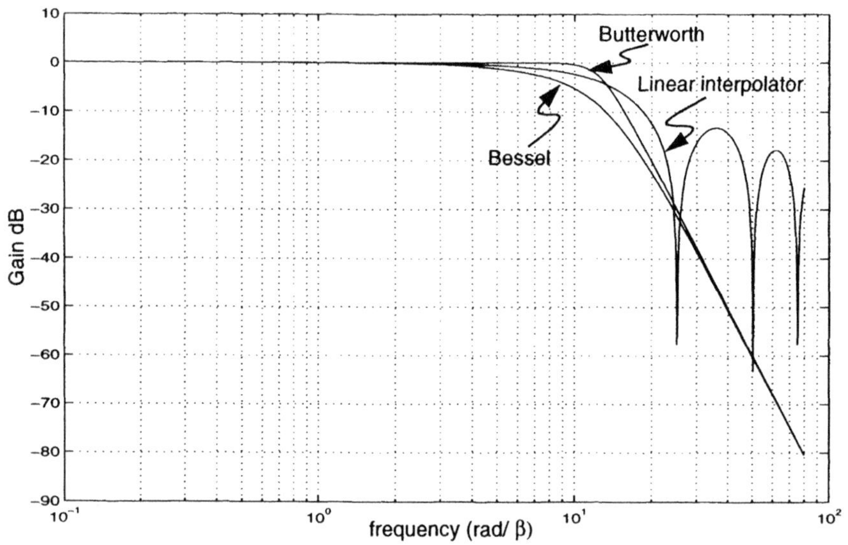

Figure 5.6.4. Magnitude responses of 5th order Butterworth, Bessel and Linear Interpolator.

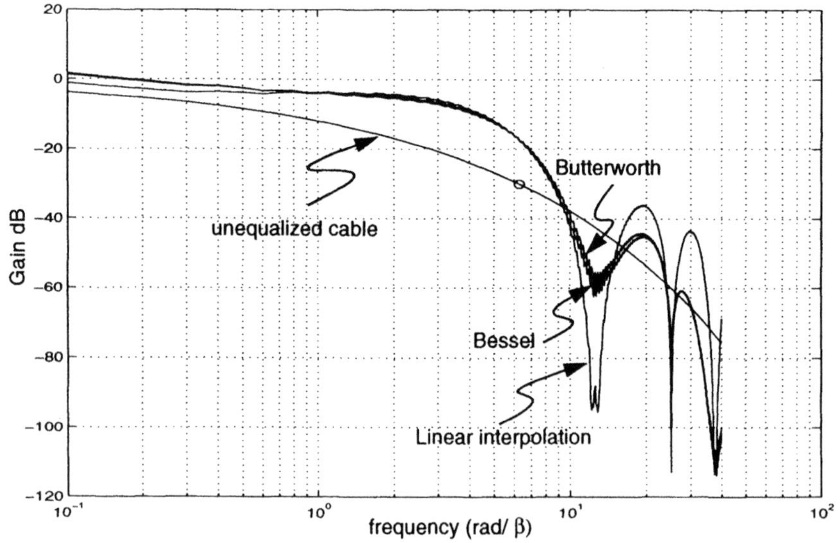

Figure 5.6.5. Magnitude responses of unequalized and equalized cable.

Chebyshev and elliptic post-filtering

For Chebyshev and elliptic filters, we also have parameters such as passband ripple R_p and stopband attenuation R_s to manipulate. Table 5.6.2 summarizes the simulation details for a Chebyshev type-1 and an elliptic post-filter. The optimum EC outputs of these post filters are almost identical as shown in Figure 5.6.6. The magnitude responses of these filters are almost identical for frequencies below $5/\beta$. However, at high frequencies (above $30/\beta$), the Chebyshev filter has better rejection (see Figure 5.6.7). This explains the similarity of the sub-optimum output (see the magnitude responses of the cable with these equalizers in Figure 5.6.8).

Although better high frequency rejection implies smaller noise enhancement of the post-filter, it is not necessarily true for the resulting hybrid EC filter. Figure 5.6.9 shows the magnitude response of an elliptic filter and two Chebyshev type-2 filters. Clearly, the Chebyshev with $R_s = 100$ has the least noise enhancement, while the elliptic with $R_s = 60$ and $R_p = 1$ has the largest noise enhancement. However, the Chebyshev with $R_s = 100$ also rejects a significant portion of the useful signal. As a result, the constraints cannot be satisfied and hence no feasible solution exists. With $R_s = 60$, sufficient signal contents are retained to permit feasibility (see Figure 5.6.10). However, the FIR-Chebyshev hybrid filter still has a larger noise gain than the FIR-elliptic as shown in Table 5.6.2.

Table 5.6.2. Simulation data for Chebyshev type 1 and elliptic post-filter.

Post-filter	Passband ripple	Stopband attenuation	Function evaluations	Search direction evaluations	Noise gain
Chebyshev I	1		417	149	0.04438
elliptic	1	100	455	167	0.04508
elliptic	1	60	690	224	0.06030
Chebyshev II		100			
Chebyshev II		60	549	173	1.2036

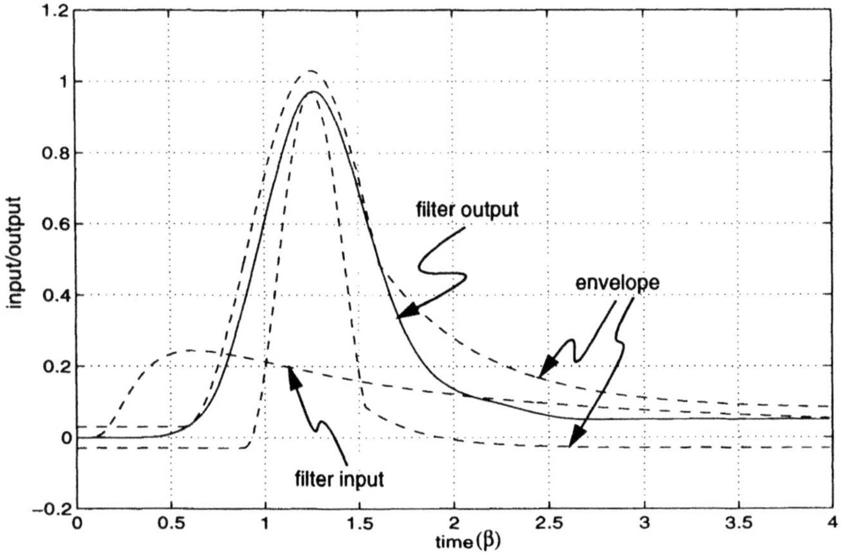

Figure 5.6.6. FIR-Chebyshev-1 and FIR-elliptic filter outputs.

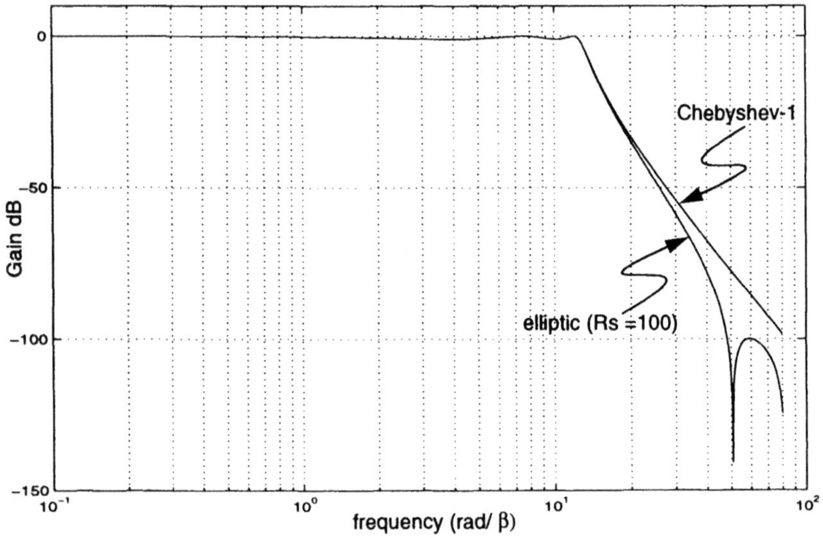

Figure 5.6.7. Magnitude responses of 5th order Chebyshev-1 and elliptic filters.

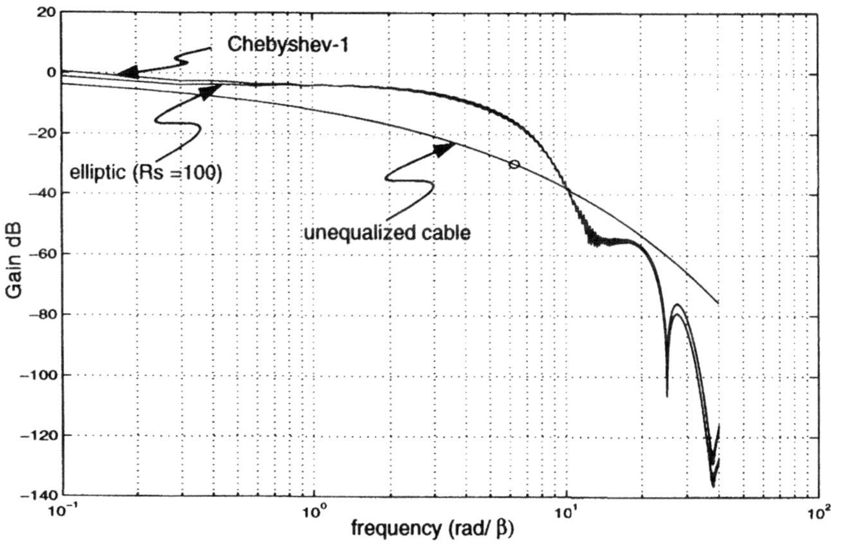

Figure 5.6.8. Magnitude responses of unequalized and equalized cable

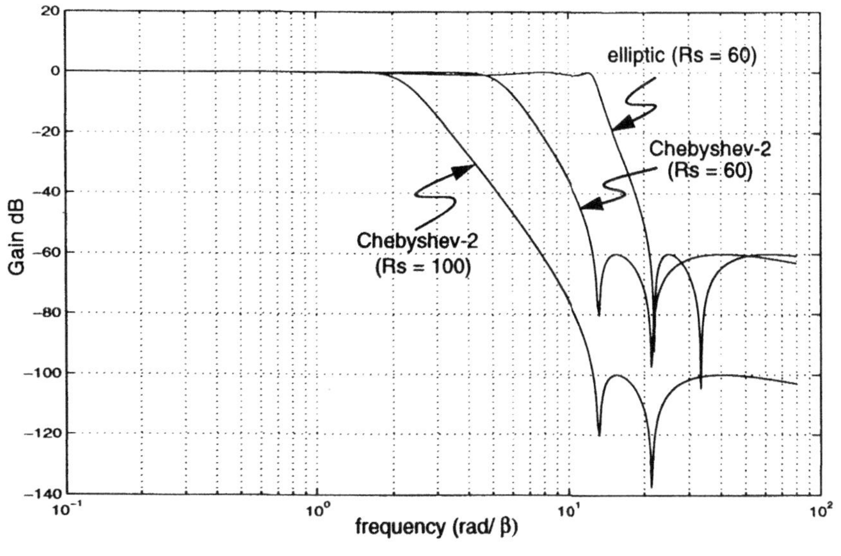

Figure 5.6.9. Magnitude responses of elliptic and Chebyshev-2 filters.

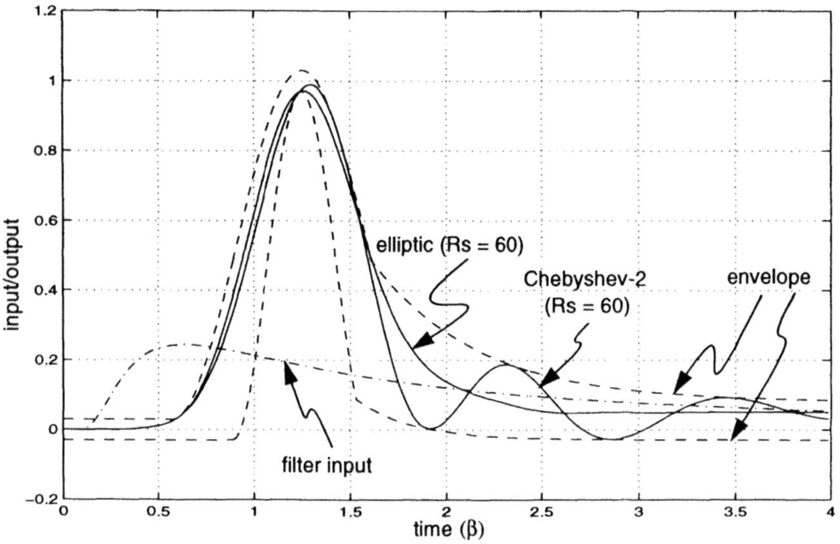

Figure 5.6.10. FIR-Chebyshev-2 and FIR-elliptic filter output.

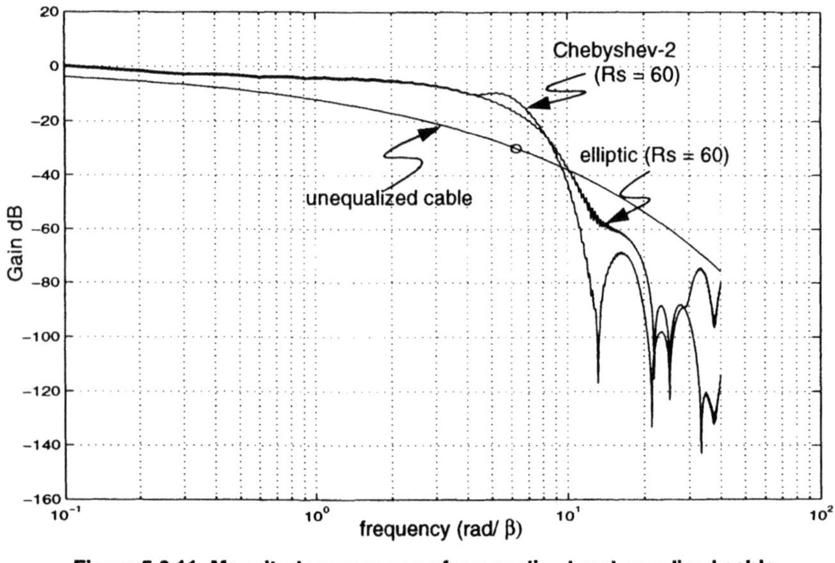

Figure 5.6.11. Magnitude responses of unequalized and equalized cable

5.7 Appendix

Lemma 5.2.1. *Suppose that q is Lipschitz continuous on a compact interval Ω, i.e.*

$$|q(\omega_2) - q(\omega_1)| \le \eta|\omega_2 - \omega_1|, \forall\omega_2, \omega_1 \in \Omega.$$

Then,

$$\exists\bar{\omega} \in \Omega \text{ such that } q(\bar{\omega}) \le -\upsilon \;\Rightarrow\; q(\omega) \le 0, \forall\omega \in [\bar{\omega} - \upsilon/\eta, \bar{\omega} + \upsilon/\eta].$$

Proof. $\forall\omega \in [\bar{\omega} - \upsilon/\eta, \bar{\omega} + \upsilon/\eta]$, the Lipschitz condition gives

$$|q(\omega) - q(\bar{\omega})| \le \eta|(\omega - \bar{\omega})| \le \eta\upsilon/\eta = \upsilon,$$

i.e. $-\upsilon \le q(\omega) - q(\bar{\omega}) \le \upsilon$ or $-\upsilon + q(\bar{\omega}) \le q(\omega) \le \upsilon + q(\bar{\omega})$. Since $q(\bar{\omega}) \le -\upsilon$, it follows that $q(\omega) \le 0$. ❐

It follows from this argument that if $\boldsymbol{a}$ satisfies the approximate constraints (5.2.1) for $\Delta \le 2\upsilon/\eta$, then

$$\phi^+(\boldsymbol{a}, t) \le 0, \;\; \phi^-(\boldsymbol{a}, t) \le 0, \forall t \in [0, \tilde{T}_c],$$

where $\tilde{T}_c = \min\{T_c, (M + 0.5)\Delta\}$.

Theorem 5.2.1. *Suppose that the output ψ and the envelope boundaries ε^+, ε^- are band limited to ω_c. If $\boldsymbol{a}$ satisfies (5.2.1), in which Δ is chosen so that*

$$\Delta^2 \sum_{k=0}^{M} [\varepsilon(j\Delta) - \upsilon] \le 2\pi\upsilon/\omega_c^2,$$

then

$$\varepsilon^-(t) \le \boldsymbol{y}^T(t)\boldsymbol{a} \le \varepsilon^+(t) \textit{ for all } t \in [0, T_c].$$

Proof: For a signal x with Fourier transform $X(\omega)$, we have

$$\sum_{k=-\infty}^{\infty} x(k\Delta)e^{-j\omega k\Delta} = \frac{1}{\Delta}\sum_{n=-\infty}^{\infty} X\left(\omega + \frac{2\pi n}{\Delta}\right).$$

If x is band limited to ω_c, and the sampling rate is above the Nyquist rate, then for all $\omega \in [-\omega_c, \omega_c]$

$$X(\omega) = \Delta \sum_{k=-\infty}^{\infty} x(k\Delta)e^{-j\omega k\Delta}.$$

Hence,

$$\begin{aligned}
|x'(t)| &= \frac{1}{2\pi}\left|\int_{-\omega_c}^{\omega_c} j\omega X(\omega)e^{j\omega t}d\omega\right| \\
&= \frac{1}{2\pi}\left|\int_{-\omega_c}^{\omega_c} j\omega\left(\Delta\sum_{k=-\infty}^{\infty} x(k\Delta)e^{-j\omega k\Delta}\right)e^{j\omega t}d\omega\right| \\
&= \frac{\Delta}{2\pi}\left|j\sum_{k=-\infty}^{\infty} x(k\Delta)\int_{-\omega_c}^{\omega_c} \omega e^{-j\omega k\Delta}e^{j\omega t}d\omega\right| \\
&\le \frac{\Delta}{2\pi}\sum_{k=-\infty}^{\infty} |x(k\Delta)|\int_{-\omega_c}^{\omega_c} |\omega|d\omega \\
&= \frac{2\Delta}{2\pi}\sum_{k=-\infty}^{\infty} |x(k\Delta)|\int_{0}^{\omega_c} \omega d\omega = \frac{\omega_c^2}{2\pi}\Delta\sum_{k=-\infty}^{\infty} |x(k\Delta)|.
\end{aligned}$$

Since the signals $\psi = \boldsymbol{y}^T\boldsymbol{a}$, ε^+, ε^- are band limited to ω_c, it follows that $\phi^+ = \boldsymbol{y}^T\boldsymbol{a} - \varepsilon^+$ and $\phi^- = \varepsilon^- - \boldsymbol{y}^T\boldsymbol{a}$ are also band limited to ω_c. Hence,

$$|\phi^{+\prime}(\boldsymbol{a}, t)| \le \frac{\omega_c^2}{2\pi}\Delta\sum_{k=-\infty}^{\infty} |\phi^+(\boldsymbol{a}, k\Delta)|$$

$$|\phi^{-\prime}(\boldsymbol{a}, t)| \le \frac{\omega_c^2}{2\pi}\Delta\sum_{k=-\infty}^{\infty} |\phi^-(\boldsymbol{a}, k\Delta)|$$

Now since $\varepsilon = 0.5(\varepsilon^+ - \varepsilon^-)$ and $\boldsymbol{a}$ satisfies (5.2.1), i.e.

$$\varepsilon^-(j\Delta) + \upsilon \le \boldsymbol{y}^T(j\Delta)\boldsymbol{a} \le \varepsilon^+(j\Delta) - \upsilon, \; j = 0, 1, \ldots, M,$$

we have

$$\max\{|\phi^+(a, j\Delta)|, |\phi^-(a, j\Delta)|\} \le 2[\varepsilon(j\Delta) - \upsilon],\ j = 0, 1, \ldots, M.$$

Because we are not concerned about the constraints outside of $\{0, 1, \ldots, M\}$, we can always choose $\varepsilon(j\Delta)$ so that

$$\max\{|\phi^+(a, j\Delta)|, |\phi^-(a, j\Delta)|\} = 0,\ \forall j \in \{0, 1, \ldots, M\}.$$

Hence,

$$\max\{|\phi^{+\prime}(a, t)|, |\phi^{-\prime}(a, t)|\} \le \frac{\omega_c^2}{\pi}\Delta \sum_{k=0}^{M} [\varepsilon(j\Delta) - \upsilon] = \frac{\omega_c^2}{\pi}K(\Delta),$$

where

$$K(\Delta) = \Delta \sum_{k=0}^{M} [\varepsilon(j\Delta) - \upsilon].$$

Note also that (5.2.1) is equivalent to $\phi^+(a, j\Delta) \le -\upsilon$ and $\phi^-(a, j\Delta) \le -\upsilon$, $j = 0, 1, \ldots, M$.

Let $\eta = (\omega_c^2/\pi)K(\Delta)$. If a point a satisfies:

$$\phi^+(a, j\Delta) \le -\upsilon \text{ and } \phi^-(a, j\Delta) \le -\upsilon,$$

then it follows from Lemma 5.2.1 that

$$\phi^+(a, t) \le 0 \text{ and } \phi^-(a, t) \le 0,\ \forall t \in [j\Delta - \upsilon/\eta, j\Delta + \upsilon/\eta].$$

Hence, if $\Delta \le 2\upsilon/\eta$, or equivalently $\Delta(\omega_c^2/\pi)K(\Delta) \le 2\upsilon$, then $\phi^+(a, t) \le 0$ and $\phi^-(a, t) \le 0$ for all $t \in [0, T_c]$. ❐

Lemma 5.3.1. *To every $\delta > 0$, there exists an $N(\delta)$ such that for $n \ge N(\delta)$, $M_n[t_0, t_0 + T_c]$ contains an element p^n for which*

$$\left|[\hat{\alpha}(p^n)](t) - [\hat{\alpha}(p^o)](t)\right| < \delta,\ t \in [t_0, t_0 + T_c],$$

$$\eta(p^o) - \eta(p^n) < \delta,$$

where p^o is a minimizer of the dual function (5.3.1), and

$$[\hat{\alpha}(p^n)](t) = \int_{t_0}^{t_0 + T_c} s(\tau - t)dp^n(\tau),$$

$$[\hat{\alpha}(p^o)](t) = \int_{t_0}^{t_0 + T_c} s(\tau - t)dp^o(\tau).$$

Proof: The integral

$$[\hat{\alpha}(p^n)](t) = \int_{t_0}^{t_0 + T_c} s(\tau - t)dp^n(\tau)$$

may be expressed for any fixed t as a limit of integrals of a sequence of simple functions (see Appendix A.8) which approaches s_τ (recall the notation $s_\tau(t) = s(\tau - t)$). Since s is continuous, such a sequence of simple functions $\{s_\tau^n\}_{n=1}^{\infty}$ is defined by setting

$$s_\tau^n(t) = s^n(\tau, t) = \sum_{j=1}^{n}\left[\int_{\tau_{j-1}}^{\tau_j} s(\xi - t)d\xi\right]\chi_{(\tau_{j-1}, \tau_j]}(\tau),$$

where $\chi_{(\tau_{j-1}, \tau_j]}$ is the indicator function on $(\tau_{j-1}, \tau_j]$. Note that

$$\int_{t_0}^{t_0 + T_c} s^n(\tau, t)dp^o(\tau) = \sum_{j=1}^{n} p^o((\tau_{j-1}, \tau_j])\int_{\tau_{j-1}}^{\tau_j} s(\xi - t)d\xi. \tag{5.7.1}$$

On the other hand, we can define a measure $p^n \in M_n[t_0, t_0 + T_c]$ such that

$$\int_{t_0}^{t_0 + T_c} x(\tau)dp^n(\tau) = \sum_{j=1}^{n} p_j^n\int_{t_{j-1}}^{t_j} x(\tau)d\tau,\ x \in C[t_0, t_0 + T_c], \tag{5.7.2}$$

where $p_j^n = p^o((\tau_{j-1}, \tau_j]), j = 1, \ldots, n$. Since for any fixed t, $s_t \in C[t_0, t_0 + T_c]$, it follows from (5.7.1) and (5.7.2) that

$$\int_{t_0}^{t_0+T_c} s^n(\tau, t)dp^o(\tau) = \int_{t_0}^{t_0+T_c} s(\tau - t)dp^n(\tau).$$

Thus, for any $\delta > 0$ there must exist an $N(\delta)$ (independent of t because of the continuity of s) such that, for $n \geq N(\delta)$

$$\left|\int_{t_0}^{t_0+T_c} s(\tau - t)dp^o(\tau) - \int_{t_0}^{t_0+T_c} s^n(\tau, t)dp^o(\tau)\right| < \delta, \; t \in [t_0, t_0 + T_c].$$

Hence, for any $\delta > 0$, there exists an $N(\delta)$ such that for $n \geq N(\delta)$,

$$\left|[\hat{\alpha}(p^n)](t) - [\hat{\alpha}(p^o)](t)\right| < \delta, \; t \in [t_0, t_0 + T_c]$$

It remains to show that for $n \geq N(\delta)$

$$\eta(p^o) - \eta(p^n) < \delta,$$

where the dual functional η is given by (5.3.1). By the same token as above, we can show that

$$\left|\int_{t_0}^{t_0+T_c} d(\tau)dp^o(\tau) - \int_{t_0}^{t_0+T_c} d(\tau)dp^n(\tau)\right| \leq \delta.$$

Further, from the definition of the ε-weighted norm (see (3.1.5)) in $M[t_0, t_0 + T_c]$, it is clear that the measure p^n specified above satisfies:

$$\sum_{j=1}^{n} \varepsilon_j |p_j^n| = \|p^n\|_\varepsilon \leq \|p^o\|_\varepsilon$$

If equality is not approached in the above with increasing n then $\|p^n\|_\varepsilon$ can be increased for example by adding to p^n a measure whose weight is γ in $(\tau_k, \tau_{k+1}]$ and zero elsewhere for some integer $k < n - 2$. Such a perturbation can be used to make

$$\left|\|p^n\|_\varepsilon - \|p^o\|_\varepsilon\right| < \delta$$

Combining the above, we see that the lemma is established. ❒

Lemma 5.7.1. *Suppose p is an integrable, non-negative and non-decreasing function and q is Lipschitz continuous on a compact interval Ω, i.e.*

$$|q(\omega_2) - q(\omega_1)| \le \eta|\omega_2 - \omega_1|,\ \forall\omega_2, \omega_1 \in \Omega.$$

Then

$$\exists\ \text{an}\ \bar{\omega} \in \Omega\ \text{such that}\ q(\bar{\omega}) > 0 \Rightarrow \forall\xi > 0, \int_{\Omega} p(q(\omega))d\omega \ge p(-\xi)\min\left(|\Omega|, \frac{\xi}{\eta}\right).$$

Proof. Supposing that there exists an $\bar{\omega} \in \Omega$ such that $q(\bar{\omega}) > 0$, then, for any $\xi > 0$, let $I \subset \Omega$ be the largest interval containing $\bar{\omega}$ such that $q(\omega) \ge -\xi$, for all $\omega \in I$. First, let's establish that $|I| \ge \min(|\Omega|, \xi/\eta)$ (see Figure 5.7.1. for illustration).

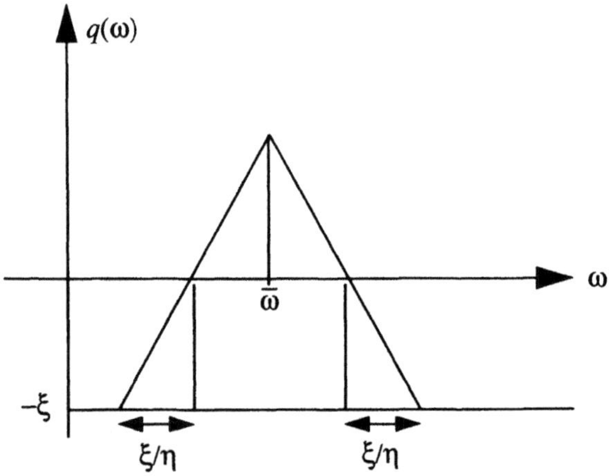

Figure 5.7.1. Illustration for lemma 5.7.1

Assume that there exists an $\omega_0 \in I$ such that $q(\omega_0) = -\xi$ (otherwise, from the definition of I and the continuity of q, $I = \Omega$). The Lipschitz condition gives

$$|q(\omega_0) - q(\bar{\omega})| = \eta|(\omega_0 - \bar{\omega})| \le \eta|I| \tag{5.7.3}$$

Since $q(\bar{\omega}) > 0$, it follows from (5.7.3) that $-\xi > q(\omega_0) - q(\bar{\omega}) \geq -\eta|I|$. Consequently, $\xi/\eta < |I| \leq |\Omega|$. Now, since p is non negative,

$$\int_{\Omega} p(q(\omega))d\omega \geq \int_{I} p(q(\omega))d\omega \geq \min\{p(q(\omega)): \omega \in I\}|I|$$

Furthermore, since p is non-decreasing and $q(\omega) \geq -\xi$, for all $\omega \in I$, it follows that

$$\min\{p(q(\omega)): \omega \in I\} \geq p(-\xi).$$

Thus,

$$\int_{\Omega} p(q(\omega))d\omega \geq p(-\xi)\min(|\Omega|, \xi/\eta). \square$$

Theorem 5.4.1. *Let*

$$\tau_{\upsilon} \equiv \sup_{\xi \in (0, \upsilon)} \{g_{\upsilon}(-\xi)\min(|\Omega_c|, \xi/\eta)\},$$

where η *is a number satisfying*

$$\eta|t_2 - t_1| \geq |\phi_j(x)(t_2) - \phi_j(x)(t_1)|$$

for all $t_2, t_1 \in \Omega_c, x \in \Theta, j \in \{1, \ldots, m\}$. *Then,* $0 < \tau < \tau_{\upsilon} \Rightarrow \mathcal{F}_{\upsilon,\tau} \subset \mathcal{F}$

Proof. Since g_{υ} is integrable, non-negative and non-decreasing, and for each $x \in \Theta$, $j \in \{1, \ldots, m\}$, $\phi_j(x)$ is Lipschitz continuous on Ω_c, it follows from Lemma 5.7.1 that

if $\exists$ a $j \in \{1, \ldots, m\}$ and a $\bar{t} \in \Omega_c$ such that $\phi_j(x)(\bar{t}) > 0$, then

$$G_{j,\upsilon}(x) \equiv \int_{\Omega_c} g_{\upsilon}(\phi_j(x)(t))dt \geq \sup_{\xi \in (0, \upsilon)} \{g_{\upsilon}(-\xi)\min(|\Omega_c|, \xi/\eta)\} \equiv \tau_{\upsilon}.$$

This is equivalent to $G_{j,\upsilon}(x) \leq \tau < \tau_{\upsilon} \Rightarrow \phi_j(x)(t) \leq 0, \forall t \in \Omega, j = 1, \ldots, m$. Thus, from the definitions of $\mathcal{F}_{\upsilon,\tau}$ and $\mathcal{F}$, it follows that $x \in \mathcal{F}_{\upsilon,\tau} \Rightarrow x \in \mathcal{F}$. $\square$

***Lemma** 5.7.2. For any $\upsilon > 0$, let*

$$\mathcal{F}_{\upsilon} = \{x \in \Theta : G_{j,\upsilon}(x) = 0, j = 1, \ldots, m\}. \tag{5.7.4}$$

Then,

(i) $\mathcal{F}_{\upsilon} \subset \mathcal{F}_{\upsilon,\tau}$

(ii) $0 < \mu < \upsilon \Rightarrow \mathcal{F}_{\upsilon} \subset \mathcal{F}_{\mu} \subset \mathcal{F}$

(iii) Under assumption A-5.4.2, for any $\alpha \in (0, 1]$,

$$\exists \delta > 0, x_{\alpha} \equiv \alpha \bar{x} + (1 - \alpha) x^{o} \in \mathcal{F}_{\delta}.$$

***Proof**.* (i) $\mathcal{F}_{\upsilon} \subset \mathcal{F}_{\upsilon,\tau}$ follows readily from the definition of $\mathcal{F}_{\upsilon,\tau}$.

(ii) To prove parts (ii) and (iii), consider an alternative expression for $\mathcal{F}_{\upsilon}$ given by

$$\mathcal{F}_{\upsilon} = \{x \in \Theta : \phi_j(x)(t) \le -\upsilon, \forall t \in \Omega_c, j = 1, \ldots, m\}.$$

Let $0 < \mu < \upsilon$, and suppose that $x \in \mathcal{F}_{\upsilon}$. Then,

$$\phi_j(x)(t) \le -\upsilon, \forall t \in \Omega_c, j = 1, \ldots, m,$$

since $-\upsilon < -\mu < 0$, $x \in \mathcal{F}_{\mu}$. As $x \in \mathcal{F}_{\mu}$ also implies $x \in \mathcal{F}$, we have $\mathcal{F}_{\upsilon} \subset \mathcal{F}_{\mu} \subset \mathcal{F}$.

(iii) By Assumption A-5.4.2, for each $\alpha \in (0, 1]$, $x_{\alpha} \equiv \alpha \bar{x} + (1 - \alpha) x^{o} \in \mathcal{F}^{o}$, i.e.

$$\phi_j(x_{\alpha})(t) < 0, \forall t \in \Omega_c, j = 1, \ldots, m.$$

Hence, there is a $\delta > 0$ such that

$$\phi_j(x_{\alpha})(t) \le -\delta, \forall t \in \Omega_c, j = 1, \ldots, m.$$

Thus, $x_{\alpha} \in \mathcal{F}_{\delta}$. ❐

***Theorem 5.4.2**. If $\tau < \tau_\upsilon$, then, $\lim_{\upsilon \to 0} f(x^o_{\upsilon,\tau}) = f(x^o)$, and any accumulation point of $\{x^o_{\upsilon(n),\tau}\}_{n=1}^\infty$ is a solution of $(\boldsymbol{P})$, where $\{\upsilon(n)\}_{n=1}^\infty$ is any positive sequence that converges to zero.*

Proof. From Assumption A-5.4.2, $\exists$ an $\bar{x} \in \mathcal{F}^o$ such that, for all $\beta \in (0, 1]$,

$$x_\beta = \beta\bar{x} + (1-\beta)x^o \in \mathcal{F}^o.$$

Moreover, since f is continuous, it follows that

$$\forall\varepsilon > 0, \exists\theta(\varepsilon) \in (0, 1] \text{ such that } 0 < \beta < \theta(\varepsilon) \Rightarrow 0 \le f(x_\beta) - f(x^o) \le \varepsilon \tag{5.7.5}$$

For a fixed value of β, there exists, from Lemma 5.7.2 (iii), a $\delta > 0$ such that $x_\beta \in \mathcal{F}_\delta$. It follows from Lemma 5.7.2 (ii) and (i) that

$$0 < \upsilon \le \delta \Rightarrow x_\beta \in \mathcal{F}_\delta \subset \mathcal{F}_\upsilon \subset \mathcal{F}_{\upsilon,\tau}.$$

Moreover, by Theorem 5.4.1, $\mathcal{F}_{\upsilon,\tau} \subset \mathcal{F}$. Thus,

$$f(x^o) \le f(x^o_{\upsilon,\tau}) \le f(x_\beta). \tag{5.7.6}$$

Combining (5.7.6) with (5.7.5) yields

$$\forall\varepsilon > 0, \exists\delta(\varepsilon) > 0 \text{ such that } 0 < \upsilon \le \delta(\varepsilon) \Rightarrow 0 \le f(x^o_{\upsilon,\tau}) - f(x^o) \le \varepsilon.$$

That is, $\lim_{\upsilon \to 0} f(x^o_{\upsilon,\tau}) = f(x^o)$.

Note that $\{x^o_{\upsilon(n),\tau}\}_{n=1}^\infty \subset \Theta$. Moreover, as Θ is compact, the sequence has an accumulation point $\hat{x} \in \Theta$, and a subsequence, again denoted by $\{x^o_{\upsilon(n),\tau}\}_{n=1}^\infty$, such that

$$\lim_{n \to \infty} \left\| x^o_{\upsilon(n),\tau} - \hat{x} \right\| = 0.$$

Thus, by the continuity of f,

$$\lim_{n \to \infty} f(x^o_{\upsilon(n),\tau}) = f(\hat{x}).$$

Combining this with the first part of the theorem gives

$$f(\hat{x}) = f(x^o).$$

From Theorem 5.4.1, $x^o_{\upsilon(n),\tau} \in \mathcal{F}_{\upsilon(n),\tau} \subset \mathcal{F}$, and since $\mathcal{F}$ is closed, $\hat{x} \in \mathcal{F}$. Hence, the accumulation point $\hat{x}$ is a solution to $(\boldsymbol{P})$. ❒

Theorem 5.4.3. *Let $\tilde{x} \in \Theta$ be such that $f(\tilde{x}) \le f(x)$, $\forall x \in \Theta$, let υ be such that $\mathcal{F}_\upsilon$ is non-empty and let $x_\upsilon \in \mathcal{F}_\upsilon$. If $\gamma > \gamma_\upsilon \equiv (f(x_\upsilon) - f(\tilde{x}))/\tau_\upsilon$, then $x^o_{\upsilon,\gamma} \in \mathcal{F}$.*

Proof. Since Θ is compact and f is continuous, $\tilde{x}$ exists. On the other hand, from the definition of $\mathcal{F}_\upsilon$, it follows that $\forall x \in \mathcal{F}_\upsilon$, $G_{j,\upsilon}(x) = 0$, consequently

$$f_{\upsilon,\gamma}(x^o_{\upsilon,\gamma}) \le f_{\upsilon,\gamma}(x_\upsilon) = f(x_\upsilon).$$

Hence, using the definition of $f_{\upsilon,\gamma}(x)$ and rearranging yields:

$$\gamma \sum_{j=1}^{m} G_{j,\upsilon}(x^o_{\upsilon,\gamma}) \le f(x_\upsilon) - f(x^o_{\upsilon,\gamma}) \le f(x_\upsilon) - f(\tilde{x})$$

$$\Rightarrow \sum_{j=1}^{m} G_{j,\upsilon}(x^o_{\upsilon,\gamma}) \le \frac{f(x_\upsilon) - f(\tilde{x})}{\gamma} < \frac{f(x_\upsilon) - f(\tilde{x})}{\gamma_\upsilon} = \tau_\upsilon.$$

Let $\tau = (f(x_\upsilon) - f(\tilde{x}))/\gamma$. Then, from the definition of $\mathcal{F}_{\upsilon,\tau}$, we have $x^o_{\upsilon,\gamma} \in \mathcal{F}_{\upsilon,\tau}$. Furthermore, since $\tau < \tau_\upsilon$, it follows from Theorem 5.4.1 that $\mathcal{F}_{\upsilon,\tau} \subset \mathcal{F}$. Thus, $x^o_{\upsilon,\gamma} \in \mathcal{F}$. ❒

Theorem 5.4.4. *If $\gamma > \gamma_\upsilon$, then, $\lim_{\upsilon \to 0} f(x^o_{\upsilon,\gamma}) = f(x^o)$, and any accumulation point of $\{x^o_{\upsilon(n),\gamma}\}_{n=1}^{\infty}$ is a solution of $(\boldsymbol{P})$, where $\{\upsilon(n)\}_{n=1}^{\infty}$ is any positive sequence that converges to zero.*

Proof. From Assumption A-5.4.2 and the continuity of f, we have (5.7.5), which is restated here,

$$\forall \varepsilon > 0, \exists\ \theta(\varepsilon) \in (0, 1] \text{ such that } 0 < \beta < \theta \Rightarrow 0 \le f(x_\beta) - f(x^o) \le \varepsilon.$$

For a fixed β, from Lemma 5.7.2 (ii) and (iii), there exists a $\delta > 0$ such that $0 < \upsilon \leq \delta \Rightarrow x_\beta \in \mathcal{F}_\upsilon$. Note from the definition of $\mathcal{F}_\upsilon$ that $\forall x \in \mathcal{F}_\upsilon, f_{\upsilon,\gamma}(x) = f(x)$. Hence,

$$f_{\upsilon,\gamma}(x^o_{\upsilon,\gamma}) \leq f_{\upsilon,\gamma}(x_\beta) = f(x_\beta).$$

Since $G_{j,\upsilon}(x^o_{\upsilon,\gamma}) \geq 0$, the penalty term is non-negative. Thus, $f(x^o_{\upsilon,\gamma}) \leq f(x_\beta)$, and hence, the first part of the theorem follows, i.e.

$$\forall \varepsilon > 0, \exists\ \delta(\varepsilon) > 0 \text{ such that } 0 < \upsilon \leq \delta \Rightarrow 0 \leq f(x^o_{\upsilon,\gamma}) - f(x^o) \leq \varepsilon.$$

From Theorem 5.4.3, $\{x^o_{\upsilon(n),\gamma}\}_{n=1}^{\infty} \subset \mathcal{F} \subset \Theta$. Consequently, the conclusion to the second part follows in the same line of reasoning as that given for Theorem 5.4.2. ❐

Theorem 5.4.5. *For each $\upsilon > 0$ such that $\mathcal{F}_\upsilon \neq \varnothing$, if $\gamma > f(x_\upsilon)/\tau_\upsilon$ for some $x_\upsilon \in \mathcal{F}_\upsilon$, then $x^o_{\upsilon,\gamma} \in \mathcal{F}$. Moreover, $\lim_{\upsilon \to 0} x^o_{\upsilon,\gamma} = x^o$.*

Proof. Since $f(x) \geq 0, \forall x \in \Theta$, it follows from Theorem 5.4.3 that, if $\gamma > f(x_\upsilon)/\tau_\upsilon$, then $x^o_{\upsilon,\gamma} \in \mathcal{F}$.

Suppose that $\lim_{\upsilon \to 0} x^o_{\upsilon,\gamma} \neq x^o$, i.e.

$$\exists \varepsilon > 0 : \forall \delta > 0, \exists \upsilon \in (0, \delta) : \left\| x^o_{\upsilon,\gamma} - x^o \right\| \geq \varepsilon.$$

Then, a real positive sequence $\{\upsilon(n)\}_{n=1}^{\infty}$, which converges to zero can be constructed, by letting

$$\upsilon(n) \in \{\upsilon : 0 < \upsilon < 1/n, \left\| x^o_{\upsilon,\gamma} - x^o \right\| \geq \varepsilon\}.$$

Clearly, the sequence $\{x^o_{\upsilon(n),\gamma}\}_{n=1}^{\infty} \subset \Theta$. Since Θ is compact, the sequence has an accumulation point y. Theorem 5.4.4 asserts that any accumulation point of the sequence $\{x^o_{\upsilon(n),\gamma}\}_{n=1}^{\infty}$ is a solution to $(\boldsymbol{P})$. Hence, y is a solution to $(\boldsymbol{P})$, but $y \neq x^o$, because x^o is not an accumulation point of $\{x^o_{\upsilon(n),\gamma}\}_{n=1}^{\infty}$ (since $\left\| x^o_{\upsilon(n),\gamma} - x^o \right\| \geq \varepsilon$ for all $n \in \{1, 2, \ldots\}$). This contradicts the uniqueness of the solution to $(\boldsymbol{P})$. ❐

Theorem 5.4.6. *For any $\bar{x} \in \mathcal{F}^o$, let*

$$\delta(\bar{x}) \equiv \min_j \min_{t \in \Omega_c} \{-[\phi_j(\bar{x})](t)/2\|\bar{x}\|_L^2\}.$$

If $\upsilon > 0$ is such that $\mathcal{F}_\upsilon$ is non-empty, and $\gamma > \|x_\upsilon\|_L^2/\tau_\upsilon$ for some $x_\upsilon \in \mathcal{F}_\upsilon$, then,

$$0 < \upsilon \le \delta(\bar{x})\min\{\varepsilon, 2\|\bar{x}\|_L^2\} \Rightarrow 0 \le \|x^o_{\upsilon,\gamma}\|_L^2 - \|x^o\|_L^2 \le \varepsilon.$$

Proof. Let $\beta = \min\{\varepsilon, 2\|\bar{x}\|_L^2\}/2\|\bar{x}\|_L^2$. Clearly, $0 < \beta \le 1$. Then,

$$\begin{aligned} 0 < \upsilon \le \delta(\bar{x})\min\{\varepsilon, 2\|\bar{x}\|_L^2\} &\Leftrightarrow 0 < \upsilon \le -\beta\phi_j(\bar{x})(t),\ \forall t \in \Omega_c, j \in \{1, \ldots, m\} \\ &\Leftrightarrow \beta\phi_j(\bar{x})(t) \le -\upsilon < 0,\ \forall t \in \Omega_c, j \in \{1, \ldots, m\}. \end{aligned} \tag{5.7.7}$$

Let $x_\beta \equiv \beta\bar{x} + (1-\beta)x^o$ and note that each ϕ_j is a convex function of x. Then, we have

$$\phi_j(x_\beta) = \phi_j(\beta\bar{x} + (1-\beta)x^o) \le \beta\phi_j(\bar{x}) + (1-\beta)\phi_j(x^o). \tag{5.7.8}$$

Since $\phi_j(x^o) \le 0,\ \forall j \in \{1, \ldots, m\}$, and $0 < \beta \le 1$, we have $(1-\beta)\phi_j(x^o) \le 0$. Thus,

$$\beta\phi_j(\bar{x}) + (1-\beta)\phi_j(x^o) \le \beta\phi_j(\bar{x}).$$

From (5.7.8) we have

$$\phi_j(x_\beta) \le \beta\phi_j(\bar{x}),\ \forall j \in \{1, \ldots, m\}.$$

Thus, by (5.7.7), it follows that

$$\begin{aligned} &\phi_j(x_\beta)(t) \le \beta\phi_j(\bar{x})(t) \le -\upsilon < 0,\ \forall t \in \Omega_c, j \in \{1, \ldots, m\} \\ &\Rightarrow G_{j,\upsilon}(x_\beta) = \int_\Omega g_\upsilon(\phi_j(x_\beta)(t))dt = 0 \\ &\Rightarrow f_{\upsilon,\gamma}(x_\beta) = \|x_\beta\|_L^2. \end{aligned} \tag{5.7.9}$$

Since $x^o_{\upsilon,\gamma}$ minimizes $f_{\upsilon,\gamma}$, we have

$$\|x^o_{\upsilon,\gamma}\|_L^2 \le f_{\upsilon,\gamma}(x^o_{\upsilon,\gamma}) \le f_{\upsilon,\gamma}(x_\beta). \tag{5.7.10}$$

Therefore, by (5.7.9) and (5.7.10),

$$\|x^o_{\upsilon,\gamma}\|_L^2 \le \|x_\beta\|_L^2 .$$

Also, as $\gamma > \|x_\upsilon\|_L^2/\tau_\upsilon$, by Theorem 5.4.3, $x^o_{\upsilon,\gamma} \in \mathcal{F}$, and hence,

$$\begin{aligned}
0 \le \|x^o_{\upsilon,\gamma}\|_L^2 - \|x^o\|_L^2 \le \|x_\beta\|_L^2 - \|x^o\|_L^2 &= (\|x_\beta\|_L - \|x^o\|_L)(\|x_\beta\|_L + \|x^o\|_L) \\
&= [\|\beta\bar{x} + (1-\beta)x^o\|_L - \|x^o\|_L][\|\beta\bar{x} + (1-\beta)x^o\|_L + \|x^o\|_L] \\
&\le \beta[\|\bar{x}\|_L - \|x^o\|_L][\beta\|\bar{x}\|_L + (2-\beta)\|x^o\|_L] \\
&= \beta(2-\beta)[\|\bar{x}\|_L - \|x^o\|_L]\left[\frac{\beta}{(2-\beta)}\|\bar{x}\|_L + \|x^o\|_L\right] \\
&< \beta(2-\beta)[\|\bar{x}\|_L - \|x^o\|_L][\|\bar{x}\|_L + \|x^o\|_L] \\
&< 2\beta\|\bar{x}\|_L^2 = \min\{\varepsilon, 2\|\bar{x}\|_L^2\} \le \varepsilon. \ \square
\end{aligned}$$

Proof of Lemma 5.4.2.

(i) Recall that from (5.4.3) that $G_\upsilon : W \to \boldsymbol{R}$ is given by

$$G_\upsilon(w) = \int_{\Omega_c} g_\upsilon(w(t))dt,$$

and that the derivative of G_υ at w, i.e. $G'_\upsilon(w): W \to \boldsymbol{R}$ is defined by (5.4.3)

$$G'_\upsilon(w)(v) = \int_{\Omega_c} g'_\upsilon(w(t))v(t)dt$$

Since $\langle Lx, x\rangle' = 2\langle Lx, \bullet\rangle$, and $\phi'_j(x) = A_j$, it follows from the chain rule that

$$\nabla f_{\upsilon,\gamma}(x) = 2\langle Lx, \bullet\rangle + \gamma\sum_{j=1}^{m} G'_\upsilon(\phi_j(x)(t))A_j . \tag{5.7.11}$$

Hence,

$$\begin{aligned}
\nabla f_{\upsilon,\gamma}(x)h &= 2\langle Lx, h\rangle + \gamma\sum_{j=1}^{m} G'_\upsilon(\phi_j(x)(t))(A_j h) \\
&= 2\langle Lx, h\rangle + \gamma\sum_{j=1}^{m}\int_{\Omega_c} g'_\upsilon(\phi_j(x)(t))(A_j h)(t)dt
\end{aligned} \tag{5.7.12}$$

(ii) Differentiating $G'_\upsilon(w)\colon W \to \boldsymbol{R}$ gives $G''_\upsilon(w)\colon W \to W^*$ defined by

$$G''_\upsilon(w)(u)(v) = \int_{\Omega_c} g''_\upsilon(w(t))u(t)v(t)dt$$

Using the product rule for differentiation

$$(FG)'(x)(h) = F'(x)hG(x) + F(x)G'(x)(h)$$

on $G'_\upsilon(\phi_j(x)(t))A_j$ gives:

$$(G'_\upsilon(\phi_j(x)(t))A_j)'(h) = G''_\upsilon(\phi_j(x)(t))A_j(h)A_j$$

Hence,

$$\nabla^2 f_{\upsilon,\gamma}(x)h = 2\langle Lh, \bullet\rangle + \gamma\sum_{j=1}^{m} G''_\upsilon(\phi_j(x)(t))A_jhA_j\,. \tag{5.7.13}$$

Consequently,

$$\begin{aligned}\langle\nabla^2 f_{\upsilon,\gamma}(x)h, k\rangle &= \langle 2Lh, k\rangle + \gamma\sum_{j=1}^{m} G''_\upsilon(\phi_j(x)(t))A_jhA_jk \\ &= \langle 2Lh, k\rangle + \gamma\sum_{j=1}^{m}\int_{\Omega_c} g''_\upsilon(\phi_j(x)(t))(A_jh)(t)(A_jk)(t)dt\end{aligned} \tag{5.7.14}$$

(iii) Using (5.7.14), we have

$$\langle\nabla^2 f_{\upsilon,\gamma}(x)h, h\rangle = \langle 2Lh, h\rangle + \gamma\sum_{j=1}^{m}\int_{\Omega_c} g''_\upsilon(\phi_j(x)(t))(A_jh)(t)(A_jh)(t)dt$$

Note that $\langle 2Lh, h\rangle \le 2\|L\|\|h\|^2$. Further, assuming A_jh is square integrable, we have

$$\begin{aligned}\int_{\Omega_c} g''_\upsilon(\phi_j(x)(t))[(A_jh)(t)]^2dt &\le h_\upsilon\int_{\Omega_c}[(A_jh)(t)]^2dt \\ &\le h_\upsilon\int_{\boldsymbol{R}}[(A_jh)(t)]^2dt = h_\upsilon\|(A_jh)\|^2 \\ &\le h_\upsilon\|A_j\|^2\|h\|^2\end{aligned}$$

Hence,

$$\langle \nabla^2 f_{\upsilon,\gamma}(x)h, h\rangle \le \left(2\|L\| + \gamma h_\upsilon \sum_{j=1}^{m} \|A_j\|^2\right)\|h\|^2.$$

Now since for all h,

$$\gamma \sum_{j=1}^{m} \int_{\Omega_c} g_\upsilon''(\phi_j(x)(t))(A_j h)(t)(A_j h)(t)dt \ge 0,$$

we have

$$\langle \nabla^2 f_{\upsilon,\gamma}(x)h, h\rangle \ge \langle 2Lh, h\rangle$$

$$\sup_{h \ne 0} \frac{\langle \nabla^2 f_{\upsilon,\gamma}(x)h, h\rangle}{\|h\|^2} \ge \sup_{h \ne 0} \frac{\langle 2Lh, h\rangle}{\|h\|^2} \tag{5.7.15}$$

Thus, the conclusion follows.

Proof of extra results for finite dimensional case.

(iv) Again using (ii), we have

$$\begin{aligned}\nabla^2 f_{\upsilon,\gamma}(\boldsymbol{u}) - \nabla^2 f_{\upsilon,\gamma}(\boldsymbol{v}) &= \gamma \sum_{j=1}^{m} \int_{\Omega_c} g_\upsilon''(\phi_j(\boldsymbol{u}, t))\boldsymbol{a}_j(t)\boldsymbol{a}_j^T(t)dt - \\ &\quad \gamma \sum_{j=1}^{m} \int_{\Omega_c} g_\upsilon''(\phi_j(\boldsymbol{v}, t))\boldsymbol{a}_j(t)\boldsymbol{a}_j^T(t)dt\end{aligned}$$

Hence, using the triangle inequality and the Lipschitz condition

$$|g_\upsilon''(x) - g_\upsilon''(y)| \le l_\upsilon |x - y|$$

gives

$$\|\nabla^2 f_{\upsilon,\gamma}(\boldsymbol{u}) - \nabla^2 f_{\upsilon,\gamma}(\boldsymbol{v})\| \le \gamma l_\upsilon \sum_{j=1}^{m} \int_{\Omega_c} |\phi_j(\boldsymbol{u}, t) - \phi_j(\boldsymbol{v}, t)| \|\boldsymbol{a}_j(t)\|^2 dt.$$

Since

$$\phi_j(\boldsymbol{u}, t) - \phi_j(\boldsymbol{v}, t) = \boldsymbol{a}_j^T(t)(\boldsymbol{u} - \boldsymbol{v}),$$

it is clear that

$$\left\|\nabla^2 f_{\upsilon,\gamma}(\boldsymbol{u}) - \nabla^2 f_{\upsilon,\gamma}(\boldsymbol{v})\right\| \le \gamma l_\upsilon \sum_{j=1}^{m} \int_{\Omega_c} \left|\boldsymbol{a}_j^T(t)(\boldsymbol{u}-\boldsymbol{v})\right| \left\|\boldsymbol{a}_j(t)\right\|^2 dt$$

$$\le \gamma l_\upsilon \|\boldsymbol{u}-\boldsymbol{v}\| \sum_{j=1}^{m} \int_{\Omega_c} \left\|\boldsymbol{a}_j(t)\right\|^3 dt. \quad \square$$

CHAPTER 6 ROBUST ENVELOPE CONSTRAINED FILTERING

In Chapters 4 and 5, we studied numerical methods for finding a filter whose response to a specified signal fits into a given envelope. Assuming that the set of feasible filters does not contain the origin, i.e. no trivial solution, and since we seek a feasible filter with the smallest possible norm, it follows that the optimum filter always lies on the boundary of the feasible set. This means the response of the optimum filter to the prescribed input touches the output envelope at some points. Consequently, perturbations of the prescribed input or filter implementation errors could result in a filter output that violates the output constraints. The design of filters that are robust to such perturbations is very important in practice. Of course, one expects that there is some penalty for achieving robustness, for example, the noise gain of the optimal filter would increase in the robust case.

We addresse the issue of designing optimal filters that are robust by introducing three new formulations of the EC filtering problem. The first incorporates input uncertainty into the constraints. This problem has been formulated as a variation of the EC filtering problem earlier in Section 2.2 of Chapter 2, and is treated in detail in this chapter. The EC with uncertain input (ECUI) filtering problem takes on a form which is intractable numerically. This major difficulty can be alleviated as shown in Section 6.1, where an analytic expression for the constraint is derived. However, the resulting constraint function is non-differentiable. Sections 6.2 and 6.3 are devoted to converting this problem into a positive definite QP problem with affine (hence differentiable) constraints. In discrete-time, results follow easily from the analog counterparts and Section 6.4 presents a summary of these results. Section 6.5 formulates the ECUI filtering problem for hybrid filters. As in the continuous-time case, the problem also takes on a numerically intractable form. A similar technique is used to convert it into a positive definite QP problem with smooth constraints. In Section 6.6, we introduce the second approach which does not consider the source of the disturbances, the key idea is to force the filter output to stay away from the envelope boundaries. In Section 6.7, we introduce the third approach which incorporates implementation uncertainty into the constraints. This problem takes on the same form as the EC filtering problem and so the algorithms of Chapter 5 can be applied. Proofs of results are given in Section 6.8 at the end of the chapter unless otherwise stated.

6.1 Constraint Function

The EC with uncertain input (ECUI) filtering problem formulated earlier in Chapter 2 addresses the robustness to input disturbances by allowing for uncertainty in the input pulse. Here the input s is not specified exactly, but is known to lie within an input envelope described by upper and lower boundaries s^- and s^+. The filter is designed so that the response of all excitations within the boundaries s^- and s^+ fit into the output mask. Of course, the penalty for robustness is the increased noise gain of the filter. For example, in the radar sidelobe reduction problem, it is desirable to design a sidelobe reduction filter which is robust in the sense that its output sidelobes remain small even if the input signal is slightly different from the nominal input. In channel equalization, the ECUI approach can handle the case where the channel characteristics are not known exactly but are known to stay within an envelope.

Recall from Section 2.2 of Chapter 2 that the ECUI problem can be posed as

$$\min_{u \in \mathcal{F}} \langle u, Lu \rangle$$

where the feasible set $\mathcal{F}$ is given by

$$\mathcal{F} = \{u \in L_2(\Omega_u) : |x * u - d| \preceq \varepsilon, \ \forall x \in L_2(\Omega_s) : |x - s| \preceq \theta\}.$$

Although the same notation is used for both partial orderings, there should be no ambiguity if the specific signal space is taken into consideration. In the excitation space $L_2(\Omega_s)$, the inequality is required to hold on the entire support Ω_s, while in the response space $L_\infty(\Omega_\psi)$, the inequality only holds on Ω_c rather than Ω_ψ.

Since the ECUI filtering problem is a CCR problem, the feasible region $\mathcal{F}$ is closed and convex if the set of all possible excitations, given by (2.2.5), i.e.

$$\mathcal{S}(s, \theta) = \{x \in L_2(\Omega_s) : |x - s| \preceq \theta\}$$

is bounded (see Lemma 2.1.2). It is assumed that $s^- \preceq s^+$ and s^+, s^- are finite energy signals so that $\mathcal{S}(s, \theta)$ is non-empty and bounded. This problem has a unique solution if the operator L is positive definite.

The description of the feasible region in terms of the set $\mathcal{S}(s, \theta)$ of all possible inputs is adequate for characterizing the problem. In particular, we have established (in Chapter 2) that it is a convex optimization problem. However, it is not helpful for computational purposes. To determine if a filter is feasible one would need to compute its response to every signal in the set $\mathcal{S}(s, \theta)$. There is no standard numerical techniques for handling problems with constraints of the form

$$|x * u - d| \preceq \varepsilon, \forall x : |x - s| \preceq \theta.$$

To solve the ECUI filtering problem, it would be necessary to have an equivalent but more explicit expression that does not contain the perturbed input x. Since the set $\mathcal{S}(s, \theta)$ of possible inputs is completely characterized by s and θ, we want an expression that would involve only d, ε, s, θ and the filter's impulse response u. This can be achieved with the help of the following lemma.

Lemma 6.1.1. *Suppose X_1 and X_2 are Banach spaces and for each X_j the subspace consisting of the real-valued members is equipped with a partial ordering*

$$x \preceq y \Leftrightarrow x(t) \le y(t),\ \forall t \in \Omega_j .$$

Let $F : X_1 \to X_2$ and $|F| : X_1 \to X_2$ be two linear operators defined by

$$(Fx)(t) \equiv \int_{\Omega_1} x(\tau) f(t, \tau) d\tau$$

and

$$(|F|x)(t) \equiv \int_{\Omega_1} x(\tau) |f(t, \tau)| d\tau,$$

respectively. Then, for a given positive $\theta \in X_1$,

$$|Fx - d| \preceq \varepsilon,\ \forall x{:}|x - s| \preceq \theta \quad \Leftrightarrow \quad |Fs - d| + |F|\theta \preceq \varepsilon .$$

Lemma 6.1.1 allows the constraints to be expressed in a form which does not involve the set $\mathcal{S}(s, \theta)$ of possible inputs. This is applicable to complex valued functions, however, for EC filtering, we only consider real valued functions. For analog ECUI, by identifying $Fx = x * u$, we see that F is linear with kernel given by $f(t, \tau) = u(t - \tau)$. Hence, by Lemma 6.1.1, we have the following result.

Theorem 6.1.1. $|x * u - d| \preceq \varepsilon, \forall x \in L_2(\Omega_s)$: $|x - s| \preceq \theta$ *if and only if*

$$|s * u - d| + \theta * |u| \preceq \varepsilon . \tag{6.1.1}$$

Inspecting the proof of Lemma 6.1.1, it is observed that for each $t \in \Omega_c$, the distance between the desired response and the output corresponding to a signal $x \in \mathcal{S}(s, \theta)$, i.e. $|(x * u)(t) - d(t)|$, is bounded above by $|(s * u)(t) - d(t)| + (\theta * |u|)(t)$. Moreover, there exists a signal $x_t \in \mathcal{S}(s, \theta)$ which achieves this bound. More concisely, for each $t \in \Omega_c$, there exists an x_t such that

$$\begin{aligned} \max_{x \in \mathcal{S}(s, \theta)} |(x * u)(t) - d(t)| &= |(x_t * u)(t) - d(t)| \\ &= |(s * u)(t) - d(t)| + (\theta * |u|)(t). \end{aligned} \tag{6.1.2}$$

We refer to the input x_t which achieves this maximum as the worst case input at t. This signal can be written as

$$x_t = s + \tilde{s}_t$$

where $\tilde{s}_t$ is the worst case perturbation given by

$$\tilde{s}_t(\tau) = \exp\{j\arg[(s * u)(t) - d(t)] - j\arg[u(t-\tau)]\}\theta(\tau).$$

Thus, a filter u is feasible if and only if for each instant $t \in \Omega_c$, its response at t to the worst case signal x_t stays inside the output envelope. Theorem 6.1.1 also implies that the uncertainty θ in the input effectively reduces the output mask by $\theta * |u|$. Note that even though for different t there exists different worst case perturbations $\tilde{s}_t$, the absolute values of these perturbations are independent of t and all are equal to the input uncertainty θ, i.e. $|\tilde{s}_t(\tau)| = \theta(\tau)$ for all τ. Hence, the norm of the worst case perturbation is, in fact, the norm of the input uncertainty θ.

Using the convolution operator notation $\Xi_x u = x * u$, the constraint function for the analog ECUI problem $G : L_2(\Omega_u) \to L_\infty(\Omega_\psi)$ can be defined as

$$G(u) = |\Xi_s u - d| + \Xi_\theta |u| - \varepsilon. \tag{6.1.3}$$

Although the constraints are applied only on Ω_c, we still require d and ε to be defined on Ω_ψ so that the above definition of G is meaningful for the cases when $\Omega_\psi \neq \Omega_c$. With this in mind, it can be easily shown that:

Lemma 6.1.2. *The constraint function G is continuous and convex.*

Theorem 6.1.1 eliminates the perturbed input x from the original constraint expression and allows the analog ECUI filtering problem to be stated as follows:

$$\begin{gathered}\min f(u),\ u \in L_2(\Omega_u) \\ \text{subject to } G(u) = |\Xi_s u - d| + \Xi_\theta |u| - \varepsilon \preceq 0.\end{gathered} \tag{6.1.4}$$

The cost and the constraint are continuous and convex functions of the filter impulse response u. The constraint function only involves the known signals d, ε, s and θ.

Clearly, (6.1.1) is far more easier to handle than the original expression. Note that G is not always everywhere differentiable with respect to u.

For the special case with $\theta = 0$, the constraint reduces to

$$|\Xi_s u - d| - \varepsilon \preceq 0.$$

This is the EC filtering problem and the constraints can be written as

$$\varepsilon^- \preceq \Xi_s u \preceq \varepsilon^+ .$$

Observe that there are now two constraint functions, each of which is differentiable with respect to the variable u.

6.2 Transformation to Smooth Problem

In the previous section, it has been shown that constraint of the ECUI filtering problem can be expressed as a continuous and convex function of the filter impulse response, see (6.1.3). This section investigates methods for solving this type of convex optimization problem with positive definite quadratic cost functional. In particular, we study a generalized form that can later be applied to the discrete-time and hybrid ECUI filters.

Problem conversion

Let H denote the Hilbert space of filter impulse responses. The space H can be $L_2(\Omega_u)$ in the case of analog filters or $l_2(\tilde{\Omega}_u)$ in the case of digital filters or $\boldsymbol{R}^n$ for finite-dimensional filters. Let Y denote the Banach space of filter outputs which can either be $L_\infty(\Omega_\psi)$ in the case of continuous-time output, or $l_\infty(\tilde{\Omega}_\psi)$ in the case of discrete-time output. The ECUI filtering problem can be posed as follows:

$$\begin{aligned} &\min \|u\|_L^2 \equiv \langle u, Lu \rangle, \; u \in H \\ &\text{subject to } |Au - d| + B|u| - \varepsilon \preceq 0, \end{aligned} \tag{6.2.1}$$

where L is a positive definite linear operator on H, $A : H \to Y$ and $B : H \to Y$ are linear operators with $0 \preceq Bu$ for all $0 \preceq u$. The partial ordering $\preceq$ is defined by

$$x \preceq y \Leftrightarrow x(t) \le y(t), \forall t \in \Omega,$$

where Ω is the support of x, y for $x, y \in H$, and $\Omega = \Omega_c \subseteq \Omega_\psi$ for $x, y \in Y$. Note that the cost functional $f(u) \equiv \|u\|_L^2$ of Problem (6.2.1) is an inner product induced norm since L is positive definite.

Problem (6.2.1) as stated is a non-smooth optimization problem because the constraint function

$$G(u) \equiv |Au - d| + B|u| - \varepsilon$$

is not everywhere differentiable with respect to the variable u. However, it can be converted into a smooth quadratic programming problem. The rationale is to find an equivalent expression for the constraint which does not involve the absolute value operator $|\bullet|$. The first absolute value operator in G is not a problem, as it is quite straightforward to show that

$$|Au - d| + B|u| - \varepsilon \preceq 0 \Leftrightarrow \begin{cases} Au + B|u| - \varepsilon^+ \preceq 0 \\ -Au + B|u| + \varepsilon^- \preceq 0 \end{cases}, \tag{6.2.2}$$

where $\varepsilon^+ = d + \varepsilon$, and $\varepsilon^- = d - \varepsilon$ as given earlier in Section 6.1.

The major difficulty is caused by the term $B|u|$. The decomposition of u into its positive and negative parts

$$\begin{aligned} u^+(t) &= [p(u)](t) = \max(u(t), 0), \\ u^-(t) &= [q(u)](t) = \max(-u(t), 0), \end{aligned} \tag{6.2.3}$$

so that $u = u^+ - u^-$ and $|u| = u^+ + u^-$ suggests that Problem (6.2.1) can be converted into the following smooth problem:

$$\begin{aligned} \min \quad & \|x-y\|_L^2 + 2\rho\langle x, y\rangle, \quad x, y \in H \\ \text{subject to} \quad & A(x-y) + B(x+y) - \varepsilon^+ \preceq 0, \\ & -A(x-y) + B(x+y) + \varepsilon^- \preceq 0, \\ & -x \preceq 0, \\ & -y \preceq 0, \end{aligned} \tag{6.2.4}$$

where $\rho \in (0, 2\lambda_{min}(L))$, and $\lambda_{min}(L)$ is the minimum eigenvalue of L.

The critical question that needs to be addressed is: how does a solution of Problem (6.2.4) relates to the solution of Problem (6.2.1) and vice versa? Before answering this question, let us establish some useful properties of the decomposition.

***Lemma 6.2.1**: Let $p: H \to H$ and $q: H \to H$ be mappings on a Hilbert space H defined by (6.2.3). Then,*

(i) $0 \preceq p(u), 0 \preceq q(u)$,

(ii) $p(u)q(u) = 0$,

(iii) $p(u) - q(u) = u$,

(iv) $p(u) + q(u) = |u|$,

(v) for each $u, v \in H$ with $0 \preceq u, 0 \preceq v$, there exists a unique non-negative function $r: H \times H \to H$ such that

$$|u - v| = u + v - r(u, v).$$

Another important issue is the uniqueness of the optimum solution to Problem (6.2.4). The following result shows that the cost functional of Problem (6.2.4) is positive definite, that is $f(x, y) \geq 0$ for all $x, y \in H$ and $f(x, y) = 0$ if and only if $x = 0$ and $y = 0$.

Lemma 6.2.2. *Let L be a positive definite linear operator on a Hilbert space H and ρ be a real number such that $\rho \in (0, 2\lambda_{min}(L))$. Then, the mapping $f: H \times H \to \mathbf{R}$ defined by*

$$f(x, y) = \|x - y\|_L^2 + 2\rho\langle x, y\rangle$$

is positive definite.

This result implies that Problem (6.2.4) has a unique solution (x^o, y^o) because the cost functional is positive definite. Furthermore, positive definiteness of the cost functional makes the numerical task much easier. Now we are ready to address the equivalence of Problems (6.2.1) and (6.2.4).

Theorem 6.2.1. *If (x^o, y^o) is the optimal solution of Problem (6.2.4), then $x^o - y^o$ is the optimal solution of Problem (6.2.1). Conversely, if u^o is the optimal solution of Problem (6.2.1), then $(p(u^o), q(u^o))$ is the optimal solution of Problem (6.2.4).*

Theorem 6.2.1 shows that the ordered pair consisting of positive and negative parts of the solution u^o to Problem (6.2.1) is, in fact, a solution to Problem (6.2.4). More importantly, if a solution (x^o, y^o) to Problem (6.2.4) is found, then we can obtain the solution u^o by the equation

$$u^o = x^o - y^o.$$

In effect, we have converted a positive definite QP problem with non-differentiable constraint into one with differentiable constraints.

Remark 6.2.1: Theorem 6.2.1 still holds for any positive value of ρ, since the proof does not make use of the condition $\rho < 2\lambda_{min}(L)$. In the case where $\rho \geq 2\lambda_{min}(L)$, the cost functional may not be strictly convex. This means that the solution (x^o, y^o) may not be unique. Nevertheless, since Problem (6.2.1) has a unique solution, the difference $x^o - y^o$ is unique.

For the special case where L is an identity operator, we can choose ρ to be unity since the eigenvalues of L are all equal to one. This deduces the cost functional of Problem (6.2.4) to the form

$$f(x, y) = \|x\|^2 + \|y\|^2.$$

This special case has been established in [115] by a different line of argument.

6.3 Application to Analog ECUI Filtering Problem

We have seen how a positive definite QP problem with non-differentiable cost of the form (6.2.1) can be converted into another positive definite QP problem of the form (6.2.4) where the constraints are differentiable. Combining the two primal variables x, y into a single variable $z = [x, y]^T$, Theorem 6.2.1 allows the analog ECUI filtering problem to be written as follows:

$$\begin{aligned}
\min \quad & \langle z, Hz\rangle, z \in L_2^2(\Omega_u) \equiv L_2(\Omega_u) \times L_2(\Omega_u) \\
\text{subject to} \quad & [\Xi_s, -\Xi_s]z + [\Xi_\theta, \Xi_\theta]z - \varepsilon^+ \preceq 0, \\
& [-\Xi_s, \Xi_s]z + [\Xi_\theta, \Xi_\theta]z + \varepsilon^- \preceq 0, \\
& [-I, 0]z \preceq 0, \\
& [0, -I]z \preceq 0,
\end{aligned} \tag{6.3.1}$$

where

$$H = \begin{bmatrix} L & \rho I - L \\ \rho I - L & L \end{bmatrix} \tag{6.3.2}$$

is a positive definite operator on $L_2^2(\Omega_u)$, $\rho \in (0, 2\lambda_{min}(L))$, and I is the identity operator on $L_2(\Omega_u)$. This is a quadratic programming problem with affine inequality constraint of the form shown in Chapter 3 where the decision variables are in $L_2(\Omega_u)$. Nonetheless, extension to $L_2^2(\Omega_u)$ is trivial. The solution to the ECUI filtering problem is obtained by using the transform

$$u = [I, -I]z.$$

Following the duality development of Chapter 3, we examine the Lagrangian $\mathcal{L} : L_2^2(\Omega_u) \times M^2(\Omega_\psi) \times L_2^2(\Omega_u) \to \boldsymbol{R}$ of Problem (6.3.1), defined by

$$\begin{aligned}\mathcal{L}(z, p_1, p_2, q_1, q_2) &= \langle z, Hz\rangle + p_1 \bullet [\Xi_s, -\Xi_s]z + p_1 \bullet [\Xi_\theta, \Xi_\theta]z - p_1 \bullet \varepsilon^+ \\ &\quad + p_2 \bullet [-\Xi_s, \Xi_s]z + p_2 \bullet [\Xi_\theta, \Xi_\theta]z + p_2 \bullet \varepsilon^- \\ &\quad + \langle q_1, [-I, 0]z\rangle + \langle q_2, [0, -I]z\rangle,\end{aligned} \tag{6.3.3}$$

where $\bullet$ denotes

$$p \bullet x = \int_{\Omega_\psi} x dp, \quad x \in C(\Omega_\psi).$$

Let $\alpha(p_1, p_2, q_1, q_2) \in L_2^2(\Omega_u)$ be defined by

$$[\alpha(p_1, p_2, q_1, q_2)](\tau) = \begin{bmatrix} [\mu_s(p_1 - p_2)](\tau) + [\mu_\theta(p_1 + p_2)](\tau) - q_1(\tau) \\ -[\mu_s(p_1 - p_2)](\tau) + [\mu_\theta(p_1 + p_2)](\tau) - q_2(\tau) \end{bmatrix},$$

where

$$[\mu_x(p)](\tau) = \int_{\Omega_c} x(t - \tau) dp(t).$$

Then, it can be verified that

$$\begin{aligned}\langle \alpha^T(p_1, p_2, q_1, q_2), z\rangle &= (p_1 - p_2) \bullet [\Xi_s, -\Xi_s]z \\ &\quad + (p_1 + p_2) \bullet [\Xi_\theta, \Xi_\theta]z - \langle [q_1, q_2], z\rangle\end{aligned} \tag{6.3.4}$$

by expanding both sides of (6.3.4) and then applying Fubini's Theorem (Appendix A.8, Theorem A.8.10). Using (6.3.4) and rearranging the Lagrangian (6.3.3), we obtain the following quadratic function of z:

$$\begin{aligned}\mathcal{L}(z, p_1, p_2, q_1, q_2) &= \langle z, Hz\rangle + (p_1 - p_2) \bullet ([\Xi_s, -\Xi_s]z - d) \\ &\quad + (p_1 + p_2) \bullet ([\Xi_\theta, \Xi_\theta]z - \varepsilon) - \langle [q_1, q_2], z\rangle \\ &= \langle z, Hz\rangle + \langle \alpha^T(p_1, p_2, q_1, q_2), z\rangle \\ &\quad - (p_1 - p_2) \bullet d - (p_1 + p_2) \bullet \varepsilon.\end{aligned} \tag{6.3.5}$$

Differentiating (6.3.5) with respect to z and then equating to zero yields

$$z = -\frac{1}{2} H^{-1} \alpha(p_1, p_2, q_1, q_2), \tag{6.3.6}$$

where the inverse of H is given by

$$H^{-1} = \begin{bmatrix} \rho^{-1}LZ & (\rho^{-1}L-I)Z \\ (\rho^{-1}L-I)Z & \rho^{-1}LZ \end{bmatrix}$$

$$= \begin{bmatrix} \rho^{-1}L & \rho^{-1}L-I \\ \rho^{-1}L-I & \rho^{-1}L \end{bmatrix}\begin{bmatrix} Z & 0 \\ 0 & Z \end{bmatrix} = \begin{bmatrix} Z & 0 \\ 0 & Z \end{bmatrix}\begin{bmatrix} \rho^{-1}L & \rho^{-1}L-I \\ \rho^{-1}L-I & \rho^{-1}L \end{bmatrix},$$

in which

$$Z = (2L-\rho I)^{-1},$$

is a symmetric operator on $L_2(\Omega_u)$. Note that since $\rho \in (0, 2\lambda_{min}(L))$, $(2L-\rho I)^{-1}$ is still positive definite. It can be easily shown that

$$LZ = ZL = (2I-\rho L^{-1})^{-1},$$

and consequently

$$H^{-1}H = HH^{-1} = \begin{bmatrix} I & 0 \\ 0 & I \end{bmatrix}.$$

Substituting (6.3.6) into the Lagrangian (6.3.5) yields the following dual functional:

$$\begin{aligned} \phi(p_1, p_2, q_1, q_2) = & -\frac{1}{4}\langle\alpha(p_1, p_2, q_1, q_2), H^{-1}\alpha(p_1, p_2, q_1, q_2)\rangle \\ & -(p_1-p_2)\bullet d-(p_1+p_2)\bullet\varepsilon. \end{aligned} \tag{6.3.7}$$

Note that

$$\langle [x, y], H^{-1}\begin{bmatrix} x \\ y \end{bmatrix}\rangle = \frac{1}{\rho}\langle ZL(x+y), (x+y)\rangle - 2\langle Zx, y\rangle.$$

Then, using (6.3.4), the dual cost functional (6.3.7) becomes:

$$\begin{aligned} \phi(p_1, p_2, q_1, q_2) = & -\frac{1}{4\rho}\langle ZL[2\mu_\theta(p_1+p_2)-q_1-q_2], [2\mu_\theta(p_1+p_2)-q_1-q_2]\rangle \\ & +\frac{1}{2}\langle Z[\mu_s(p_1-p_2)+\mu_\theta(p_1+p_2)-q_1], -\mu_s(p_1-p_2)+\mu_\theta(p_1+p_2)-q_2\rangle \\ & -(p_1-p_2)\bullet d-(p_1+p_2)\bullet\varepsilon \end{aligned}$$

By Theorem 3.1.1, once $(p_1^o, p_2^o, q_1^o, q_2^o)$ is found to maximize the dual cost functional (6.3.7) over $P_M^{\oplus} \times P_M^{\oplus} \times P_L^{\oplus} \times P_L^{\oplus}$, the primal solution is given by

$$z^o = -\frac{1}{2}H^{-1}\alpha(p_1^o, p_2^o, q_1^o, q_2^o).$$

Thus, the optimal solution u^o to Problem (6.2.1) can be obtained from Theorem 6.2.1 as follows:

$$\begin{aligned} u^o &= -\frac{1}{2}[I, -I]H^{-1}\alpha(p_1^o, p_2^o, q_1^o, q_2^o) \\ &= -\frac{1}{2}Z[I, -I]\alpha(p_1^o, p_2^o, q_1^o, q_2^o) \\ &= -\frac{1}{2}(2L - \rho I)^{-1}[2\mu_s(p_1^o - p_2^o) - q_1^o + q_2^o]. \end{aligned}$$

Note that $p_1^o, p_2^o, q_1^o, q_2^o$ depend on the value of ρ. Similar to the EC filtering Problem, the $\mu_s(p_1^o - p_2^o)$ can be interpreted as a filter matched to the input signal s, followed by a pulse shaping filter determined by $p_1^o - p_2^o$. Since $\rho \in (0, 2\lambda_{min}(L))$, taking the limiting case where it is infinitesimally small, the term $(2L - \rho I)^{-1}$ becomes $L^{-1}/2$. Hence, the solution can be interpreted as a whitening filter followed by the filter

$$[2\mu_s(p_1^o - p_2^o) - q_1^o + q_2^o].$$

Finite dimensional filter for analog problem

We now consider a finite dimensional filter with impulse response of the form

$$u = K\boldsymbol{a} = \boldsymbol{v}^T\boldsymbol{a}$$

as given in Section 2.4, in which $\boldsymbol{v}^T = [v_0, v_1, \ldots, v_{n-1}]$ is a vector of n linearly independent elements of $L_2(\Omega_u)$ and $\boldsymbol{a} \in \boldsymbol{R}^n$ is the vector of filter parameters. The feasible region $\mathcal{F} \cap [\{v_i\}_{i=0}^{n-1}]$ is embedded in an n-dimensional subspace and can be characterized by the set of feasible filter coefficients, which is defined in $\boldsymbol{R}^n$ as

$$\mathcal{F}^n = \{\boldsymbol{a} \in \boldsymbol{R}^n : |x * (K\boldsymbol{a}) - d| \preceq \varepsilon, \forall x : |x - s| \preceq \theta\}.$$

Using Theorem 6.1.1, the ECUI problem for finite dimensional filters is

$$\begin{aligned} &\min\ \boldsymbol{a}^T K^\dagger L K \boldsymbol{a}, \boldsymbol{a} \in \boldsymbol{R}^n \\ &\text{subject to } \left|\Xi_s K\boldsymbol{a} - d\right| + \Xi_\theta |K\boldsymbol{a}| - \varepsilon \preceq 0. \end{aligned} \tag{6.3.8}$$

Note that the transform technique of Section 6.2 cannot be applied to the finite dimensional case, because the image $K(\boldsymbol{R}^n)$ is the subspace $[\{v_i\}_{i=0}^{n-1}]$ of $L_2(\Omega_u)$, which may not be closed under the mappings p and q. That is, there may not be any $\boldsymbol{x}, \boldsymbol{y} \in \boldsymbol{R}^n$ for which $K\boldsymbol{x} = p(K\boldsymbol{a})$ and $K\boldsymbol{y} = q(K\boldsymbol{a})$.

Suppose that the image $K(\boldsymbol{R}^n)$ is closed under p and q, for example the Walsh functions [33]. Then, for each $\boldsymbol{a} \in \boldsymbol{R}^n$, there exist unique $\boldsymbol{x}, \boldsymbol{y} \in \boldsymbol{R}^n$ such that

$$p(K\boldsymbol{a}) = K\boldsymbol{x} \text{ and } q(K\boldsymbol{a}) = K\boldsymbol{y}.$$

Hence, the conditions of Theorem 6.2.1 remain valid when we replace x and y in Problem (6.2.4) by $K\boldsymbol{x}$ and $K\boldsymbol{y}$, respectively.

The cost functional can be expressed in terms of the vectors $\boldsymbol{x}, \boldsymbol{y} \in \boldsymbol{R}^n$ as follows:

$$\|x - y\|_L^2 + 2\rho\langle x, y\rangle = (\boldsymbol{x} - \boldsymbol{y})^T K^\dagger L K(\boldsymbol{x} - \boldsymbol{y}) + 2\rho \boldsymbol{x}^T K^\dagger K \boldsymbol{y}, \tag{6.3.9}$$

where $K^\dagger$ is the adjoint of K and $K^\dagger L K$ is the Gram matrix (see (2.4.5) in Chapter 2).

Combining the two $\boldsymbol{R}^n$ vectors $\boldsymbol{x}, \boldsymbol{y}$ into one $\boldsymbol{R}^{2n}$ vector z by defining

$$z^T = [\boldsymbol{x}^T, \boldsymbol{y}^T],$$

so that $\boldsymbol{x} = [I_n, 0_n]z$, $\boldsymbol{y} = [0_n, I_n]z$, $\boldsymbol{x} - \boldsymbol{y} = [I_n, -I_n]z$ and $\boldsymbol{x} + \boldsymbol{y} = [I_n, I_n]z$, we can rewrite (6.3.9) as

$$z^T H z$$

where H is a positive definite matrix given by

$$\begin{aligned} H &= [I_n, -I_n]^T K^\dagger L K [I_n, -I_n] + 2\rho [I_n, 0_n]^T K^\dagger K [0_n, I_n] \\ &= \begin{bmatrix} K^\dagger L K & 2\rho K^\dagger K - K^\dagger L K \\ -K^\dagger L K & K^\dagger L K \end{bmatrix}. \end{aligned} \tag{6.3.10}$$

or alternatively, in a symmetric form

$$\begin{aligned} H &= [I_n, -I_n]^T K^{\dagger} L K [I_n, -I_n] \\ &\quad + \rho [I_n, 0_n]^T K^{\dagger} K [0_n, I_n] + \rho [0_n, I_n]^T K^{\dagger} K [I_n, 0_n] \\ &= \begin{bmatrix} K^{\dagger} L K & \rho K^{\dagger} K - K^{\dagger} L K \\ \rho K^{\dagger} K - K^{\dagger} L K & K^{\dagger} L K \end{bmatrix}. \end{aligned} \tag{6.3.11}$$

The constraints can be written as

$$\begin{aligned} \Xi_s K [I_n, -I_n] z + \Xi_\theta K [I_n, I_n] z - \varepsilon^+ &\preceq 0 \\ -\Xi_s K [I_n, -I_n] z + \Xi_\theta K [I_n, I_n] z + \varepsilon^- &\preceq 0 \\ -K [I_n, 0_n] z &\preceq 0 \\ -K [0_n, I_n] z &\preceq 0 \end{aligned}$$

Consequently, recalling from (3.3.4) in Chapter 3 that the composite operator $\Xi_s K$ can be expressed in an alternative form:

$$(\Xi_s K \boldsymbol{a})(t) = \boldsymbol{y}_s^T(t) \boldsymbol{a},$$

where

$$\boldsymbol{y}_s^T(t) = [(\Xi_s v_0)(t), \ldots, (\Xi_s v_{n-1})(t)],$$

the finite-dimensional analog ECUI filtering problem can be posed as the following finite-dimensional QP problem:

$$\begin{aligned} \min \quad & z^T H z, z \in \boldsymbol{R}^{2n} \\ \text{subject to} \quad & [\boldsymbol{y}_s^T(t), -\boldsymbol{y}_s^T(t)] z + [\boldsymbol{y}_\theta^T(t), \boldsymbol{y}_\theta^T(t)] z - \varepsilon^+ \le 0, \forall t \in \Omega_c \\ & -[\boldsymbol{y}_s^T(t), -\boldsymbol{y}_s^T(t)] z + [\boldsymbol{y}_\theta^T(t), \boldsymbol{y}_\theta^T(t)] z + \varepsilon^- \le 0, \forall t \in \Omega_c \\ & -\boldsymbol{v}^T(t) [I_n, 0_n] z \le 0, \forall t \in \Omega_u \\ & -\boldsymbol{v}^T(t) [0_n, I_n] z \le 0, \forall t \in \Omega_u \end{aligned} \tag{6.3.12}$$

If $z^o \in \boldsymbol{R}^{2n}$ is the solution to Problem (6.3.12), then $\boldsymbol{a}^o = [I_n, -I_n] z^o$ is the solution to the finite-dimensional analog ECUI filtering problem.

Example using Walsh functions

The basis functions used here are the Walsh functions [33]:

$$v_0(t) = 1, t \in [0, T],$$

$$v_1^0(t) = \begin{cases} 1, t \in [0, 0.5T) \\ -1, t \in (0.5T, T] \end{cases},$$

and for $k = 1, \ldots, 2^{m-1}$, $m = 1, 2, \ldots$

$$v_{m+1}^{(2k-1)}(t) = \begin{cases} v_m^{(k)}(t), & t \in [0, 0.5T) \\ (-1)^{k+1} v_m^{(k)}(2t-T), & t \in (0.5T, T] \end{cases},$$

$$v_{m+1}^{(2k)}(t) = \begin{cases} v_m^{(k)}(t), & t \in [0, 0.5T) \\ (-1)^k v_m^{(k)}(2t-T), & t \in (0.5T, T] \end{cases}.$$

Consider the compression of a 13-bit Barker coded signal with an allowable error of 3% as shown in Figure 6.3.1 (a unit on the time axis corresponds to one bit interval β). We require all responses to signals within this input mask to fit into an output mask with allowable sidelobes of 0.03 and a mainlobe peak of 0.69 ± 0.075 as shown in Figure 6.3.2.

Without loss of generality, we consider the case where the input noise is white. In this example, we use 32 Walsh functions, i.e. $m = 5$, $k = 16$. To match the bit interval, we use $T = 32\beta$. For computational purpose, the continuum of constraints is discretized to 64 constraints per bit interval.

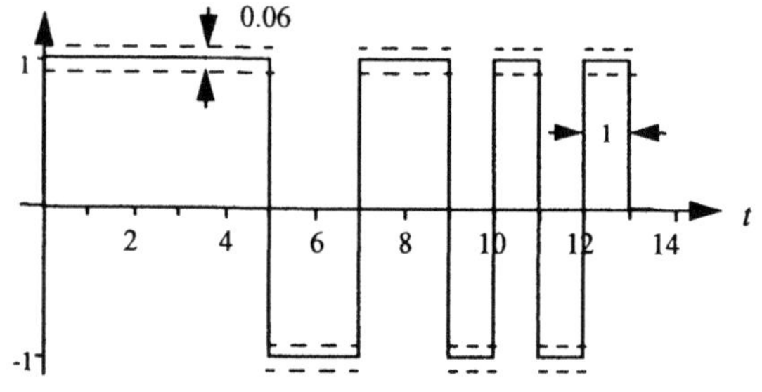

Figure 6.3.1. 13 -bit Barker-coded signal with input mask

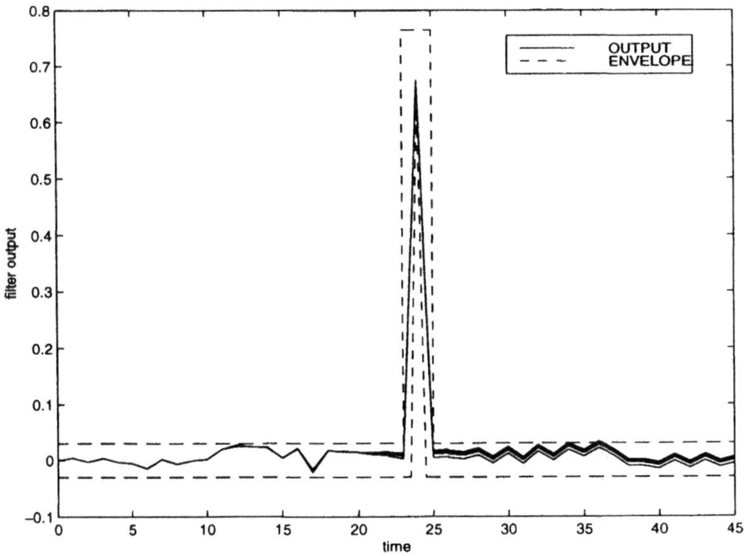

Figure 6.3.2. Filter outputs

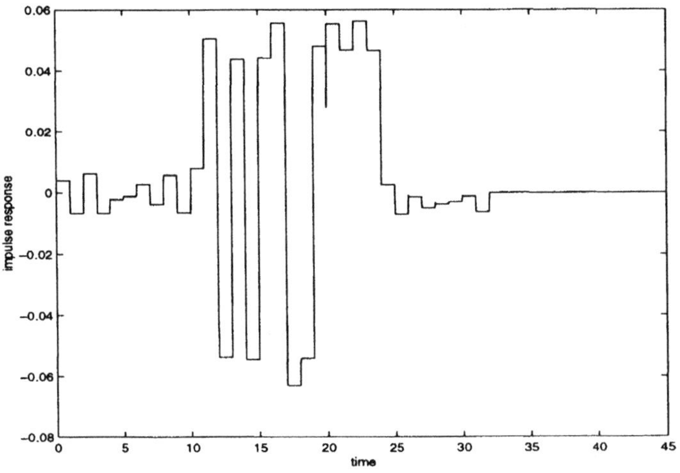

Figure 6.3.3. Impulse response of optimal ECUI filter

This type of approximation results in a finite dimensional quadratic programming problem with a finite number of affine inequality constraints which can be easily dealt with by MATLAB routines. Figure 6.3.2 also shows the filter's responses to signals that were randomly perturbed about the nominal input, but still staying within the input mask. Note that these responses stay within the output envelope. Figure 6.3.3 plots the impulse response of the optimal ECUI which has a noise gain of $1.113 \times 10^{-3}T$, i.e. 0.0356.

The next example concerns the equalization of a digital transmission channel consisting of a coaxial cable operating at the DX3 rate (45Mb/s) [5], [12]. The coaxial cable has a 30dB attenuation at a normalized frequency of $2\pi/\beta$, where β is the baud period (22.35×10^{-9}s). An equalizing filter is required to shape all signals within a 2% tolerance of the cable's impulse response so that they fit into the envelope given by the DSX3 pulse template (see Figure 6.3.4). Note that the gap in the output mask is 0.1.

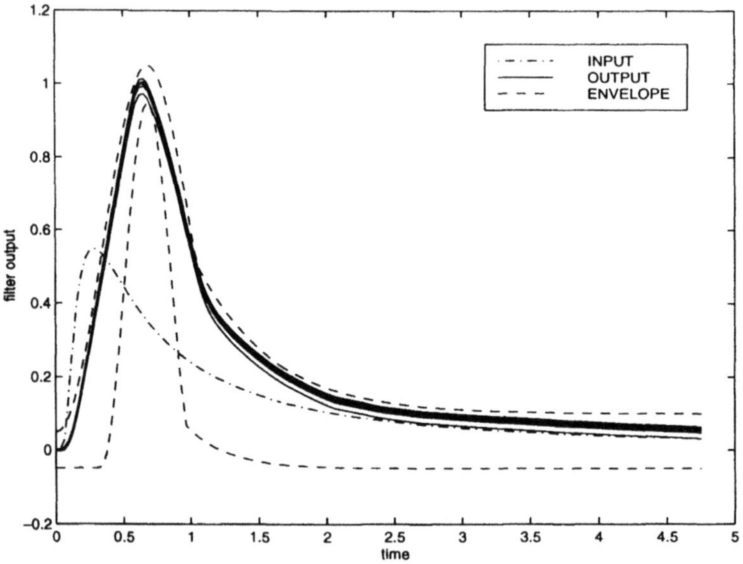

Figure 6.3.4. DSX3 example

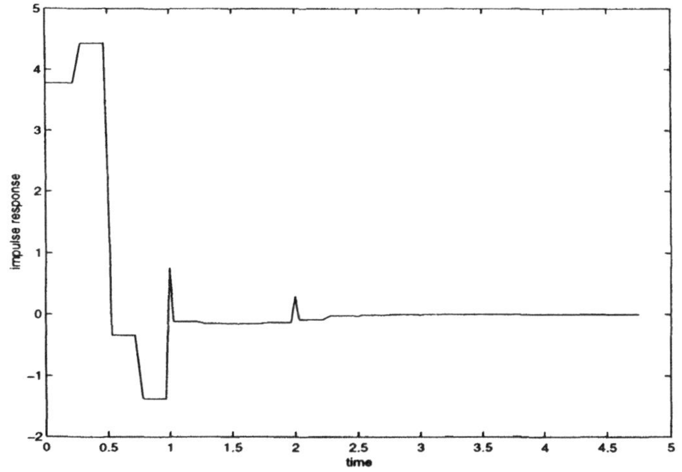

Figure 6.3.5. Impulse response of optimal ECUI filter

In this example, we use $m = 4$, $k = 8$ and $T = 4\beta$. The effective duration of the input is taken to be 20β. For computational purpose, the continuum of constraints is discretized to 32 constraints per Baud interval. Figure 6.3.4 shows the filter's response to signals that were randomly perturbed about the nominal input, but still remain within the input mask. Note that these responses all stay within the prescribed output mask. Figure 6.3.5 shows the impulse response of the optimal filter which has noise gain of $2.25T$, i.e. 22.35×10^{-8}.

Approximations for finite dimensional filter

When the image $K(\boldsymbol{R}^n)$ is not closed under p and q, we can apply the smooth transformation on an approximate version of the finite dimensional problem. If we approximate the constraint of the finite dimensional problem by

$$|\Xi_s K\boldsymbol{a} - d| + \Xi_\theta |K||\boldsymbol{a}| - \varepsilon \preceq 0, \tag{6.3.13}$$

where $|K| : \boldsymbol{R}^n \to L_2(\Omega_u)$ is a linear operator defined by $|K|\boldsymbol{a} = |\boldsymbol{v}^T|\boldsymbol{a}$, then it is clear that any feasible point of this approximate problem is a feasible point of the original finite dimensional problem, i.e. Problem (6.3.8). This approximate problem is equivalent to the following problem by Theorem 6.2.1

$$\begin{aligned}
&\min \ z^T H z, z \in \mathbf{R}^{2n} \\
&\text{subject to } \ [y_s^T(t), -y_s^T(t)]z + [\tilde{y}_\theta^T(t), \tilde{y}_\theta^T(t)]z - \varepsilon^+(t) \le 0, \forall t \in \Omega_c \\
&\qquad -[y_s^T(t), -y_s^T(t)]z + [\tilde{y}_\theta^T(t), \tilde{y}_\theta^T(t)]z + \varepsilon^-(t) \le 0, \forall t \in \Omega_c \\
&\qquad -z \le 0,
\end{aligned} \tag{6.3.14}$$

where

$$\tilde{y}_\theta^T(t) = [\tilde{y}_{\theta,0}(t), \ldots, \tilde{y}_{\theta,n-1}(t)], \ \tilde{y}_{\theta,j}(t) = (\Xi_\theta|v_j|)(t).$$

6.4 Discrete-time

The ECUI filtering problem can be formulated in discrete-time domain in the same way as that for continuous-time. Lemma 6.1.1 and Theorem 6.1.1 are readily applicable to the discrete-time scenario. In this section, we summarize these results for the discrete-time ECUI filtering problem, which can be posed as follows:

$$\min_{u \in \mathcal{F}} \langle u, Lu \rangle,$$

where

$$\mathcal{F} = \{u \in l_2(\tilde{\Omega}_u) : |x * u - d| \preceq \varepsilon, \forall x : |x - s| \preceq \theta\}$$

is closed and convex with appropriate partial orderings on the excitation and response spaces. Using Lemma 6.1.1, a filter $u \in l_2(\tilde{\Omega}_u)$ is feasible, i.e. $u \in \mathcal{F}$, if and only if

$$|s * u - d| + \theta * |u| \preceq \varepsilon.$$

As a result, the ECUI filtering problem can be rewritten as

$$\begin{aligned}
&\min \ \langle u, Lu \rangle, u \in l_2(\tilde{\Omega}_u) \\
&\text{subject to } |\Xi_s u - d| + \Xi_\theta |u| - \varepsilon \preceq 0.
\end{aligned} \tag{6.4.1}$$

The difficulties associated with the non-differentiable constraints can be avoided if we convert this problem into the following positive definite QP problem:

$$\begin{aligned} &\min \ \langle z, Hz\rangle, z \in l_2^2(\tilde{\Omega}_u) \\ &\text{subject to } [\Xi_s, -\Xi_s]z + [\Xi_\theta, \Xi_\theta]z - \varepsilon^+ \preceq 0 \\ &\qquad [-\Xi_s, \Xi_s]z + [\Xi_\theta, \Xi_\theta]z + \varepsilon^- \preceq 0 \\ &\qquad [-I, 0]z \preceq 0 \\ &\qquad [0, -I]z \preceq 0 \end{aligned} \tag{6.4.2}$$

where $\rho \in (0, 2\lambda_{min}(L))$. Let z^o denote the solution Problem (6.4.2). Then, Theorem 6.2.1 shows that the optimal ECUI filter u^o is given by

$$u^o = [I, -I]z^o.$$

The dual cost functional for Problem (6.4.2) is given by

$$\begin{aligned} \phi(p_1, p_2, q_1, q_2) = &-\frac{1}{4\rho}\langle ZL[2\mu_\theta(p_1+p_2)-q_1-q_2], [2\mu_\theta(p_1+p_2)-q_1-q_2]\rangle \\ &+\frac{1}{2}\langle Z[\mu_s(p_1-p_2)+\mu_\theta(p_1+p_2)-q_1], -\mu_s(p_1-p_2)+\mu_\theta(p_1+p_2)-q_2\rangle \\ &-\sum_{i\in\Omega_c}(p_1-p_2)(i)d(i)-\sum_{i\in\Omega_c}(p_1+p_2)(i)\varepsilon(i), \end{aligned}$$

where

$$\begin{aligned} Z &= (2L-\rho I)^{-1}, \\ [\mu_x(p)](k) &= \sum_{i\in\Omega_c} x(k-i)p(i). \end{aligned}$$

If $(p_1^o, p_2^o, q_1^o, q_2^o)$ maximizes the dual functional over $P_M^{\oplus}\times P_M^{\oplus}\times P_l^{\oplus}\times P_l^{\oplus}$, then the optimal ECUI filter u^o is given by

$$u^o = -\frac{1}{2}(2L-\rho I)^{-1}[2\mu_s(p_1^o-p_2^o)-q_1^o+q_2^o].$$

6.5 Hybrid ECUI Filtering

Recall from (2.6.5) of Section 2.6 that when a hybrid filter with discrete-time impulse response $u \in l_2(\tilde{\Omega}_u)$ is excited by a signal $x \in L_2(\Omega_s)$, the output is given by

$$(\Xi_x u)(t) = (\tilde{\Xi}_x\Psi u)(t) = \sum_{k\in\Omega_s}\sum_{l\in\Omega_u}\Lambda(t-kT_s-lT_s)u(l)x(kT_s).$$

where $\Psi: l_2(\tilde{\Omega}_u) \to L_\infty(\Omega_\psi)$ and $\tilde{\Xi}_x : L_\infty(\Omega_\psi) \to L_\infty(\Omega_\psi)$ are linear mappings given, respectively, by

$$(\Psi u)(\varsigma) \equiv \sum_{l = \tilde{\Omega}_u} \Lambda(\varsigma - lT_s)u(l),$$

$$(\tilde{\Xi}_x v)(t) \equiv \sum_{k \in \tilde{\Omega}_s} v(t - kT_s)x(kT_s).$$

The feasible region can thus be defined as

$$\mathcal{F} = \{u \in l_2(\tilde{\Omega}_u) : |\Xi_x u - d| \preceq \varepsilon, \ \forall x \in L_2(\Omega_s) : |x - s| \preceq \theta\}$$

As in the previous formulation for analog filters, the above description of the feasible region is again not very useful computationally. Identifying the mapping $F : L_2(\Omega_s) \to C(\Omega_\psi)$ by

$$Fx = \tilde{\Xi}_x \Psi u,$$

we see that F is linear with kernel

$$f(t, k) = (\Psi u)(t - kT_s).$$

Hence, similar to the case of analog filters, Lemma 6.1.1 allows the constraints to be expressed in a form that does not involve the input perturbations as shown in the following theorem.

Theorem 6.5.1. $|\tilde{\Xi}_x \Psi u - d| \preceq \varepsilon, \forall x \in L_2(\Omega_s) : |x - s| \preceq \theta$ *if and only if*

$$|\tilde{\Xi}_s \Psi u - d| + \tilde{\Xi}_\theta |\Psi u| \preceq \varepsilon.$$

Similar to the analog case, for each instant $t \in \Omega_c$, the distance between the desired response and the output corresponding to a signal $x \in \mathcal{S}(s, \theta)$, i.e. $|(\Xi_x u)(t) - d(t)|$, is bounded above by $|(\Xi_s u)(t) - d(t)| + [\tilde{\Xi}_\theta |\Psi u|\ ](t)$. Moreover, there exists a signal $x_t \in \mathcal{S}(s, \theta)$ which achieves this bound. More concisely, for each $t \in \Omega_c$, there exists an x_t such that

$$\begin{aligned} \max_{x \in \mathcal{S}(s, \theta)} |(\Xi_x u)(t) - d(t)| &= |(\Xi_{x_t} u)(t) - d(t)| \\ &= |(\Xi_s u)(t) - d(t)| + (\tilde{\Xi}_\theta |\Psi u|)(t) \end{aligned} \tag{6.5.1}$$

We refer to the input x_t which achieves this maximum as the worst case input at t. This signal can be written as

$$x_t = s + \tilde{s}_t,$$

where $\tilde{s}_t$ is the worst case perturbation given by

$$\tilde{s}_t(\tau) = \exp\{j\arg[(\Xi_s u)(t) - d(t)] - j\arg[(\Psi u)(t-\tau)]\}\theta(\tau).$$

Thus, a filter u is feasible if and only if, for each instance $t \in \Omega_c$, its response at t to the worst case signal x_t stays inside the output envelope. Theorem 6.5.1 also implies that the uncertainty θ in the input effectively reduces the output mask by $\tilde{\Xi}_\theta|\Psi u|$. Note that even though for different t there exist different worst case perturbations, the absolute values of these perturbations are independent of t and all are equal to the input uncertainty θ, i.e. $|\tilde{s}_t(\tau)| = \theta(\tau)$ for all τ.

Theorem 6.5.1 yields an explicit constraint function $G : l_2(\tilde{\Omega}_u) \to L_\infty(\Omega_\psi)$ given by

$$G(u) = |\tilde{\Xi}_s \Psi u - d| + \tilde{\Xi}_\theta|\Psi u| - \varepsilon. \tag{6.5.2}$$

The ECUI filtering problem for hybrid filter can then be stated as

$$\begin{gathered} \min \langle u, Lu\rangle, u \in l_2(\tilde{\Omega}_u) \\ \text{subject to } G(u) \preceq 0. \end{gathered} \tag{6.5.3}$$

Similar to the continuous-time case, the constraint function G is continuous and convex. Again, this is a non-smooth optimization problem. However, this problem cannot be transformed into a smooth form using the techniques of Section 6.2, because the image $\Psi(l_2(\tilde{\Omega}_u))$, may not be closed under the mappings p and q. That is, there may not be any $x, y \in l_2(\tilde{\Omega}_u)$ for which $\Psi x = p(\Psi u)$ and $\Psi y = q(\Psi u)$.

Approximations for hybrid filters

The above technique puts too much restriction on post-filter's pulse response Λ. Alternatively, if we use the following approximation,

$$[\tilde{\Xi}_\theta|\Psi u|](t) \le \sum_{i=0}^{\infty} \sum_{k \in \tilde{\Omega}_u} \theta(iT_s)\,|\Lambda(t - iT_s - kT_s)||u(k)| \equiv [\tilde{\Xi}_\theta|\Psi||u|](t),$$

then the approximate hybrid ECUI problem can be written as

$$\begin{aligned} &\min\ f(u),\ u \in l_2(\tilde{\Omega}_u) \\ &\text{subject to } |\Xi_s u - d| + \tilde{\Xi}_\theta|\Psi||u| - \varepsilon \preceq 0. \end{aligned} \tag{6.5.4}$$

That is, we approximate the hybrid ECUI problem by using a smaller feasible region, so as not to violate the constraints, at the expense of raising the cost. Then, we apply Theorem 6.2.1 to transform Problem (6.5.4) into the following smooth problem.

$$\begin{aligned} \min\ & \|x - y\|_L^2 + 2\rho\langle x, y\rangle,\ x, y \in l_2(\tilde{\Omega}_u) \\ \text{subject to}\ & \Xi_s(x - y) + \tilde{\Xi}_\theta|\Psi|(x + y) - \varepsilon^+ \preceq 0, \\ & -\Xi_s(x - y) + \tilde{\Xi}_\theta|\Psi|(x + y) + \varepsilon^- \preceq 0, \\ & -x \preceq 0, \\ & -y \preceq 0. \end{aligned}$$

Finite dimensional hybrid filters

Consider the case where the discrete-time filter in the hybrid configuration has finite structures with impulse response of the form

$$u = Ka = \boldsymbol{v}^T \boldsymbol{a}$$

as given in Section 2.7, in which $\boldsymbol{v}^T = [v_0, v_1, \ldots, v_{n-1}]$ is a vector of n linearly independent elements of $l_2(\tilde{\Omega}_u)$ and $\boldsymbol{a} \in \boldsymbol{R}^n$ is the vector of filter parameters. The feasible region $\mathcal{F} \cap [\{v_i\}_{i=0}^{n-1}]$ is embedded in an n-dimensional subspace and can be characterized by the set of feasible filter coefficients, which is defined in $\boldsymbol{R}^n$ as

$$\mathcal{F}^n = \{\boldsymbol{a} \in \boldsymbol{R}^n : |\Xi_x K\boldsymbol{a} - d| \preceq \varepsilon,\ \forall x : |x - s| \preceq \theta\}.$$

It follows from Theorem 6.5.1 that the ECUI problem can be restated for hybrid filters with finite structures as follows:

$$\min \boldsymbol{a}^T K^{\dagger} L K \boldsymbol{a}, \boldsymbol{a} \in \boldsymbol{R}^n$$
$$\text{subject to } \left|\bar{\Xi}_s \Psi K \boldsymbol{a} - d\right| + \bar{\Xi}_\theta |\Psi K \boldsymbol{a}| - \varepsilon \preceq 0. \tag{6.5.5}$$

Using the approximation presented in Problem (6.5.4), the approximate problem for finite dimensional hybrid filters can be written as

$$\min \boldsymbol{a}^T K^{\dagger} L K \boldsymbol{a}, \boldsymbol{a} \in \boldsymbol{R}^n$$
$$\text{subject to } \left|\bar{\Xi}_s \Psi K \boldsymbol{a} - d\right| + \bar{\Xi}_\theta |\Psi||K \boldsymbol{a}| - \varepsilon \preceq 0. \tag{6.5.6}$$

For finite analog filter structures, Theorem 6.2.1 does not necessarily transform the non-smooth problem into a smooth one. The same goes for the approximate hybrid ECUI filtering Problem (6.5.6). However, a set of linearly independent functions $\{v_i\}_{i=0}^{n-1}$, whose span is closed under the mappings p and q, can be realized by using the well known FIR filters.

Recall from (3.3.4) in Chapter 3 that the composite operator $\Xi_s K$ can be expressed in an alternative form as

$$(\Xi_s K \boldsymbol{a})(t) = \boldsymbol{y}_s^T(t)\boldsymbol{a},$$

where

$$\boldsymbol{y}_s^T(t) = [y_{s,0}(t), \ldots, y_{s,n-1}(t)], \; y_{s,j} = \Xi_s v_j.$$

In particular, for FIR filters, we have

$$y_{s,j}(t) = \sum_{i=0}^{\infty} \Lambda(t - iT_s) s[(i-j)T_s].$$

Furthermore,

$$\bar{\Xi}_\theta |\Psi||K \boldsymbol{a}| = z_\theta^T(t) |\boldsymbol{a}|,$$

where

$$z_{\theta,j}(t) = \sum_{i=0}^{\infty} |\Lambda(t - iT_s)| \theta[(i-j)T_s].$$

Hence, the approximate constraint simplifies to

$$\left|y_s^T(t)a - d(t)\right| + z_\theta^T(t)|a| \le \varepsilon(t),\ \forall t \in \Omega_c.$$

Suppose further that the excitation is of finite duration or finite support. Then,

$$y_s^T(t) = [\Lambda(t), \ldots, \Lambda(t-(n+m-1)\tau)]\begin{bmatrix} s(0) & 0 & \cdots & 0 \\ \vdots & s(0) & & \vdots \\ s((m-1)\tau) & \vdots & \ddots & 0 \\ 0 & s((m-1)\tau) & & s(0) \\ \vdots & & \ddots & \vdots \\ 0 & \cdots & 0 & s((m-1)\tau) \end{bmatrix}$$

and

$$z_\theta^T(t) = [|\Lambda(t)|, \ldots, |\Lambda(t-(n+m-1)\tau)|]\begin{bmatrix} \theta(0) & 0 & \cdots & 0 \\ \vdots & \theta(0) & & \vdots \\ \theta((m-1)\tau) & \vdots & \ddots & 0 \\ 0 & \theta((m-1)\tau) & & \theta(0) \\ \vdots & & \ddots & \vdots \\ 0 & \cdots & 0 & \theta((m-1)\tau) \end{bmatrix}.$$

As an example, consider the compression of a 13-bit Barker coded signal in Section 6.3. For this example, we use a 27-tap FIR filter and a Bessel post-filter of 3th order with cut-off frequency $\omega_c = 2\pi/\beta$. To obtain the approximate solution, we have constrained the output at every $t_i = i\beta/8$. Figures 6.5.1 and 6.5.2 show, respectively, the filter's responses to the nominal input and signals that were randomly perturbed about the nominal input but still fit inside the input mask. Observe that these responses stay within the boundary of the output envelope. The noise gain of the ECUI filter is 0.027.

The next example is the equalization application presented in the analog case. The input is sampled by the A/D converter at 0.25 of the Baud rate. The effective duration of the input is taken to be 20β. The order of digital filter used in this example is 16. The interpolating filter is a 5th order Bessel filter with a cut-off frequency of $1.1\pi/\beta$. We discretize the continuum of constraints at 8 constraints per sampling period. Figures 6.5.3 and 6.5.4 show, respectively, the filter's responses to the nominal input and signals that were randomly perturbed about the nominal input but still fit inside the input mask. The noise gain of the optimal filter is 0.608.

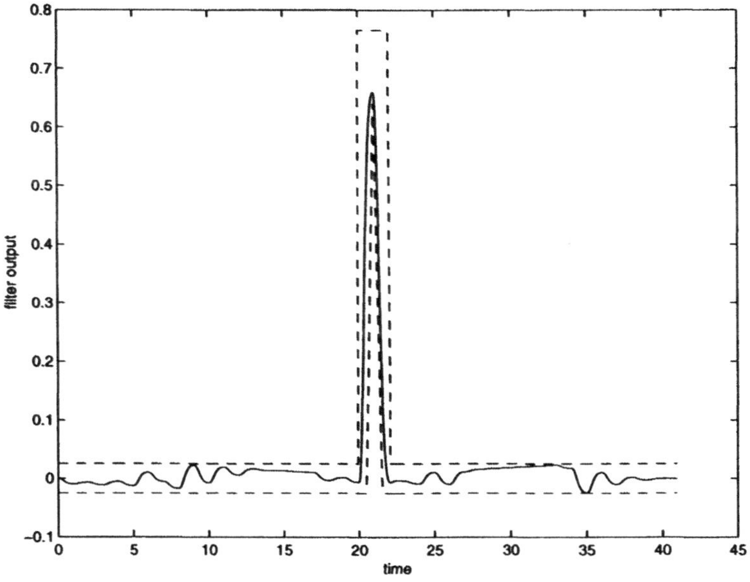

Figure 6.5.1. Response of filter to nominal Barker Coded input

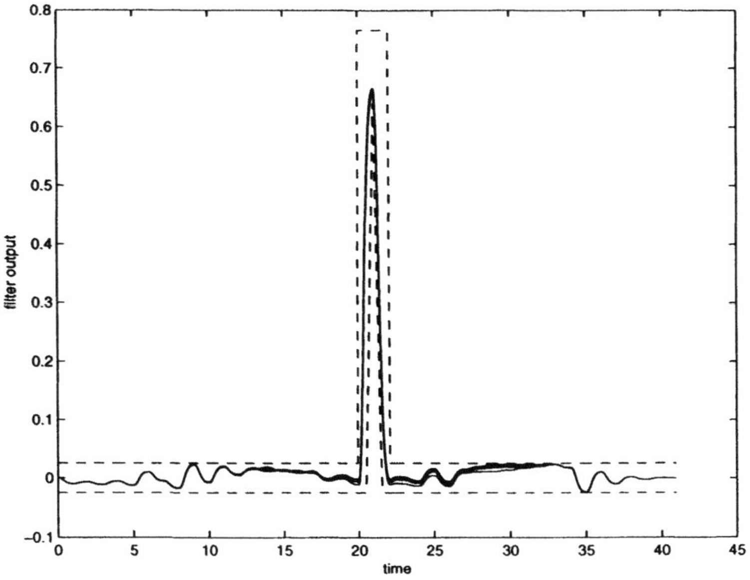

Figure 6.5.2. Response of filter to perturbed inputs

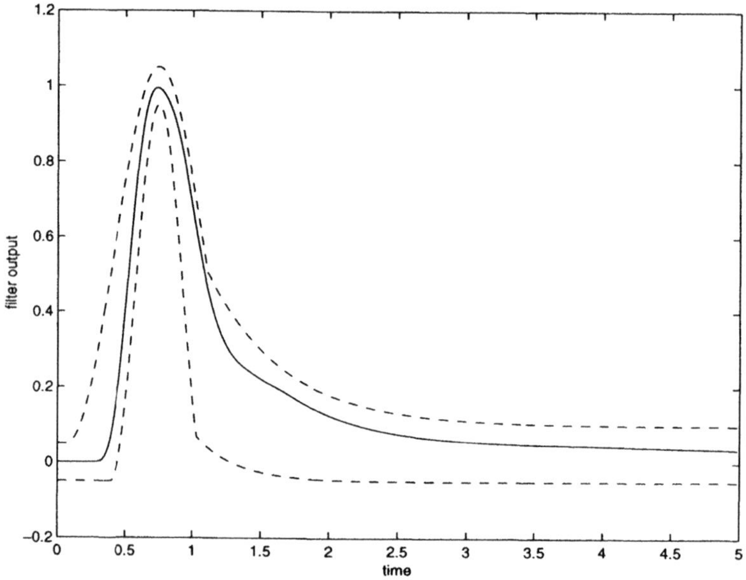

Figure 6.5.3. Filter's response to nominal input

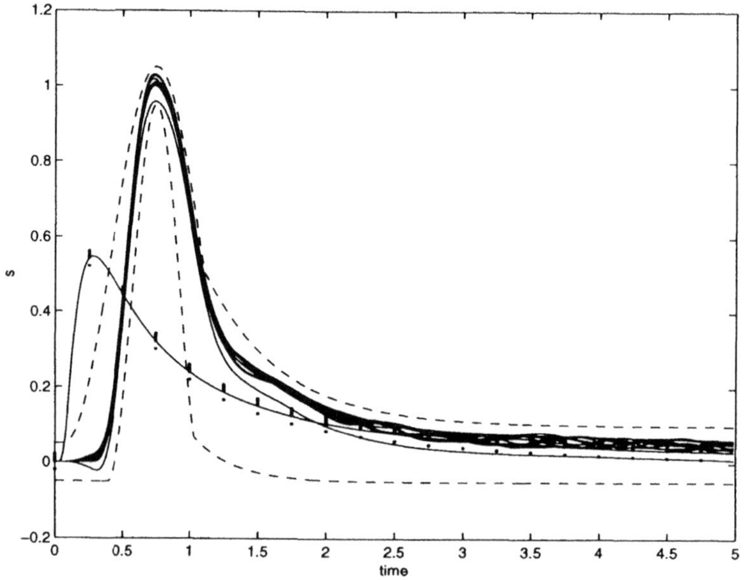

Figure 6.5.4. Filter's response to perturbed inputs

6.6 Constraint Robustness

In this section, we present a technique for providing a guard band on the output mask and investigate how the trade-off between the noise gain and the constraint robustness can be achieved by proposing a new optimization problem. This is a generic approach which does not take into consideration the source or cause of the perturbation. Instead, we try to force the filter output to stay as far away from the mask as possible subject to an allowable increase on the noise gain.

Recall the EC filtering Problem (5.1.1):

$$\min \langle u, Lu \rangle$$
$$\text{subject to } \varepsilon^- \preceq \Xi_s u \preceq \varepsilon^+,$$

where Ξ_s is defined by (2.1.3), and L is defined by (2.1.5). For a given filter u (which may or may not satisfy the envelope constraints), consider the difference between its response $\Xi_s u$ and upper or lower mask boundaries:

$$\begin{aligned} \phi^+(u) &= \Xi_s u - \varepsilon^+ \\ \phi^-(u) &= \varepsilon^- - \Xi_s u \end{aligned} \tag{6.6.1}$$

It is clear that u is feasible if and only if $\phi^+(u)$ and $\phi^-(u)$ are in the negative cone

$$P_C^- = \{x \in C(\Omega_\psi): x(t) \le 0, \forall t \in \Omega_c\}.$$

That is, $[\phi^+(u)](t)$ and $[\phi^-(u)](t)$ are non-positive for all $t \in \Omega_c$. To quantify the notion of robustness for a filter u, let us define its constraint robustness margin as:

$$\rho(u) \equiv \min\left\{ \min_{t \in \Omega_c} [-\phi^+(u)](t), \min_{t \in \Omega_c} [-\phi^-(u)](t) \right\}, \tag{6.6.2}$$

which measures how far the filter output is from the envelope. Clearly, u is a feasible point of the EC filtering Problem (5.1.1) if and only if $\rho(u) \ge 0$ (hence u is not feasible if and only if $\rho(u) < 0$). The feasible region of the EC filtering problem can be expressed in terms of the robustness margin as

$$\mathcal{F} = \{u \in L_2(\Omega_u) \colon \rho(u) \geq 0\}.$$

If $\rho(u) \geq 0$, then $\rho(u)$ is the smallest distance of the output $\Xi_s u$ from the boundaries of the mask. Therefore, we may say that the filter u is robust with constraint robustness margin $\rho(u)$. On the other hand, if $\rho(u)$ is negative, then u is infeasible and $|\rho(u)|$ gives the largest distance of the output $\Xi_s u$ from the mask boundaries.

Let u^o denote the optimal solution of the EC filtering Problem (5.1.1). Note that $\rho(u^o) = 0$, since the optimal filter output touches the mask at some point. Hence, any filter with a positive robustness margin has a noise gain not less than u^o (otherwise there would be a contradiction to the optimality of u^o). To achieve robustness, we must allow for the increase in noise gain and maximize the robustness margin as shown in the following constrained optimization problem:

$$\begin{aligned} &\max \ \rho(u) \\ &\text{subject to } \ \|u\|_L^2 \leq (1+\delta)\|u^o\|_L^2 \end{aligned} \tag{6.6.3}$$

where $\delta > 0$ is a constant that specifies the allowable amount of increase of the output noise power. For a given δ, the feasible region $\mathcal{F}_{1,\delta}$ of Problem (6.6.3) is:

$$\mathcal{F}_{1,\delta} = \{u \in L_2(\Omega_u) : \|u\|_L^2 \leq (1+\delta)\|u^o\|_L^2\}.$$

Let u^* denote an optimal solution of Problem (6.6.3). Then, it is clear that $\rho(u^*) \geq 0$, since $u^o \in \mathcal{F}_{1,\delta}$ and $\rho(u^o) = 0$. Note that for $\delta = 0$, $u^* = u^o$.

Define $f(u, \sigma) = \sigma$, the constrained maximization Problem (6.6.3) can be converted into the following standard maximization problem:

$$\begin{aligned} &\max f(u,\sigma),\ (u,\sigma) \in L_2(\Omega_u) \times \boldsymbol{R} \\ &\text{subject to } \ g_1(u,\sigma) \equiv \sigma + \phi^+(u) \preceq 0, \\ &\qquad g_2(u,\sigma) \equiv \sigma + \phi^-(u) \preceq 0, \\ &\qquad g_3(u,\sigma) \equiv \|u\|_L^2 - (1+\delta)\|u^o\|_L^2 \leq 0. \end{aligned} \tag{6.6.4}$$

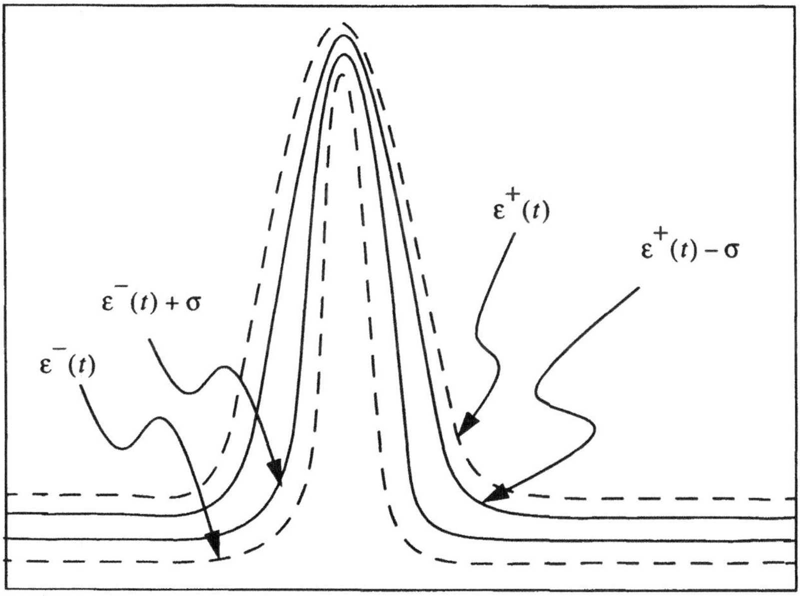

Figure 6.6.1. Weighted constraint robustness margin

Note that g_3 maps $L_2(\Omega_u) \times \boldsymbol{R}$ to $\boldsymbol{R}$, g_1 and g_2 map $L_2(\Omega_u) \times \boldsymbol{R}$ to $L_\infty(\Omega_\psi)$ and the partial ordering $\preceq$ is defined as before in Section 6.1. Define the feasible region for Problem (6.6.4) as

$$\mathcal{F}_{2,\delta} = \{(u, \sigma) \in L_2(\Omega_u) \times \boldsymbol{R} : g_i(u, \sigma) \preceq 0, i = 1, 2, g_3(u, \sigma) \leq 0\}.$$

Intuitively, for positive σ, the first two constraints of Problem (6.6.4) effectively shrinks the mask about the "desired" response d as shown in Figure 6.6.1. The larger we make σ, the tighter the mask becomes. Conversely, for negative σ, the mask is expanded about d. Thus, by maximizing σ, we force the filter output away from the boundary. Note that if (u^*, σ^*) is the solution of Problem (6.6.4), then similar to Problem (6.6.3), $\sigma^* \geq 0$ and the distance between the output waveform $[\Xi_s u^*](t)$ and the output mask is at least equal to $\rho(u^*)$. In fact, Problems (6.6.3) and (6.6.4) are equivalent as shown in the following theorem.

Theorem 6.6.1. *If (u^*, σ^*) is a solution to Problem (6.6.4), then u^* is a solution to Problem (6.6.3) and $\rho(u^*) = \sigma^*$. Conversely, if u^* is a solution to Problem (6.6.3), then $(u^*, \rho(u^*))$ is a solution of Problem (6.6.4).*

From convex optimization theory, we know that if a solution to this problem exists, then any relative minimum is a global minimum.

In practice, it may be important to require a larger constraint robustness margin on certain intervals. In this case, a weighting function β can be used to specify the weightings in different time intervals. Define the weighted constraint robustness as:

$$g_1(u, \sigma) \equiv \sigma\beta + \phi^{+}(u) \preceq 0$$
$$g_2(u, \sigma) \equiv \sigma\beta + \phi^{-}(u) \preceq 0$$

where β is a non-negative piecewise continuous weighting function. We normalize the weighting function so that $\max_{t \in \Omega_c} \beta(t) = 1$. In this form, we can see that the envelope is also tightened, however, not by a constant amount but by a variable amount specified by the weighting function β. In other words, β is used to specify the weightings in different time intervals.

All previous results follow if we define the cost of Problem (6.6.3) to be:

$$\rho_{\beta}(u) \equiv \min\left\{ \min_{t \in \Omega_c} \frac{[-\phi^{+}(u)](t)}{\beta(t)}, \min_{t \in \Omega_c} \frac{[-\phi^{-}(u)](t)}{\beta(t)} \right\}$$

Characterization of filter structure

For convenience, we write Problem (6.6.4) as an equivalent minimization problem, i.e. we minimize $-\sigma$ instead of maximizing σ. The Lagrangian for this problem is then given by

$$\begin{aligned} \mathcal{L}(u, \sigma, p_1, p_2, p_3) = & -\sigma + p_1 \bullet (\Xi_s u + \sigma\beta - \varepsilon^{+}) \\ & + p_2 \bullet (\varepsilon^{-} + \sigma\beta - \Xi_s u) \\ & + p_3(\|u\|_L^2 - (1 + \delta)\|u^o\|_L^2) \end{aligned} \tag{6.6.5}$$

Recall that $\varepsilon^{+} = d + \varepsilon$ and $\varepsilon^{-} = d - \varepsilon$. Then, rearranging (6.6.5) gives

$$\begin{aligned} \mathcal{L}(u, \sigma, p_1, p_2, p_3) = & \ p_3\langle u, Lu\rangle + (p_1 - p_2) \bullet \Xi_s u \\ & -(p_1 - p_2) \bullet d + (p_1 + p_2) \bullet (\sigma\beta - \varepsilon) - \sigma - p_3(1 + \delta)\|u^o\|_L^2 . \end{aligned}$$

Using the same argument as in Section 3.2, we can combine the two measures p_1 and p_2 into one measure p and the Lagrangian (6.6.5) can be written as

$$\begin{aligned}\mathcal{L}(u, \sigma, p_1, p_2, p_3) &= p_3\langle u, Lu\rangle + \langle \hat{\alpha}(p), u\rangle \\ &\quad - p \bullet d + \sigma\|p\|_\beta - \|p\|_\varepsilon - \sigma - p_3(1+\delta)\|u^o\|_L^2,\end{aligned} \tag{6.6.6}$$

where $\hat{\alpha}(p) \in L_2(\Omega_u)$ for analog filters with

$$[\hat{\alpha}(p)](\tau) = \int_{\Omega_c} s(t-\tau)dp(t),$$

and $\hat{\alpha}(p) \in l_2(\tilde{\Omega}_u)$ for hybrid filters with

$$[\hat{\alpha}(p)](j) = \int_{\Omega_c} \sum_{i=0}^{\infty} s[(i-j)T_s]\Lambda(t-iT_s)dp(t).$$

We assume the constraint g_3 is active and non-degenerate so that $p_3 > 0$. Differentiating (6.6.6) with respect to u and then equating to zero gives

$$u = -\frac{1}{2p_3}L^{-1}\hat{\alpha}(p).$$

Hence, the dual functional is

$$\begin{aligned}\phi(p, p_3) &= \min_{(u,\sigma)} \mathcal{L}(u, \sigma, p, p_3) \\ &= \min_{\sigma}\left\{\sigma(\|p\|_\beta - 1) - \frac{1}{4p_3}\langle\hat{\alpha}(p), L^{-1}\hat{\alpha}(p)\rangle - p \bullet d - \|p\|_\varepsilon - p_3(1+\delta)\|u^o\|^2\right\} \\ &= \begin{cases} -\frac{1}{4p_3}\langle\hat{\alpha}(p), L^{-1}\hat{\alpha}(p)\rangle - p \bullet d - \|p\|_\varepsilon - p_3(1+\delta)\|u^o\|_L^2, & \text{for } \|p\|_\beta - 1 \ge 0 \\ -\infty, & \text{otherwise}\end{cases}\end{aligned}$$

If the constraint g_3 is active and non-degenerate, the solution is unique and can be expressed as follows:

$$u^* = -\frac{1}{2p_3^*}L^{-1}\hat{\alpha}(p^*) \tag{6.6.7}$$

This gives a structural characterization of the optimum robustness EC filter. The filter in (6.6.7) can be interpreted as consisting of a matched filter followed by a second filter which is determined by the Lagrange multipliers p_3^* and p^* (see also Section 3.2 for the interpretation of $\hat{\alpha}(p^*)$ for different filter configurations).

Finite dimensional filter

Consider a finite dimensional filter with impulse response described by

$$u = Ka = \boldsymbol{v}^T\boldsymbol{a},$$

as given in Section 2.4. Recall that the EC filtering problem for these filters can be posed in terms of the vector of filter parameter as problem (5.1.3):

$$\min \boldsymbol{a}^T Q\boldsymbol{a}, \boldsymbol{a} \in \boldsymbol{R}^n$$
$$\text{subject to } \varepsilon^-(t) \le \boldsymbol{y}_s^T(t)\boldsymbol{a} \le \varepsilon^+(t), t \in \Omega_c$$

where $Q = K^{\dagger}LK$, and $\boldsymbol{y}_s^T(t)$ is given by (3.3.4) in Chapter 3.

Thus, the constraint robustness problem for finite filter structure is:

$$\max f(\boldsymbol{a}, \sigma) \equiv \sigma, (\boldsymbol{a}, \sigma) \in \boldsymbol{R}^{n+1}$$

subject to

$$[g_1(\boldsymbol{a}, \sigma)](t) \equiv \sigma\beta(t) - \varepsilon^+(t) + \boldsymbol{y}_s^T(t)\boldsymbol{a} \le 0, t \in \Omega_c, \tag{6.6.8}$$

$$[g_2(\boldsymbol{a}, \sigma)](t) \equiv \sigma\beta(t) + \varepsilon^-(t) - \boldsymbol{y}_s^T(t)\boldsymbol{a} \le 0, t \in \Omega_c, \tag{6.6.9}$$

$$g_3(\boldsymbol{a}, \sigma) \equiv \|\boldsymbol{a}\|_Q^2 - (1+\delta)\|\boldsymbol{a}^o\|_Q^2 \le 0. \tag{6.6.10}$$

The Lagrangian for this problem is

$$\begin{aligned}\mathcal{L}(\boldsymbol{a}, \sigma, p_1, p_2, p_3) &= -\sigma + p_1 \bullet (\Xi_s K\boldsymbol{a} + \sigma\beta - \varepsilon^+) + p_2 \bullet (\varepsilon^- + \sigma\beta - \Xi_s K\boldsymbol{a}) \\ &\quad + p_3(\|K\boldsymbol{a}\|^2 - (1+\delta)\|K\boldsymbol{a}^o\|^2)\end{aligned}$$

Rearranging and combining the two measures p_1 and p_2 into one measure p, the Lagrangian can be written as

$$\mathcal{L}(\boldsymbol{a}, \sigma, p, p_3) = p_3\|\boldsymbol{a}\|_Q^2 + \langle K^\dagger\hat{\alpha}(p), \boldsymbol{a}\rangle \\ -p \bullet d + \sigma\|p\|_\beta - \|p\|_\varepsilon - \sigma - p_3(1+\delta)\|\boldsymbol{a}^o\|_Q^2 \tag{6.6.11}$$

in which

$$K^\dagger\hat{\alpha}(p) = [p \bullet \boldsymbol{y}_s] = \int_{\Omega_c} \boldsymbol{y}_s dp$$

(see Section 3.2). Differentiating the Lagrangian (6.6.11) with respect to $\boldsymbol{a}$ yields

$$2p_3 Q\boldsymbol{a} + p \bullet \boldsymbol{y}_s = 0$$

Substituting this back into (6.6.11), we get the dual functional

$$\begin{aligned} \phi(p, p_3) &= \min_{(\boldsymbol{a}, \sigma)} \mathcal{L}(\boldsymbol{a}, \sigma, p, p_3) \\ &= \min_{\sigma}\left\{\sigma(\|p\|_\beta - 1) - \frac{1}{4p_3}\langle p \bullet \boldsymbol{y}_s, Q^{-1} p \bullet \boldsymbol{y}_s\rangle - p \bullet d - \|p\|_\varepsilon - p_3(1+\delta)\|\boldsymbol{a}^o\|^2\right\} \\ &= \begin{cases} -\dfrac{1}{4p_3}\langle p \bullet \boldsymbol{y}_s, Q^{-1} p \bullet \boldsymbol{y}_s\rangle - p \bullet d - \|p\|_\varepsilon - p_3(1+\delta)\|\boldsymbol{a}^o\|_Q^2, & \text{for } \|p\|_\beta - 1 \geq 0 \\ -\infty, & \text{otherwise} \end{cases} \end{aligned}$$

If the constraint g_3 is active and non-degenerate, the solution is unique and can be expressed as follows:

$$\boldsymbol{a}^* = -\frac{1}{2p_3^*} Q^{-1} K^\dagger\hat{\alpha}(p^*) = -\frac{1}{2p_3^*} Q^{-1} p^* \bullet \boldsymbol{y}_s$$

Discrete-time filters follow without any change. In practice, the constrained optimization Problem (6.6.4) can be solved using standard optimization softwares such as the MATLAB optimization toolbox.

Numerical example with Laguerre filter

Consider again the design of an equalizer for a digital transmission channel consisting of a coaxial cable on which data is transmitted according to the DX3 standard [5], [12]. The design objective is to find an equalizer which takes the impulse response of a coaxial cable with a 30dB loss at the Baud frequency and produces an output that lies within the DSX3 pulse template. For computational purpose, we consider the time domain constraints at a finite number of points rather than the entire continuum.

In our numerical studies, we use 1024 points over the interval $[0, 32\beta]$. Using a Laguerre filter with 14 coefficients and $p = 12$, we first solve the optimal EC filtering problem for which the optimal noise gain is: $\|u^o\|^2 = \|a^o\|^2 = 54.2$. The output mask, input signal and the output are shown in Figure 6.6.2. Figure 6.6.3 shows the frequency responses of the unequalized cable and the equalized cable. To achieve maximum constraint robustness margin with $\delta = 1.5$, we solve the robust EC filtering problem for

$$\beta(t) = \begin{cases} 3\varepsilon(t), & |\varepsilon(t)| \geq 0.05 \\ 0.035, & |\varepsilon(t)| \leq 0.03 \\ \varepsilon(t), & \text{elsewhere} \end{cases} .$$

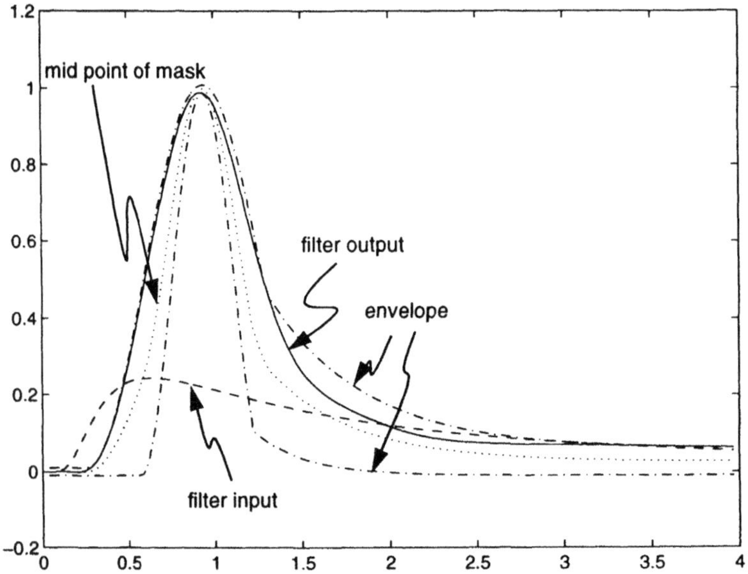

Figure 6.6.2. Laguerre filter - EC approach

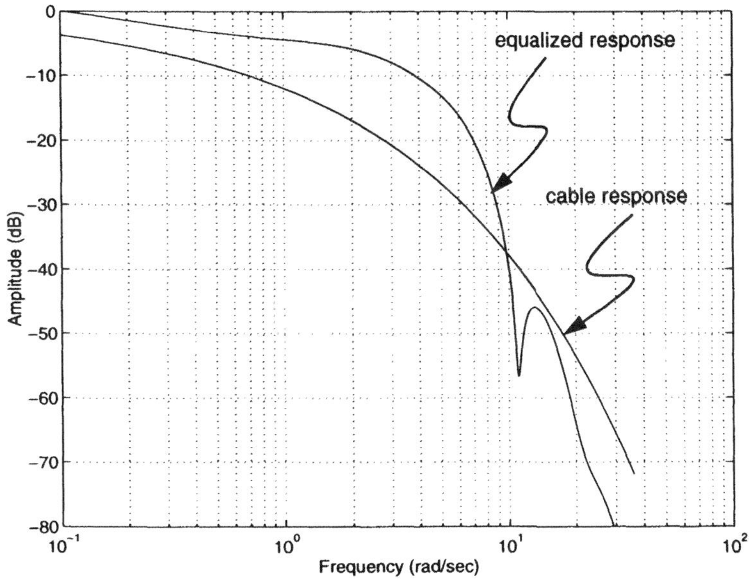

Figure 6.6.3. Magnitude response EC approach

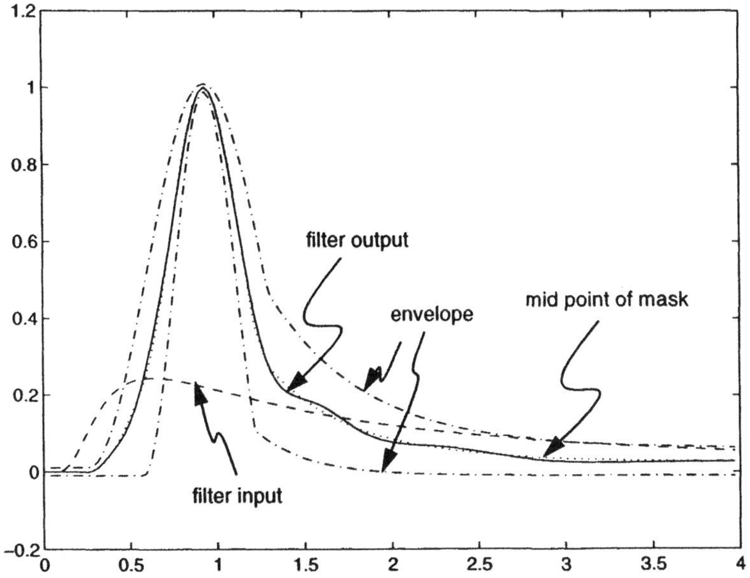

Figure 6.6.4. Laguerre filter - EC approach with constraint robustness

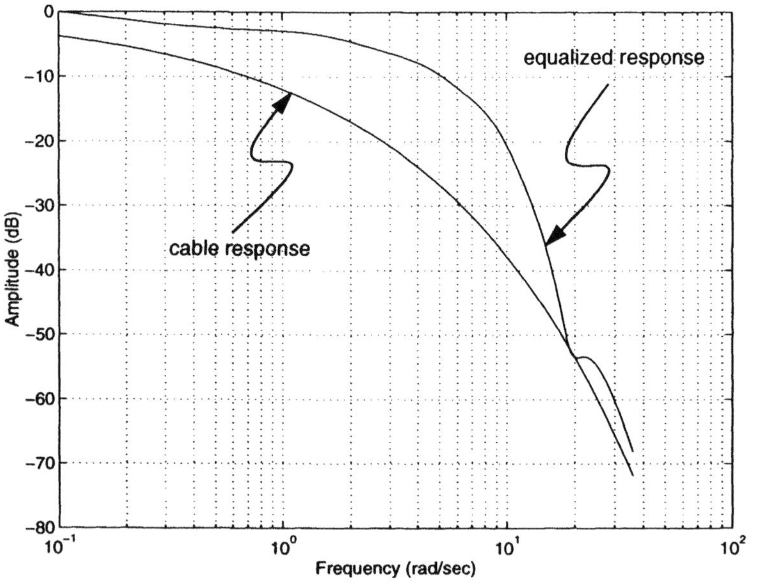

Figure 6.6.5. Magnitude response with constraint robustness

The corresponding output is shown in Figure 6.6.4 and the frequency response is shown in Figure 6.6.5. From Figure 6.6.4, we see that the output is very close to the center of the output mask. The robustness to constraint violation of this filter in comparison to the optimal EC filter is self-evident. Note also that frequency response of the robust-EC equalized channel is flatter than that of the EC equalized channel.

6.7 EC Filtering with Uncertain Implementation.

Apart from input uncertainty, which has been addressed in Section 6.1, there is also implementation uncertainty. In practice, it is often the case that a designed filter or system impulse response cannot be implemented exactly, there is always implementation errors. These errors can arise from component mismatch or quantization of filter coefficients. For example, when digital techniques are used, the actual impulse response achieved differs from the desired value because of finite word lengths. As a result, the constraints will invariably be violated. In this section, the robustness to implementation uncertainty is addressed. Our aim is to determine filters which ensure that the constraints are satisfied even in the presence implementation errors. Such filters may or may not exist depending on the magnitude of the implementation errors.

The robust margin technique of Section 6.6 tries to push the filter output further away from the mask boundaries by modifying the shape of the mask. However, there is no guarantee that the actual implementation of the resulting solution satisfies the constraints. Without knowledge of the errors, the robust margin is a sensible formulation. However, given knowledge of the implementation errors, for example the magnitude bound, it is not clear how the mask should be modified so that the constraints are met. In the following, the EC filtering problem with bounded implementation errors is formulated. In particular, we are able to establish the relationship with the robust margin formulation by showing how the output mask can be modified so that the original constraints are met in the presence of implementation errors.

It is assumed that the implementation errors are bounded in the sense that for a given filter u, we can achieve actual implementations that are known to be within δ of u, where δ is a bounded function. In other words, any implementation of u is known to belong to the set $\mathcal{V}(u, \delta)$ defined by

$$\mathcal{V}(u, \delta) \equiv \{v \in L_2(\Omega_u) : |v - u| \preceq \delta\}.$$

If δ is a constant, then the region $\mathcal{V}(u, \delta)$ corresponds to a ball centered at u with radius δ. If we use the sup-norm metric

$$\|v - u\| = \sup_{t \in \Omega_u} |(v - u)(t)|,$$

$\mathcal{V}(u, \delta)$ is, in this case, a hyper-cube. A non constant δ reflects non-uniform weightings. Regions with large values of δ mean large uncertainty and vice-versa.

In the EC filtering problem, we seek a filter u such that the response to a prescribed signal s stays within some specified output envelope. With uncertainty in the implementations, we are interested in those u such that the response of any element of $\mathcal{V}(u)$ to the signal s stays inside the output envelope. The set $\mathcal{F}$ of feasible filters is defined by

$$\begin{aligned}\mathcal{F} &= \{u \in L_2(\Omega_u) \colon \Xi_s v \in \Psi(d, \varepsilon),\ \forall v \in \mathcal{V}(u, \delta)\} \\ &= \{u \in L_2(\Omega_u) \colon |\Xi_s v - d| \preceq \varepsilon,\ \forall v \in L_2(\Omega_u) \colon |v - u| \preceq \delta\}.\end{aligned} \tag{6.7.1}$$

where Ξ_s is defined by (2.1.3). Hence, the EC filtering problem with implementation uncertainty is formulated as

$$\min_{u \in \mathcal{F}} \langle u, Lu \rangle \tag{6.7.2}$$

where L is defined by (2.1.5). The u that solves this problem is such that any implementation v that is within δ of u will still force the filter output to be within the prescribed output mask. The problem of envelope constrained filters with uncertain implementation is related to the ECUI problem as shown in the following.

Similar to the uncertain input problem, the description of the constraint given by (6.7.1) is not very helpful for computational purposes. Making the identification $F = \Xi_s$ and using Lemma 6.1.1, we obtain

$$|\Xi_s v - d| \preceq \varepsilon,\ \forall v \colon |v - u| \preceq \delta \text{ iff } |\Xi_s u - d| + |\Xi_s|\delta \preceq \varepsilon. \tag{6.7.3}$$

Thus, to ensure robustness to implementation errors, we effectively tighten the output mask by reducing ε to $\varepsilon - |\Xi_s|\delta$. In the robust constraint formulation of Section 6.6, the function β is used to specify the relative reduction to the mask for different times. However, it is not clear how β should be chosen. As a consequence of (6.7.3), we can use the normalized $|\Xi_s|\delta$ as the weighting function β in the robust margin formulation. Of course, if the uncertainty δ in the implementation is too large, the resulting output mask would not admit a feasible solution. We can determine from the implementation uncertainty δ whether robustness can be achieved. For instance, if there exists an interval where $(|\Xi_s|\delta)(t) \geq \varepsilon(t)$, then there is no feasible solution.

Using (6.7.3), Problem (6.7.2) can be written as

$$\begin{aligned}&\min\ \langle u, Lu \rangle \\ &\text{subject to } |\Xi_s u - d| + |\Xi_s|\delta \preceq \varepsilon.\end{aligned} \tag{6.7.4}$$

The same applies to discrete-time filters and hybrid filters. Note that for hybrid filters $F = \Xi_s : l_2(\tilde{\Omega}_u) \to C(\Omega_\psi)$, in which the kernel is

$$f(t, l) = \sum_{k \in \Omega_s} \Lambda(t - kT_s - lT_s)s(kT_s).$$

Problem (6.7.4) has the same form as the EC filtering Problem (5.1.1). Here, we are minimizing the cost (e.g. noise gain) of the theoretical filter u (i.e. without implementation error). The actual implementation $\hat{u}^o$ of the optimal theoretical filter u^o is not known exactly. The cost at $\hat{u}^o$ may be larger or smaller than that of u^o, but its response to the signal s still stays inside the output envelope.

Examples with finite dimensional filters

For finite dimensional filters,

$$(F\boldsymbol{a})(t) \equiv \boldsymbol{y}_s^T(t)\boldsymbol{a},$$

it follows from Lemma 6.1.1, that

$$|y_s^T(t)\hat{\boldsymbol{a}} - d(t)| \le \varepsilon(t),\ \forall \hat{\boldsymbol{a}}: |\hat{\boldsymbol{a}} - \boldsymbol{a}| \le \delta \text{ iff } |y_s^T(t)\boldsymbol{a} - d(t)| + |y_s^T(t)|\delta \le \varepsilon(t).$$

As an example, consider a pulse compression application using a hybrid configuration with a 27 tap FIR filter and a 6th order Bessel post-filter with cut-off frequency of $2\pi/\beta$ (β is the bit interval). The response of the system to a 13-bit Barker coded input is required to fit inside the mask of Figure 6.7.1. The solution $\boldsymbol{a}^o$ has entries in the range $[-0.06, 0.06]$. For an 8-bit representation of $\boldsymbol{a}^o$ on an interval $[a, b]$, the bound ς on the quantization error is $(b-a)2^{-9}$. Using 8-bit representation of $\boldsymbol{a}^o$ on $[-0.065, 0.065]$ gives $\varsigma = 0.13 \times 2^{-9}$ as the bound on the quantization error. Figure 6.7.1 shows the responses of the perturbations $\hat{\boldsymbol{a}}$ within $\delta = \varsigma[1, \ldots, 1]^T$ of $\boldsymbol{a}^o$ in the sense that $|\hat{\boldsymbol{a}} - \boldsymbol{a}^o| \le \delta$.

Solving the problem with the implementation robustness constraint ensures that the response of each filter within δ of the optimum solution $\boldsymbol{a}^o$ also stays inside the mask. This is verified by Figure 6.7.2, which shows the responses of some of these perturbed filters. The price for robustness is a 3% increase in noise gain.

The next example considers the design of a pulse shape to be transmitted over cable with 4.5db loss at 770 kHz so that the resulting pulse fits into a DS1 mask. Figure 6.7.3 shows the optimal solution with implementation perturbation $\varsigma = 0.052$.

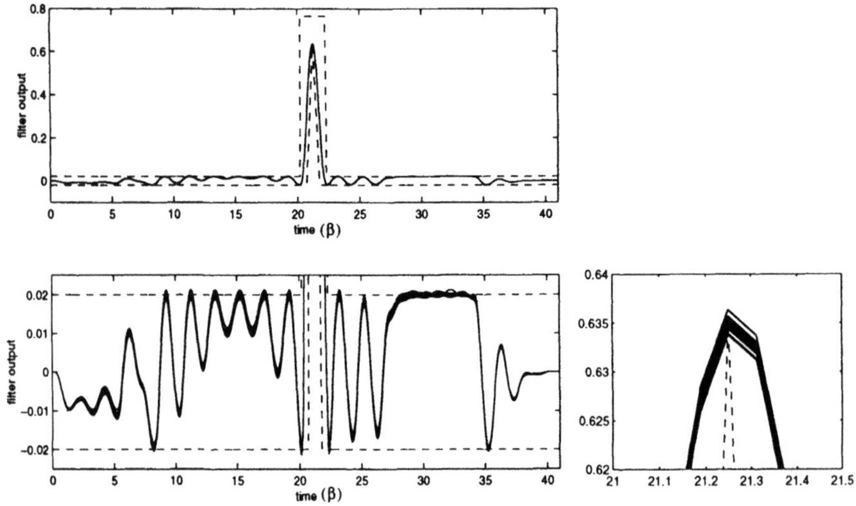

Figure 6.7.1. Response of perturbed filter

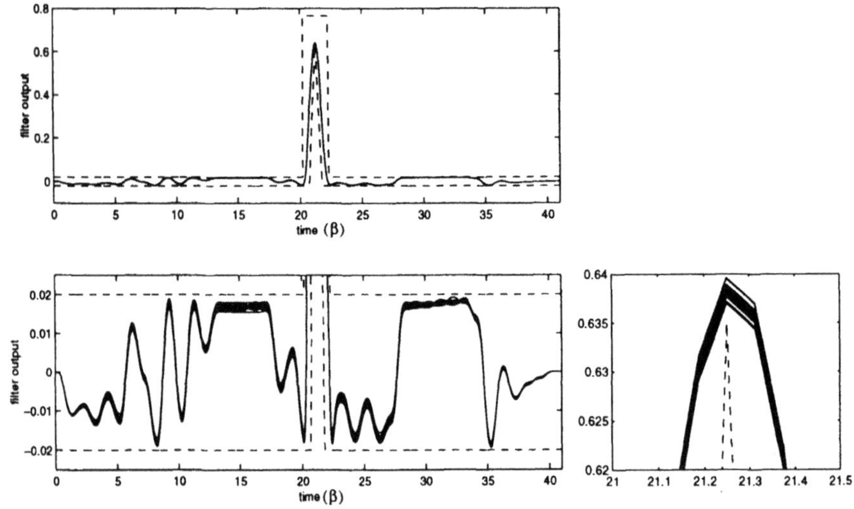

Figure 6.7.2. Response of perturbed robust filter

Observe that the output violates the mask constraints. However, the implementation robustness approach yields a solution that satisfies the mask constraints despite of implementation errors (see Figure 6.7.4). This is achieved at the expense of a 9% increase in the signal energy.

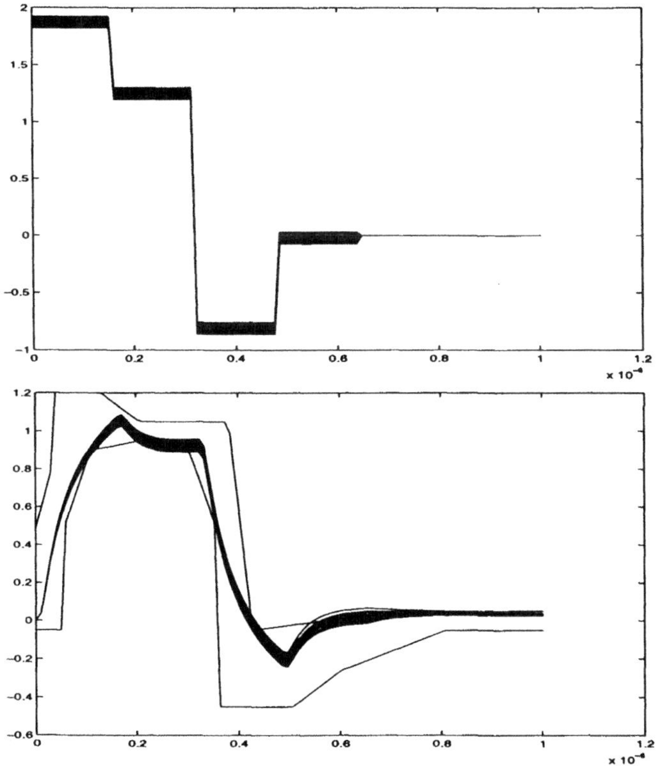

Figure 6.7.3. Signal Energy = 5.73

To illustrate the application of Lemma 6.1.1 to problems other than EC filters, we consider the design of linear phase (FIR) filters with magnitude response specifications. For linear phase filters, the filter coefficients $\tilde{h}_i$, $i = 0, \ldots, n-1$, are symmetric about the middle coefficient, assuming n is odd. The frequency response is given by [91]

$$H(e^{j\omega}) = \boldsymbol{c}^T(\omega)\boldsymbol{h}e^{-j\omega(n-1)/2}$$

where $\boldsymbol{c}(\omega) = [c_0(\omega), \ldots, c_M(\omega)]^T$ and $\boldsymbol{h} = [h_0, \ldots, h_M]^T$ are, respectively, given by

$$c_i(\omega) = \cos(i\omega), \text{ for } i = 0, \ldots, M, \; M = (n-1)/2,$$

and

$$h_0 = \tilde{h}_M, \; h_i = 2\tilde{h}_{M-i}, \text{ for } i = 1, \ldots, M.$$

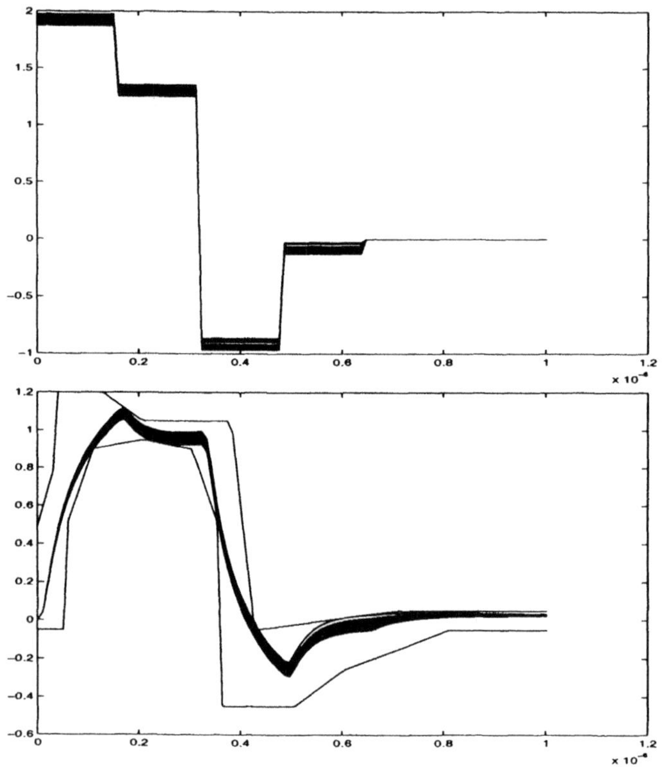

Figure 6.7.4. Signal Energy = 6.25, ~ 9% increase

This also holds for even n, in which, $h_i = 2\tilde{h}_{M-i}$ and $c_i(\omega) = \cos((i+0.5)\omega)$, $i = 0, \ldots, M$, $M = n/2-1$.

With the following desired response:

$$d(\omega) = D(\omega)e^{-j\omega(n-1)/2},$$

where $D \geq 0$, the mask constraint requires that

$$|c^T(\omega)h - D(\omega)| \leq \varepsilon(\omega), \ \ \forall\omega \in \Omega_c \subseteq [0, \pi].$$

The cost function used is the filter norm so as to minimize the sensitivity of the filter to implementation uncertainty.

For a 22-tap linear phase lowpass filter with

$$D(\omega) = \begin{cases} 1, \omega \in [0, 0.3\pi] \\ 0, \omega \in [0.45\pi, \pi] \end{cases},$$

$$\varepsilon(\omega) = \begin{cases} 0.050, \omega \in [0, 0.3\pi] \\ 0.025, \omega \in [0.45\pi, \pi] \end{cases},$$

the optimal solution $\boldsymbol{h}^o$ has entries taking on values in the range $[-0.13, 0.67]$. For convenience, we choose the interval $[-0.2, 0.8]$ and use an 8-bit representation, i.e. $\varsigma = 2^{-9}$. Figure 6.7.5 shows the magnitude responses of perturbed filters $\hat{\boldsymbol{h}}$ within $\delta = \varsigma[1, \ldots, 1]^T$ of $\boldsymbol{h}^o$ in the sense that $|\hat{\boldsymbol{h}} - \boldsymbol{h}^o| \le \delta$. Note the constraint violations near 0, 0.15 and on the stopband.

The robustness to implementation error approach ensures that the response of each perturbed filter $\hat{\boldsymbol{h}}$ within δ of the optimum solution $\boldsymbol{h}^o$, (i.e. $|\hat{\boldsymbol{h}} - \boldsymbol{h}^o| \le \delta$) also stays inside the mask. This is verified in Figure 6.7.6, which shows the magnitude responses of some of these perturbations. The price for robustness is a 4% increase in the noise gain.

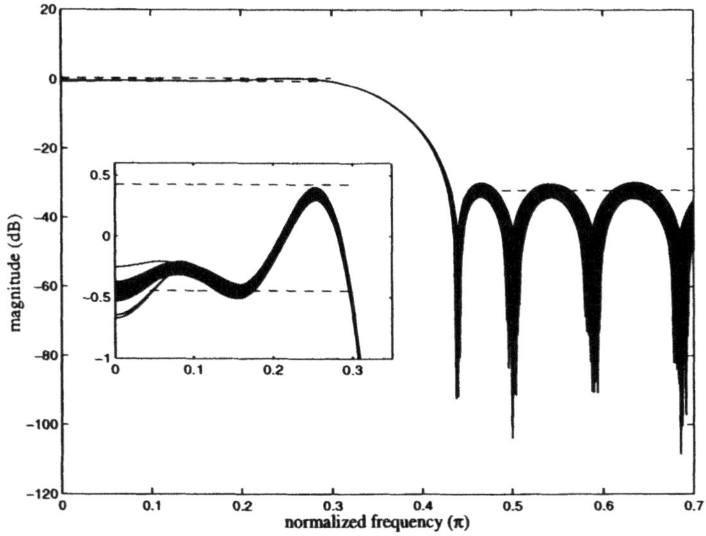

Figure 6.7.5. Magnitude response of perturbations

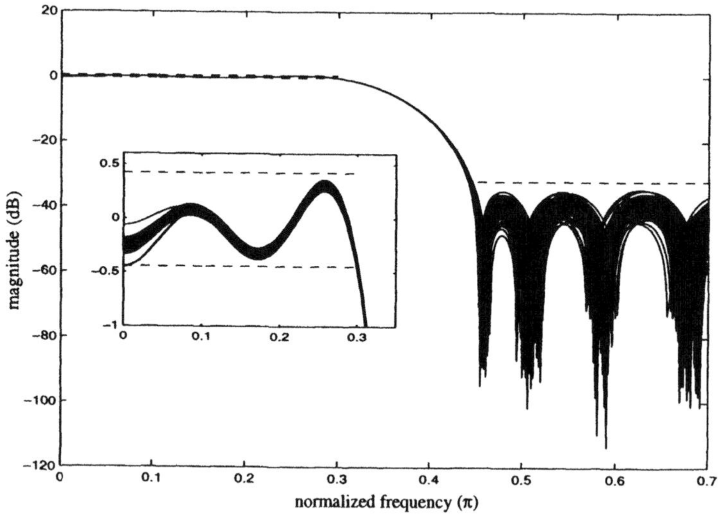

Figure 6.7.6. Magnitude response of perturbations of robust filter

6.8 Appendix

Lemma 6.1.1. *Suppose X_1 and X_2 are Banach spaces and for each X_j the subspace consisting of the real-valued members is equipped with a partial ordering*

$$x \preceq y \Leftrightarrow x(t) \le y(t), \forall t \in \Omega_j .$$

Let $F : X_1 \to X_2$ and $|F| : X_1 \to X_2$ be two linear operators defined by

$$(Fx)(t) \equiv \int_{\Omega_1} x(\tau) f(t, \tau) d\tau$$

and

$$(|F| x)(t) \equiv \int_{\Omega_1} x(\tau) |f(t, \tau)| d\tau,$$

respectively. Then, for a given positive $\theta \in X_1$,

$$|Fx - d| \preceq \varepsilon, \ \forall x : |x - s| \preceq \theta \quad \Leftrightarrow \quad |Fs - d| + |F|\theta \preceq \varepsilon .$$

Proof: Observe that

(i) Let $x \in X_1$ be such that $|x(\tau)| \le \theta(\tau)$, $\forall \tau \in \Omega_1$. Then, $\forall t \in \Omega_2$

$$|(Fx)(t)| \le \int_{\Omega_1} |x(\tau)||f(t, \tau)| d\tau \le \int_{\Omega_1} \theta(\tau) |f(t, \tau)| d\tau = (|F|\theta)(t). \tag{6.8.1}$$

(ii) For each $t \in \Omega_2$, let $x_t(\tau) = \theta(\tau) e^{-j \arg(f(t, \tau))}$. Then, $|x_t| \le \theta$, and

$$(Fx_t)(t) = \int_{\Omega_1} \theta(\tau) e^{-j \arg(f(t, \tau))} f(t, \tau) d\tau = \int_{\Omega_1} \theta(\tau) |f(t, \tau)| d\tau = (|F|\theta)(t). \tag{6.8.2}$$

Now, for any x such that $|x - s| \le \theta$, let $x = s + \tilde{s}$. Then, clearly $|\tilde{s}| \le \theta$. Suppose that $|Fs - d| + \Upsilon\theta \le \varepsilon$. Then, using linearity of F and (6.8.1) yields

$$|Fx - d| = |Fs - d + F\tilde{s}| \preceq |Fs - d| + |F\tilde{s}| \preceq |Fs - d| + |F|\theta \preceq \varepsilon .$$

That is, $|Fs - d| + |F|\theta \preceq \varepsilon$ implies $|Fx - d| \preceq \varepsilon, \forall x : |x - s| \preceq \theta$.

Conversely, suppose that $|Fx - d| \le \varepsilon, \forall x : |x - s| \le \theta$. Let $x = s + \tilde{s}$. Then,

$$\forall t \in \Omega_2, \forall \tilde{s} \in X_1 : |\tilde{s}| \le u, |(Fs)(t) - d(t) + (F\tilde{s})(t)| \le \varepsilon(t). \tag{6.8.3}$$

For each $t \in \Omega_2$, let $\alpha_t = e^{j\arg[(Fs)(t) - d(t)]}$. Using (6.8.2), there exists $x_t \in X_1$ with $|x_t| \le \theta$ such that $(Fx_t)(t) = (|F|\theta)(t)$. Since $|\alpha_t| \le 1$, it is clear that $|\alpha_t x_t| \le \theta$. Consequently, for each $t \in \Omega_2$, substituting $\tilde{s} = \alpha_t x_t$, i.e.

$$(F\tilde{s})(t) = [F(\alpha_t s_t)](t) = \alpha_t(|F|\theta)(t),$$

into (6.8.3) yields,

$$|(Fs)(t) - d(t) + \alpha_t(|F|\theta)(t)| \le \varepsilon(t). \tag{6.8.4}$$

Note that $|p + qe^{j\arg p}| = |p| + q$ for any $p \in \boldsymbol{C}$, $q \ge 0$, and as $|F|\theta \ge 0$, we have

$$|(Fs)(t) - d(t) + \alpha_t(|F|\theta)(t)| = |(Fs)(t) - d(t)| + (|F|\theta)(t), \ \forall t \in \Omega_2.$$

Thus, it follows from (6.8.4) that $|(Fs)(t) - d(t)| + (|F|\theta)(t) \le \varepsilon(t), \ \forall t \in \Omega_2$. ❐

Lemma 6.1.2. *The constraint function G is continuous and convex.*

Proof: Since $s, \theta \in L_2(\Omega_s)$, the constraint function G is continuous because it is the composition of the continuous mappings $|\bullet|$, Ξ_s, Ξ_θ (see Lemma 2.1.2). G is also convex with respect to the partial ordering $\preceq$, since for any $\alpha \in (0, 1)$,

$$\begin{aligned} |\Xi_s(\alpha u + (1-\alpha)v) - d| &= |\alpha \Xi_s u + (1-\alpha)\Xi_s v - d| \\ &= |\alpha(\Xi_s u - d) + (1-\alpha)(\Xi_s v - d)| \\ &\preceq \alpha|\Xi_s u - d| + (1-\alpha)|\Xi_s v - d| \end{aligned}$$

and

$$\begin{aligned} \Xi_\theta|\alpha u + (1-\alpha)v| &\preceq \Xi_\theta(\alpha|u| + (1-\alpha)|v|) \\ &= \alpha \Xi_\theta|u| + (1-\alpha)\Xi_\theta|v|. \text{❐} \end{aligned}$$

Lemma 6.2.1. *Let $p:H \to H$ and $q:H \to H$ be mappings on a Hilbert space H defined by (6.2.3). Then,*

(i) $0 \preceq p(u), 0 \preceq q(u)$,

(ii) $p(u)q(u) = 0$,

(iii) $p(u) - q(u) = u$,

(iv) $p(u) + q(u) = |u|$,

(v) *for each $u, v \in H$ with $0 \preceq u, 0 \preceq v$, there exists a unique non-negative function $r:H \times H \to H$ such that*

$$|u - v| = u + v - r(u, v).$$

Proof. Parts (i) and (ii) are trivial.

(iii)

$$\begin{aligned}[p(u)](t) - [q(u)](t) &= \max(u(t), 0) - \max(-u(t), 0) \\ &= \begin{cases} u(t) - 0, & u(t) \ge 0 \\ 0 - -u(t), & u(t) \le 0 \end{cases} \\ &= u(t)\end{aligned}$$

(iv)

$$\begin{aligned}[p(u)](t) + [q(u)](t) &= \max(u(t), 0) + \max(-u(t), 0) \\ &= \begin{cases} u(t) + 0, & u(t) \ge 0 \\ 0 - u(t), & u(t) \le 0 \end{cases} \\ &= |u(t)|\end{aligned}$$

(v) Let, $r(u, v)$ be defined by

$$[r(u, v)](t) = u(t)(1 - \mathrm{sgn}(u(t) - v(t))) + v(t)(1 + \mathrm{sgn}(u(t) - v(t))).$$

Then, it is clear that $r(u, v)$ is non-negative if u and v are also non-negative. Furthermore, we can rewrite $r(u, v)$ as follows:

$$[r(u, v)](t) = \begin{cases} 2v(t), & u(t) - v(t) \geq 0 \\ 2u(t), & u(t) - v(t) \leq 0 \end{cases}.$$

Hence,

$$\begin{aligned} u(t) + v(t) - [r(u, v)](t) &= \begin{cases} u(t) - v(t), & u(t) - v(t) \geq 0 \\ v(t) - u(t), & u(t) - v(t) \leq 0 \end{cases} \\ &= |u(t) - v(t)| \end{aligned}$$

This completes the proof. ❐

Theorem 6.8.1. *(Mercer's Theorem [62]). If an autocovariance $C(t, u)$ is continuous, then it can be expanded in the uniformly absolutely convergent series:*

$$C(t, u) = \sum_{k=1}^{\infty} \lambda_k \psi_k(t) \psi_k(u)$$

where $\{\lambda_k\}_{k=1}^{\infty}$ and $\{\psi_k\}_{k=1}^{\infty}$ are the eigenvalues and the corresponding orthonormal eigenvectors of the operator $\tilde{C}$ defined by

$$(\tilde{C}x)(t) = \int_0^{\infty} C(t, u) x(u) du.$$

Lemma 6.2.2. *Let L be a positive definite linear operator on a Hilbert space H and ρ be a real number such that $\rho \in (0, 2\lambda_{min}(L))$. Then, the mapping $f{:}H \times H \to \mathbf{R}$ defined by*

$$f(x, y) = \|x - y\|_L^2 + 2\rho \langle x, y \rangle$$

is positive definite.

Proof. Since

$$(Lx)(t) = \int_{\Omega_u} R_{nn}(t - \tau) x(\tau) d\tau,$$

and assuming the auto-correlation $R_{nn}(t-\tau)$ is continuous, then by Mercer's theorem (Theorem 6.8.1), we can express the kernel of L as follows:

$$R_{nn}(t-\tau) = C(t, t+\tau) = \sum_{k=1}^{\infty} \lambda_k \psi_k(t) \psi_k(t+\tau)$$

where $\{\lambda_k\}_{k=1}^{\infty}$ and $\{\psi_k\}_{k=1}^{\infty}$ are, respectively, the eigenvalues and corresponding orthonormal eigenvectors of the operator L. Since L is positive -definite, the eigenvalues $\{\lambda_k\}_{k=1}^{\infty}$ are positive. (For the discrete-time filters, the integral is replaced by sums and the continuity requirement is redundant). Hence, we can express the operator L in terms of projections into its eigenspaces:

$$Lx = \sum_{k=1}^{\infty} \lambda_k \langle \psi_k, x \rangle \psi_k = \sum_{k=1}^{\infty} \lambda_k P_k x$$

where $P_k x$ denotes the projection $\langle \psi_k, x \rangle \psi_k$ of x into the kth eigenvector ψ_k. Now, expanding $f(x, y)$, we have

$$\begin{aligned} f(x, y) &= \langle Lx, x \rangle + \langle Ly, y \rangle - 2\langle Lx, y \rangle + 2\rho \langle x, y \rangle \\ &= \sum_{k=1}^{\infty} \lambda_k [\langle P_k x, x \rangle + \langle P_k y, y \rangle - 2\langle P_k x, y \rangle] + 2\rho \langle x, y \rangle \end{aligned}$$

Since the P_j 's are orthogonal projections, it follows that

$$\langle P_j x, y \rangle = \langle P_j x, \sum_{k=1}^{\infty} P_k y \rangle = \sum_{k=1}^{\infty} \langle P_j x, P_k y \rangle = \langle P_j x, P_j y \rangle$$

Hence, we can write $f(x, y)$ as:

$$\begin{aligned} f(x, y) &= \sum_{k=1}^{\infty} \lambda_k [\|P_k x\|^2 + \|P_k y\|^2 - 2\langle P_k x, P_k y \rangle] + \sum_{k=1}^{\infty} 2\rho \langle P_k x, P_k y \rangle \\ &= \sum_{k=1}^{\infty} \lambda_k \left[\|P_k x\|^2 + \|P_k y\|^2 - 2\langle P_k x, P_k y \rangle + \frac{2\rho}{\lambda_k} \langle P_k x, P_k y \rangle\right] \qquad (6.8.5) \\ &= \sum_{k=1}^{\infty} \lambda_k [\|P_k x + q(\lambda_k) P_k y\|^2 + \|P_k y\|^2 (1 - q^2(\lambda_k))] \end{aligned}$$

where $q(\lambda_k) = \rho/\lambda_k - 1$. Now since $0 < \rho < 2\lambda_{min}(L)$, we have

$$-1 < \frac{\rho}{\lambda_k} - 1 < \frac{2\lambda_{min}(L)}{\lambda_k} - 1 \leq 1 .$$

That is, $1 - q^2(\lambda_k) > 0$ and consequently, $f(x, y) \geq 0$ for all $x, y \in H$. Moreover, since each term inside the summation of (6.8.5) is non-negative, $f(x, y) = 0$ implies that all these terms are zero. Hence,

$$\|P_k x + q(\lambda_k) P_k y\|^2 = 0 \text{ and } \|P_k y\|^2 = 0 \text{ for all } k$$

This means $P_k y = 0$ and $P_k x = 0$ for all k. Thus, $(x, y) = (0, 0)$. ❒

Lemma 6.8.1. *If (x^o, y^o) is an optimal solution of Problem (6.2.4), then $\langle x^o, y^o \rangle = 0$.*

Proof. Let

$$\begin{aligned} g &= p(x^o - y^o), \\ h &= q(x^o - y^o). \end{aligned}$$

Then, from Lemma 6.2.1 (i), $0 \preceq g, 0 \preceq h$. Using parts (iii) and (v) of Lemma 6.2.1 with the facts that (x, y) satisfies the constraints of Problem (6.2.4) and that $0 \preceq B(r(x^o, y^o))$, we obtain

$$\begin{aligned} |A(g-h) - d| + B(g+h) &= |A(x^o - y^o) - d| + B(x^o + y^o) - B(r(x^o, y^o)) \\ &\preceq \varepsilon - B(r(x^o, y^o)) \preceq \varepsilon. \end{aligned}$$

That is, (g, h) also satisfies the constraints of Problem (6.2.4). Moreover, from Lemma 6.2.1 (ii), $\langle g, h \rangle = 0$. Hence, using the optimality of (x^o, y^o), we have

$$\|x^o - y^o\|_L^2 + 2\rho \langle x^o, y^o \rangle \leq \|g - h\|_L^2 + 2\rho \langle g, h \rangle = \|g - h\|_L^2 .$$

However, from Lemma 6.2.1 (iii), $\|g - h\|_L^2 = \|x^o - y^o\|_L^2$, hence

$$\|x^o - y^o\|_L^2 + 2\rho\langle x^o, y^o\rangle \le \|x^o - y^o\|_L^2.$$

Noting that $0 \preceq x^o, 0 \preceq y^o$ and $0 < \rho$, we have $\langle x^o, y^o\rangle = 0$. ❐

Theorem 6.2.1. *If (x^o, y^o) is an optimal solution of Problem (6.2.4), then $u^o = x^o - y^o$ is an optimal solution of Problem (6.2.1). Conversely, if u^o is an optimal solution of Problem (6.2.1), then $(p(u^o), q(u^o))$ is an optimal solution of Problem (6.2.4).*

Proof. Suppose that (x^o, y^o) is an optimal solution of Problem (6.2.4). Then, by Lemma 6.2.1 (v) and $0 \preceq B(r(x^o, y^o))$, we have

$$\begin{aligned}|A(x^o - y^o) - d| + B|x^o - y^o| &= |A(x^o - y^o) - d| + B(x^o + y^o) - B(r(x^o, y^o)) \\ &\preceq |A(x^o - y^o) - d| + B(x^o + y^o) \preceq \varepsilon.\end{aligned}$$

That is, $x^o - y^o$ satisfies the constraints of (6.2.1).

Now, for any u satisfying the constraints of (6.2.1), it follows from Lemma 6.2.1 (iii), (iv) that

$$|A(p(u) - q(u)) - d| + B(p(u) + q(u)) = |Au - d| + B|u| \preceq \varepsilon,$$

i.e. $(p(u), q(u))$ satisfies the constraints of (6.2.4). From the optimality of (x^o, y^o),

$$\|x^o - y^o\|_L^2 + 2\rho\langle x^o, y^o\rangle \le \|p(u) - q(u)\|_L^2 + 2\rho\langle p(u), q(u)\rangle. \tag{6.8.6}$$

Since $\langle p(u), q(u)\rangle = 0$ (by Lemma 6.2.1 (ii)), $\langle x^o, y^o\rangle = 0$ (by Lemma 6.8.1), and $p(u) - q(u) = u$ (by Lemma 6.2.1 (iii)), we can rewrite (6.8.6) as

$$\|x^o - y^o\|_L^2 \le \|p(u) - q(u)\|_L^2 = \|u\|_L^2.$$

Hence, $x^o - y^o$ is the optimal solution of Problem (6.2.1).

Conversely, suppose that u^o is an optimal solution of Problem (6.2.1). Then, from Lemma 6.2.1 (iii) and (iv),

$$|A(p(u^o) - q(u^o)) - d| + B(p(u^o) + q(u^o)) = |Au^o - d| + B|u^o| \preceq \varepsilon .$$

That is, $(p(u^o), q(u^o))$ satisfies the constraints of Problem (6.2.4).

For any (x, y) satisfying the constraints of Problem (6.2.4), by Lemma 6.2.1 (v) and $0 \preceq B(r(x^o, y^o))$, we have

$$\begin{aligned} |A(x-y) - d| + B(|x-y|) &= |A(x-y) - d| + B(x + y - r(x, y)) \\ &\preceq |A(x-y) - d| + B(x+y) \preceq \varepsilon. \end{aligned}$$

That is, $x - y$ satisfies the constraints of (6.2.1).

By Lemma 6.2.1 (iii) and the optimality of u^o, we have, for all (x, y) satisfying the constraints of (6.2.4),

$$\|p(u^o) - q(u^o)\|_L^2 = \|u^o\|_L^2 \leq \|x - y\|_L^2 . \tag{6.8.7}$$

Since $\langle p(u^o), q(u^o)\rangle = 0$ (by Lemma 6.2.1 (ii)) and $0 \preceq x, 0 \preceq y$, $0 < \rho$, we can rewrite (6.8.7) as

$$\|p(u^o) - q(u^o)\|_L^2 + 2\rho\langle p(u^o), q(u^o)\rangle \leq \|x - y\|_L^2 + 2\rho\langle x, y\rangle .$$

Thus, $(p(u^o), q(u^o))$ is an optimal solution of (6.2.4). ❒

Lemma 6.8.2. *If u is a feasible point of Problem (6.6.3), then $(u, \rho(u))$ is a feasible point of Problem (6.6.4). On the other hand, if (u, σ) is a feasible point of Problem (6.6.4), then*

(i) $\rho(u) \geq \sigma$

(ii) *$(u, \rho(u))$ is also a feasible point of Problem (6.6.4) u is a feasible point of Problem (6.6.3).*

Proof. From the definition of $\rho(u)$ (see (6.6.2)), it is clear that for all $t \in \Omega_c$,

$$\rho(u) \le [-\phi^+(u)](t),$$

$$\rho(u) \le [-\phi^-(u)](t).$$

Hence, $(u, \rho(u))$ is a feasible point of Problem (6.6.4).

Suppose that (u, σ) is a feasible point of Problem (6.6.4). Then, for all $t \in \Omega_c$,

$$\sigma \le [-\phi^+(u)](t),$$

$$\sigma \le [-\phi^-(u)](t).$$

(i) Hence, from the definition of $\rho(u)$,

$$\rho(u) \equiv \min\left\{ \min_{t \in \Omega_c} [-\phi^+(u)](t), \min_{t \in \Omega_c} [-\phi^-(u)](t) \right\} \ge \sigma.$$

(ii) Using (i) and the definition of $\rho(u)$, it is clear that the pair $(u, \rho(u))$ is also a feasible point of Problem (6.6.4). ❒

Theorem 6.6.1. *If (u^*, σ^*) is a solution to Problem (6.6.4), then u^* is a solution to Problem (6.6.3) and $\rho(u^*) = \sigma^*$. Conversely, if u^* is a solution to Problem (6.6.3), then $(u^*, \rho(u^*))$ is a solution of Problem (6.6.4).*

Proof. Suppose that (u^*, σ^*) is a solution to Problem (6.6.4). Then, by Lemma 6.8.2 (i), $\rho(u^*) \ge \sigma^*$. On the other hand, by Lemma 6.8.2 (ii), the pair $(u^*, \rho(u^*))$ also satisfies the constraints. Hence, by the optimality of (u^*, σ^*), we have $\sigma^* \ge \rho(u^*)$. Therefore, $\rho(u^*) = \sigma^*$.

Consider any u satisfying the constraints of Problem (6.6.3). Lemma 6.8.2 asserts that $(u, \rho(u))$ satisfies the constraints of Problem (6.6.4). Hence, by the optimality of (u^*, σ^*), we have

$$\rho(u) \le \sigma^* = \rho(u^*).$$

However, from Lemma 6.8.2 (ii), u^* is a feasible point of Problem (6.6.3). Thus u^* is a solution of Problem (6.6.3).

Conversely, suppose that u^* is a solution of Problem (6.6.3). Then, by Lemma 6.8.2, $(u^*, \rho(u^*))$ is a feasible point of Problem (6.6.4). Consider any (u, σ) which satisfies the constraints of Problem (6.6.4). From Lemma 6.8.2 (ii), $(u, \rho(u))$ is a feasible point of Problem (6.6.4) and u is a feasible point of Problem (6.6.3). By the optimality of u^* and Lemma 6.8.2 (i), we have

$$\rho(u^*) \geq \rho(u) \geq \sigma,$$

i.e. $f(u^*, \rho(u^*)) \geq f(u, \sigma)$. Thus, $(u^*, \rho(u^*))$ is an optimum solution of Problem (6.6.4). ❐

APPENDIX A MATHEMATICAL BACKGROUND

This Appendix is a review of the basic ideas in functional analysis which also serves as a reference for the necessary mathematical background required in the monograph.

A.1 Topological Space

In this section, the notion of a topology is introduced by defining open sets and closed sets. Equiped with a topology of open sets, we can proceed to the notion of continuity of a mapping and convergence of sequences. The open sets also cater for other concepts such as denseness, separability, compactness and Hausdorff property.

Definition A.1.1. A *topology* on a non-empty set X is a subset $\mathcal{T}$ of the power set 2^X (i.e. a collection of all subsets of X) such that

(1) $X \in \mathcal{T}$ and $\varnothing \in \mathcal{T}$, where $\varnothing$ denotes the empty set,

(2) if $G_1, G_2 \in \mathcal{T}$, then $G_1 \cap G_2 \in \mathcal{T}$,

(3) if $\{G_\alpha : \alpha \in J\} \subseteq \mathcal{T}$, then $\bigcup_{\alpha \in J} G_\alpha \in \mathcal{T}$.

Definition A.1.2. The combination $(X, \mathcal{T})$ is called a *topological space* and the elements or members of $\mathcal{T}$ are called the *open sets* of the topological space. The *closed sets* of a topology $\mathcal{T}$ on X are complements of the members of $\mathcal{T}$, i.e. a set $S \subset X$ is said to be

(1) *open* if $S \in \mathcal{T}$, and

(2) *closed* if $S^c \in \mathcal{T}$.

Note that X and $\varnothing$ are both open and closed. The intersection of any family of closed sets is closed and the union of a finite family of closed sets is closed. When no confusion can arise the topological space $(X, \mathcal{T})$ is simply written as X.

Definition A.1.3. Let $(X, \mathcal{T})$ and $(Y, \mathcal{U})$ be two topological spaces and let $f: X \to Y$ be a function from X to Y. Then, f is said to be *continuous* if, given any open set $G \subset Y$, the inverse image of G under f, $f^{-1}(G)$, is open in X, i.e.

$$G \in \mathcal{U} \Rightarrow f^{-1}(G) \in \mathcal{T}.$$

Definition A.1.4. Let A be a subset of X, where $(X, \mathcal{T})$ is a topological space.

(1) The *interior* A° of A, is defined as the maximal open subset of A, i.e.

$$A^\circ = \bigcup_{B \subset A,\, B \in \mathcal{T}} B$$

(2) The *closure* $\bar{A}$ of A, is defined as the minimal closed superset of A, i.e.

$$\bar{A} = \bigcap_{B \supset A,\, B^c \in \mathcal{T}} B$$

(3) The *boundary* ∂A of A, is defined as $\bar{A} \backslash A^\circ$.

Definition A.1.5. A *neighborhood* of a point x in a topological space $(X, \mathcal{T})$ is any set $V \subset X$ such that its interior, V°, contains x.

A set in a topological space is open if and only if it contains a neighborhood of each of its points. If A is open, then for each point x of A, A is an open neighborhood of

x, which is contained in A. Conversely, if a set A contains a neighborhood of each of its points, then it contains an open neighborhood of each of its points. The union of all these open neighborhoods is A, and hence A is open.

Definition A.1.6. Let $A \subset X$ be non-empty. A point $x \in X$ is said to be an *accumulation point* of A if every neighborhood of x contains points of A other than x.

If A is a closed set, then it contains all its accumulation points. Let x be an accumulation point of A. Then, every open set containing x contains points of A. This means any open set not containing points of A cannot possibly contain x. Now, A^c does not contain any point of A, and since A^c is open, A^c cannot possibly contain x. Hence $x \in A$.

Definition A.1.7. Let $(X, \mathcal{T})$ be a topological space. Then, a subset A of X is said to be *dense* in X if $\bar{A} = X$.

Definition A.1.8. The function f is said to be *continuous* at a point $x_0 \in X$ if given any open neighborhood V of $f(x_0)$, there exists an open neighborhood U such that $U \subseteq f^{-1}(V)$.

If f is a continuous one-to-one mapping of X onto Y, and if the inverse mapping f^{-1} is also continuous, then f is called a *homeomorphism* or *topological isomorphism*. In this situation, the spaces X and Y are said to be *homeomorphic*.

Let $\mathcal{T}_1$ and $\mathcal{T}_2$ be two topologies on X. Then, $\mathcal{T}_2$ is said to be *stronger* than $\mathcal{T}_1$ (or $\mathcal{T}_1$ is *weaker* than $\mathcal{T}_2$) if $\mathcal{T}_1 \subset \mathcal{T}_2$.

Definition A.1.9. Let $(X, \mathcal{T})$ be a topological space, and let A be a non-empty subset of X. The family $\mathcal{T}_A = \{A \cap B : B \in \mathcal{T}\}$ forms a topology on A and is called the *relative topology* on A induced by the topology $\mathcal{T}$ on X.

Definition A.1.10. A topological space $(X, \mathcal{T})$ is said to be *separable* if X contains a countable subset that is dense in X.

Definition A.1.11. A family $\mathcal{B}$ of subsets of X is a *base* for a topology $\mathcal{T}$ on X if $\mathcal{B} \subset \mathcal{T}$ and, for each $x \in X$ and each neighborhood U of x, there is a member $V \in \mathcal{B}$ such that $x \in V \subset U$. A family $\mathcal{F}$ of subsets of X is a *sub-base* for a topology $\mathcal{T}$ on X if the family of finite intersections of members of $\mathcal{F}$ is a base for $\mathcal{T}$.

Theorem A.1.1. If a topological space has a countable base, then it is separable.

Definition A.1.12. A sequence $\{x_n\} \subset X$ is said to *converge* to a point $x \in X$, i.e. $x_n \to x$, if each neighborhood of x contains all but a finite number of elements of the sequence.

Definition A.1.13. Let $(X, \mathcal{T})$ be a topological space, and let A be a non-empty subset of X. Then, a countable collection of open sets $C = \{G_i\}$ is called an *open covering* of A, if $A \subset \bigcup G_i$. A subset $\mathcal{S}$ of C is called a (open) *sub-covering* of A if $\mathcal{S}$ is also an open covering of A.

Definition A.1.14. A topological space $(X, \mathcal{T})$ is said to be *compact* if every open covering of X has a finite subcovering; *countably compact* if every countable covering of X has a finite subcovering; *sequentially compact* if every sequence has a convergent subsequence.

A subset A in a topological space $(X, \mathcal{T})$ is *compact* if every open covering of A has a finite subcovering; *countably compact* if every countable covering of A has a finite subcovering; *sequentially compact* if every sequence has a convergent subsequence which converges in A.

Note that compactness implies countably compactness which in turn implies sequentially compactness. A closed subset of a compact space is compact. A continuous image of a compact space is compact. Hence, a real continuous function on a compact space attains its supremum and infimum.

Definition A.1.15. A topological space $(X, \mathcal{T})$, where singleton sets are closed, is called *Hausdorff* if, for every pair of distinct points $x, y \in X$, there are disjoint neighborhoods of x and y (the separation axiom). $(X, \mathcal{T})$, is called *regular* if, for every closed set A and every $x \notin A$, there are disjoint neighborhoods of A and x. $(X, \mathcal{T})$, is called *normal* if, for every pair of disjoint closed sets A and B, there are disjoint neighborhoods of A and B.

A compact subset of a Hausdorff space is closed. A compact Hausdorff space is normal.

Lemma A.1.1 (Urysohn's Lemma [19], [100]). *If A and B are disjoint closed subsets of a normal topological space X, then there exists a continuous function $\sigma : X \to [0, 1]$ that takes the value 0 everywhere on A and 1 everywhere on B.*

A.2 Metric Spaces

An important class of topological spaces is the class of metric spaces. Metric provides the concept of distance between points in a set. This notion leads to boundedness (or total boundedness) and completeness. It turns out that compactness in a metric space is equivalent to total boundedness and completeness.

Definition A.2.1. A *metric* on a set X is a function $d : X \times X \to \mathbf{R}$ such that

(1) $d(x, y) \geq 0, \forall x, y \in X$,

(2) $d(x, y) = 0 \Leftrightarrow x = y$,

(3) $d(x, y) = d(y, x), \forall x, y \in X$, and

(4) $d(x, z) = d(x, y) + d(y, z), \forall x, y, z \in X$.

The combination (X, d) is called a *metric space.*

Definition A.2.2. An *open ball* of radius r centered at x_0, denoted by $B(x_0, r)$, is defined as

$$B(x_0, r) = \{x \in X : d(x, x_0) < r\}.$$

Similarly, a closed ball of radius r centered at x_0, denoted by $\bar{B}(x_0, r)$, is defined as

$$\bar{B}(x_0, r) = \{x \in X : d(x, x_0) \leq r\}.$$

Definition A.2.3. A subset M of a metric space X is said to be open if it contains a ball about each of its points. A subset K of X is said to be closed if its complement (in X) is open.

Strictly speaking, use of the terms 'open' and 'closed' in the above definition is an abuse of notation because a topology has not yet been defined on X. However, it is not difficult to show that the collection of all 'open' subsets of a metric space X is, in fact, a topology on X and hence justifying the definition.

With this topology, the family of open balls $\{B(x, 1/n): n \in J, x \in X\}$ forms a base for X.

It is possible to define different metrics on the same set X. Let d_1, d_2 be two different metrics on X. Then, (X, d_1) and (X, d_2) are generally two different metric spaces.

Definition A.2.4. Two metrics on a set X are *equivalent* if they give rise to the same topology. Two metrics d_1, d_2 are *strongly equivalent* if there exist $\alpha, \beta > 0$ such that

$$\alpha d_1(x, y) \leq d_2(x, y) \leq \beta d_1(x, y),\ \forall x, y \in X.$$

Note that strongly equivalent implies equivalent. A metric d_1 is always equivalent to $d_2 = d_1/(1 + d_1)$. Also, if d_1 is bounded, then d_1, d_2 are strongly equivalent.

Definition A.2.5. Let (X, d) be a metric space. A sequence $\{x_n\}$ is said to be a *Cauchy* sequence if

$$\forall \varepsilon > 0, \exists N : m, n > N \Rightarrow d(x_n, x_m) < \varepsilon$$

Theorem A.2.1. Every Cauchy sequence with convergent subsequence converges.

Definition A.2.6. A metric space (X, d) is said to be *complete* if every Cauchy sequence converges in X. A subset A of a metric space (X, d) is complete if every Cauchy sequence converges in A.

A metric space is normal, and hence a convergent sequence can have only one limit.

Theorem A.2.2. A subset A of a complete metric space (X, d) is complete if and only if A is closed.

Definition A.2.7. A subset A of a metric space (X, d) is said to be *totally bounded* if, for each $\varepsilon > 0$, there exists a finite collection of open balls of radius ε that covers A.

Theorem A.2.3. A set in $\boldsymbol{R}^n$ is compact if and only if it is closed and bounded. Thus, a closed subset of a compact set in $\boldsymbol{R}^n$ is compact.

Theorem A.2.4. A subset A of a metric space (X, d) is compact if and only if A is complete and totally bounded.

Corollary A.2.1. A subset A of a complete metric space (X, d) is compact if and only if it is closed and totally bounded.

Theorem A.2.5. A set A in a metric space (X, d) is compact if and only if each infinite subset of A has an accumulation point in A.

Definition A.2.8. Let (X, d) be a metric space, and let $A \subset X$. If every sequence in A contains a subsequence that converges with respect to the metric d to a point in X, then A is said to be *conditionally sequentially compact*. The set A is *sequentially compact* if it is conditionally sequentially compact and closed (that is the limit does not leave the set A).

For a metric space, compactness, countable compactness and sequential compactness are equivalent. A compact space is separable.

Definition A.2.9. Let (X, d_X) and (Y, d_Y) be two metric spaces and let $f : X \to Y$ be a function from X to Y. Then, f is said to be *continuous* at $x_0 \in X$ if

$$\forall \varepsilon > 0, \exists \delta(\varepsilon, x_0) > 0 \text{ such that } d_X(x, x_0) < \delta(\varepsilon, x_0) \Rightarrow d_Y(f(x), f(x_0)) < \varepsilon$$

It is said to be *uniformly continuous* at $x_0 \in X$ if δ does not depend on x_0.

The function f is said to be continuous if it is continuous everywhere in X. The set of all such continuous functions is denoted by $C(X, Y)$. If the metric space X is compact, then f is uniformly continuous.

Definition A.2.10. Let (X, d) and $(\tilde{X}, \tilde{d})$ be metric spaces. Then, a mapping $T : X \rightarrow \tilde{X}$ is said to be *isometric* or an *isometry* if T preserves distances, i.e.

$$\forall x, y \in X, \tilde{d}(T(x), T(y)) = d(x, y).$$

The space X is said to be isometric to the space $\tilde{X}$ if there exists a bijective isometry of X onto $\tilde{X}$. The spaces X and $\tilde{X}$ are called isometric spaces.

A.3 Vector Spaces

Definition A.3.1. A *vector space* over a field K is a non-empty set X together with two operations called vector addition and scalar multiplication satisfying the following axioms:

(1) closed under vector addition:

$$x, y \in X \Rightarrow x + y \in X$$

(2) commutative with respect to vector addition:

$$x + y = y + x, \forall x, y \in X$$

(3) associative with respect to vector addition:

$$x + (y + z) = (x + y) + z, \forall x, y, z \in X$$

(4) zero vector: there exists a unique vector 0 (the zero vector) such that

$$x + 0 = 0 + x = x, \forall x \in X$$

(5) additive complement:

$$\forall x \in X, \exists \text{ a unique element } -x \text{ such that } -x + x = 0$$

(6) closed under scalar multiplication:

$$x \in X, \alpha \in K \Rightarrow \alpha x \in X$$

(7) associative with respect to scalar multiplication:

$$\alpha(\beta x) = (\alpha\beta)x, \forall x \in X, \forall \alpha, \beta \in K$$

(8) distributive with respect to vector addition:

$$\alpha(x+y) = \alpha x + \alpha y, \forall x, y \in X, \forall \alpha \in K$$

(9) distributive with respect to scalar addition:

$$(\alpha + \beta)x = \alpha x + \beta x, \forall x \in X, \forall \alpha, \beta \in K$$

(10) identity: $1x = x, \forall x \in X$.

If $X_1, \ldots, X_m$ are vector spaces, their direct product $X_1 \times \ldots \times X_m$ is also a vector space. Its elements are denoted by $(x_1, \ldots, x_m)$ with $x_i \in X_i$, $i = 1, \ldots, m$. The vector addition and scalar multiplication operations are defined for any $(x_1, \ldots, x_m), (y_1, \ldots, y_m) \in X_1 \times \ldots \times X_m$, and $\alpha \in K$ by

$$(x_1, \ldots, x_m) + (y_1, \ldots, y_m) = (x_1 + y_1, \ldots, x_m + y_m)$$

and

$$\alpha(x_1, \ldots, x_m) = (\alpha x_1, \ldots, \alpha x_m)$$

Definition A.3.2. A *linear combination* of the vectors $x_1, \ldots, x_m \in X$ is an expression of the form

$$\alpha_1 x_1 + \ldots + \alpha_m x_m$$

Definition A.3.3. A finite collection $\{x_1, \ldots, x_m\}$ of vectors in X is said to be *linearly dependent* if there exists scalars $\alpha_1, \ldots, \alpha_m$ not all zero, such that

$$\alpha_1 x_1 + \ldots + \alpha_m x_m = 0$$

A finite subset of X is said to be *linearly independent* if it is not linearly dependent.

Definition A.3.4. A linear combination $\alpha_1 x_1 + \ldots + \alpha_m x_m$ of vectors $x_1, \ldots, x_m \in X$ is called a *convex combination* if $\alpha_i \geq 0$, $i = 1, \ldots, m$, and $\sum_{i=1}^{m} \alpha_i = 1$

Definition A.3.5. A non-empty subset Y of a vector space X is called a *sub-space* of X if Y is closed under (induced) addition and scalar multiplication.

Let S and T be subspaces of a vector space X. Then, the intersection $S \cap T$ is also a subspace of X.

Definition A.3.6. The *span* of a set $S \subset X$, denoted by $[S]$, is defined as the set of all vectors in X which are linear combinations of vectors of S. Note that $[S]$ is a subspace of X.

Definition A.3.7. If X is a vector space, and B is a linearly independent subset of X, then B is called a *basis* (or *Hamel basis*) for X.

Every non-empty vector space has a basis.

Definition A.3.8. The *direct sum* of two subsets S and T in a vector space X, denoted by $S \oplus T$, consists of all vectors of X of the form $s + t$, where $s \in S$ and $t \in T$.

Let M and N be subspaces of a vector space X. Then, their direct sum, $M \oplus N$, is also a subspace of X.

Definition A.3.9. The *translation* of a subspace is said to be a *linear variety* or an *affine sub-space.*

An affine sub-space V can be written as $V = x_0 \oplus M$, where M is a subspace.

Definition A.3.10. Let S be a non-empty subset of a vector space X. The *affine subspace generated by* S, denoted by $v(S)$, is defined as the intersection of all affine subspaces in X that contain S, or equivalently, the minimal affine subspace containing S.

Definition A.3.11. An *isomorphism* of a vector space X onto a vector space $\tilde{X}$ over the same field is a bijective mapping which preserves vector addition and scalar multiplication.

A.4 Normed Spaces

The vector space axioms only describe algebraic properties of the elements of the space: addition and scalar multiplication. What are missing are the topological concepts such as openness, closeness, convergence and completeness. These concepts can be provided by the introduction of a metric to the vector space. To guarantee such a relation between "algebraic" and "geometric" properties, a norm is introduced which uses the algebraic properties of the space and also defines a metric.

Definition A.4.1. Let X be a vector space and a topological space. It is called a *linear topological space* if the mappings

(1) $(x, y) \to x + y$ from $X \times X$ to X, and

(2) $(\alpha, x) \to \alpha x$ from $R \times X$ to X

are continuous.

Definition A.4.2. A *norm* $\|\bullet\|$ on a vector space X over a field K is a function from X to $[0, \infty)$ such that the following axioms are satisfied:

(1) $\|x\| \geq 0, \forall x \in X, \|x\| = 0 \Leftrightarrow x = 0$,

(2) $\|x + y\| \leq \|x\| + \|y\|, \forall x, y \in X$, and

(3) $\|\alpha x\| = |\alpha| \|x\|, \forall x \in X, \forall \alpha \in K$.

The combination $(X, \|\bullet\|)$ is called a *normed (linear) space.*

A normed space $(X, \|\bullet\|)$ can be regarded as a metric space (X, d) with the metric d induced by the norm $\|\bullet\|$ as follows:

$$d(x, y) = \|x - y\|$$

A metric d induced by a norm on a normed space X has the *translational invariance* property, i.e. for all $x, y \in X$ and $\alpha \in K$

$$d(x + a, y + a) = d(x, y)$$
$$d(\alpha x, \alpha y) = |\alpha| d(x, y)$$

Definition A.4.3. A *Banach space* is a complete normed space with respect to the metric induced by the norm.

Theorem A.4.1. Let $(X, \|\bullet\|)$ be a normed space. Then, there is a Banach space $\hat{X}$ and an isometry A from X onto a subspace W of $\hat{X}$ which is dense in $\hat{X}$. The space $\hat{X}$ is unique except for isometries.

Definition A.4.4. An *isomorphism* of a normed space $(X, \|\bullet\|_X)$ onto a normed space $(\tilde{X}, \|\bullet\|_{\tilde{X}})$ is a bijective linear operator $T : X \to \tilde{X}$ such that $\forall x \in X, \|Tx\|_{\tilde{X}} = \|x\|_X$. In other words, T is a vector space isomorphism which preserves norm.

Theorem A.4.2. If the closed unit ball of a normed space $(X, \|\bullet\|)$ is compact, then X is finite dimensional.

A.5 Inner Product Spaces

In normed spaces, we can add vectors, multiply vectors by scalars, just as in elementary vector algebra. Furthermore, the norm on such a space generalizes the elementary concept of length of a vector. However, what is still missing in a general normed space is the notion of orthogonality. For this, we introduce inner product spaces.

Definition A.5.1. An *inner product* in a vector space X over a field K is a function $\langle \bullet, \bullet \rangle$ from $X \times X$ to K such that

(1) $\langle x, x \rangle \geq 0, \forall x \in X$ and $\langle x, x \rangle = 0 \Leftrightarrow x = 0$,

(2) $\langle x, y \rangle = \langle y, x \rangle^*, \forall x, y \in X$, and

(3) $\langle \alpha x + \beta y, z \rangle = \alpha \langle x, z \rangle + \beta \langle y, z \rangle, \forall \alpha, \beta \in K, \forall x, y, z \in X$.

If we let $\|x\| = \sqrt{\langle x, x \rangle}$, then X becomes a normed space.

Definition A.5.2. A *Hilbert space* is a complete inner product space with the norm induced by the inner product.

Inner product spaces are normed spaces and Hilbert spaces are Banach spaces. An inner product space and its corresponding norm satisfies the Cauchy-Schwartz inequality

$$|\langle x, y\rangle| \le \|x\|\|y\|.$$

Definition A.5.3. An element x of an inner product space X is said to be *orthogonal* to an element $y \in X$ (or x and y are orthogonal, written as $x \perp y$) if

$$\langle x, y\rangle = 0$$

Similarly, for subsets $A, B \subset X$, $x \perp A$ if $x \perp a, \forall a \in A$, and $A \perp B$ if $a \perp b, \forall a \in A, \forall b \in B$.

Definition A.5.4. An *orthogonal set* M in an inner product space is a subset $M \subset X$, whose elements are pairwise orthogonal. An *orthonormal set* $M \subset X$ is an orthogonal set whose elements have unity norm, i.e.

$$\forall x, y \in X, \langle x, y\rangle = \begin{cases} 0, & \text{if } x \ne y \\ 1, & \text{if } x = y \end{cases}.$$

An orthogonal set is linearly independent.

Theorem A.5.1 (Bessel inequality). Let $\{e_k\}_{k=1}^{\infty}$ be an orthonormal sequence in an inner product space X. Then, for every $x \in X$,

$$\sum_{k=1}^{\infty} |\langle x, e_k\rangle|^2 \le \|x\|^2$$

Definition A.5.5. An *isomorphism* T of an inner product space X onto an inner product space $\tilde{X}$ over the same scalar field is a bijective linear operator $T : X \to \tilde{X}$ which preserves inner product, i.e.

$$\forall x, y \in X, \langle Tx, Ty\rangle = \langle x, y\rangle$$

For any inner product space X, there exists a Hilbert space H and an isomorphism A from X onto a dense subspace $W \subset H$. The space H is unique except for the isomorphisms.

Definition A.5.6. Let Y be a subset of an inner product space X. Then, the set $\{z \in H : z \perp Y\}$ of all vectors orthogonal to Y is called the *orthogonal complement* of Y, and is denoted as $Y^{\perp}$.

Let H be a Hilbert space, and let Y be a closed subspace. Then, $H = Y \oplus Y^{\perp}$.

If Y is a closed subspace of a Hilbert space H, then $Y^{\perp\perp} = Y$.

Definition A.5.7. A subset A of an inner product space X is said to be *total* if the span of A is dense in X.

Theorem A.5.2. There exists a total orthonormal set (not necessarily countable) in every Hilbert space.

Definition A.5.8. A sequence of vectors $\{e_k\}_{k=1}^{\infty}$ is called an *orthonormal basis* for a Hilbert space H if it is orthonormal and total.

A Hilbert space is separable if and only if it has an orthonormal basis (Note that for a countable total orthonormal set, Gram-Schmidt procedure can be used to get an orthonormal basis).

The sequence $\{e_k\}_{k=1}^{\infty}$ is an orthonormal basis for a Hilbert space H if and only if

$$\forall x \in H, \sum_{k=1}^{\infty} \langle x, e_k \rangle e_k = x$$

Theorem A.5.3 (Parseval's equation). The sequence $\{e_k\}_{k=1}^{\infty}$ is an orthonormal basis for a Hilbert space H if and only if

$$\forall x \in H, \sum_{k=1}^{\infty} |\langle x, e_k \rangle|^2 = \|x\|^2$$

A.6 Linear Operators

In vector spaces and, in particular, normed spaces, a mapping is called an operator.

Definition A.6.1. A *linear operator* T is an operator with *domain* $\mathcal{D}(T)$ being a vector space and *range* $\mathcal{R}(T)$ lying in a vector space over the same field K, satisfying

$$\forall x, y \in \mathcal{D}(T), \alpha, \beta \in K, T(\alpha x + \beta y) = \alpha Tx + \beta Ty$$

Observe that Tx is used instead of $T(x)$. This is standard in functional analysis.

Definition A.6.2. Let T be a linear operator $T: \mathcal{D}(T) \to \mathcal{R}(T)$. Then, the *null space* or the *kernel* of T, $\mathcal{N}(T)$, is defined as the set of all $x \in \mathcal{D}(T)$ such that $Tx = 0$.

Let $T: \mathcal{D}(T) \to \mathcal{R}(T)$ be a linear operator. Then, $\mathcal{R}(T)$ and $\mathcal{N}(T)$ are vector spaces, and

$$\dim \mathcal{D}(T) < \infty \Rightarrow \dim \mathcal{R}(T) \leq \dim \mathcal{D}(T).$$

Theorem A.6.1. Let X, Y be vector spaces over the same field. Let $T: \mathcal{D}(T) \to \mathcal{R}(T)$ be a linear operator with $\mathcal{D}(T) \subset X$ and $\mathcal{R}(T) \subset Y$. Then,

(1) the inverse $T^{-1}: \mathcal{R}(T) \to \mathcal{D}(T)$ exists if and only if $Tx = 0 \Rightarrow x = 0$,

(2) if T^{-1} exists, it is a linear operator,

(3) if $\dim \mathcal{D}(T) < \infty$ and T^{-1} exists, then $\dim \mathcal{R}(T) = \dim \mathcal{D}(T)$.

Theorem A.6.2 (Banach inverse theorem [66]). Let X, Y be Banach spaces over the same field K and T be a continuous linear operator from X onto Y. Suppose that the inverse T^{-1} exists. Then, T^{-1} is continuous.

Definition A.6.3. Let $(X, \|\bullet\|_X)$, $(Y, \|\bullet\|_Y)$ be normed spaces over the same field K and $T: \mathcal{D}(T) \to \mathcal{R}(T)$ be a linear operator with $\mathcal{D}(T) \subset X$ and $\mathcal{R}(T) \subset Y$. Then, T is said to be *bounded* if there exists a real number c such that

$$\forall x \in \mathcal{D}(T), \|Tx\|_Y \leq c\|x\|_X$$

Theorem A.6.3. Let $(X, \|\bullet\|_X)$, $(Y, \|\bullet\|_Y)$ be normed spaces over the same field K and $T: \mathcal{D}(T) \to \mathcal{R}(T)$ be a linear operator with $\mathcal{D}(T) \subset X$ and $\mathcal{R}(T) \subset Y$. Then,

(1) T is continuous if and only if T is bounded,

(2) if T is continuous at any single point in $\mathcal{D}(T)$, then T is continuous.

Theorem A.6.4. Let $(X, \|\bullet\|_X)$, $(Y, \|\bullet\|_Y)$ be normed spaces over the same field K and let $\mathcal{L}(X, Y)$ denote the set of all bounded linear operators from X to Y. Then, $\mathcal{L}(X, Y)$ is a normed space with the *operator norm* defined as

$$\|T\| = \sup_{x \in \mathcal{D}(T)\setminus\{0\}} \frac{\|Tx\|_Y}{\|x\|_X} = \sup_{x \in \mathcal{D}(T),\, \|x\|_X = 1} \|Tx\|_Y .$$

Moreover, if Y is a Banach space, then $\mathcal{L}(X, Y)$ is also a Banach space.

Definition A.6.4. Let X and Y be vector spaces over the same field K. Then, a *sesquilinear form* or (*sesquilinear functional*) h on $X \times Y$ is a mapping $h : X \times Y \to K$ which is linear in the first argument and conjugate linear in the second argument.

Definition A.6.5. Let $(X, \|\bullet\|_X)$, $(Y, \|\bullet\|_Y)$ be normed spaces over the same field K. Then, a sesquilinear form h is said to be *bounded* if there is a real number c such that

$$\forall x \in X, \forall y \in Y, |h(x, y)| \le c\|x\|_X\|y\|_Y$$

and the norm of h is defined by

$$\|h\| = \sup_{x \in X\setminus\{0\},\, y \in Y\setminus\{0\}} \frac{|h(x, y)|}{\|x\|_X\|y\|_Y} = \sup_{\|x\|_X = 1,\, \|y\|_Y = 1} |h(x, y)|_Y$$

Definition A.6.6. Let X and Y be normed spaces and $T : X \to Y$ a bounded linear operator. Then, the *adjoint operator* $T^\dagger : Y' \to X'$ is defined by

$$(T^\dagger g)(x) = g(Tx),$$

where X', Y' are the dual spaces of X and Y, respectively. (See definition of dual space in Appendix A.7)

The adjoint operator $T^\dagger : X' \to Y'$ is linear and bounded with $\|T^\dagger\| = \|T\|$.

Theorem A.6.5 (Riesz representation). Let H_1 and H_2 be Hilbert spaces over the same scalar field K, and let $h : H_1 \times H_2 \to K$ be a bounded sesquilinear form. Then, h has a representation $h(x, y) = \langle Sx, y\rangle$, where $S : H_1 \to H_2$ is a bounded linear operator. S is uniquely determined by h and $\|S\| = \|h\|$.

Definition A.6.7. Let H_1 and H_2 be Hilbert spaces over the same scalar field K and let $T : H_1 \to H_2$ be a bounded linear operator. Then, the *Hilbert adjoint operator* $T^\dagger : H_2 \to H_1$ of T is the operator satisfying

$$\forall x \in H_1, \forall y \in H_2, \langle Tx, y\rangle = \langle x, T^\dagger y\rangle .$$

The Hilbert adjoint operator $T^\dagger$ of T is unique and is a bounded linear operator with norm $\|T^\dagger\| = \|T\|$.

Let H_1 and H_2 be Hilbert spaces over the same scalar field K, let $S : H_1 \to H_2$ and $T : H_1 \to H_2$ be bounded linear operators, and let $\alpha \in K$ be a scalar. Then,

(1) $\langle T^\dagger y, x\rangle = \langle y, Tx\rangle$,

(2) $(S + T)^\dagger = S^\dagger + T^\dagger$,

(3) $(\alpha T)^\dagger = \alpha^* T^\dagger$,

(4) $T^{\dagger\dagger} = T$,

(5) $\|T^\dagger T\| = \|TT^\dagger\| = \|T\|^2$,

(6) $T^\dagger T = 0 \Leftrightarrow T = 0$, and

(7) $(ST)^\dagger = T^\dagger S^\dagger$.

Definition A.6.8. A bounded linear operator $T : H \to H$ on a Hilbert space H is said to be

(1) *self-adjoint* or *Hermitian* if $T^\dagger = T$,

(2) *unitary* if T is bijective and $T^\dagger = T^{-1}$, and

(3) *normal* if $TT^\dagger = T^\dagger T$.

Note that if T is self-adjoint or unitary, then T is normal.

Theorem A.6.6. The product of two bounded self-adjoint operators S and T on a Hilbert space is self-adjoint if and only if the operators commute, i.e. $ST = TS$.

Theorem A.6.7. Let $\{T_n\}_{n=1}^{\infty}$ be a sequence of bounded self adjoint linear operators, on a Hilbert space H. Suppose $T_n \to T$, i.e. $\|T_n - T\| \to 0$ with respect to the usual operator norm. Then, T is a bounded self-adjoint linear operator on H.

Theorem A.6.8. Let $U : H \to H$ and $V : H \to H$ be unitary operators on the Hilbert space H. Then,

(1) U is isometric,

(2) $\|U\| = 1$, provided $H \neq \{0\}$,

(3) U^{-1} $(= U^{\dagger})$ is unitary, and

(4) UV is unitary.

A.7 Linear Functionals and Dual Spaces

A functional is an operator whose range lies on the scalar field.

Definition A.7.1. A *linear functional* f is a linear operator with domain in a vector space and range in the scalar field.

Definition A.7.2. The set of all linear functionals on a vector space X over a scalar field K is a vector space, denoted by X' and called the *algebraic dual space* of X, under the algebraic operations

$$\forall f, g \in X', \alpha, \beta \in K, \ (\alpha f + \beta g)(x) = \alpha f(x) + \beta g(x), \forall x \in X$$

The value of the linear functional $f \in X'$ at a point $x \in X$ is denoted by $f(x)$ or $f \bullet x$.

Of greater importance for our purpose is the subspace of the algebraic dual consisting of all bounded (i.e. continuous) linear functionals on a normed space $(X, \|\bullet\|)$.

Definition A.7.3. The subspace of the algebraic dual consisting of all continuous linear functionals on a normed space $(X, \|\bullet\|)$ is a normed space, denoted by X^* and called the *continuous dual* (or simply *dual)* space of X, with norm defined by

$$\|f\| = \sup_{x \in X\backslash\{0\}} \frac{|f(x)|}{\|x\|} = \sup_{x \in X, \|x\| = 1} |f(x)|$$

This is a special case of the operator norm defined in Section A.6.

Theorem A.7.1. The dual space X^* of a normed space X is a Banach space (regardless of whether X is a Banach space or not).

Definition A.7.4. The *second dual* X^{**} of X is defined as the dual of X^*.

Definition A.7.5 (Canonical mapping). Let $C : X \to X^{**}$ be a mapping such that $C(x) : X^* \to K$ is defined by

$$C(x)(f) = f(x), \forall f \in X^*.$$

Then, C is called the *canonical mapping* or *canonical embedding* of X into X^{**}.

Note that C is a one-to-one linear operator. Hence, it is a vector space isomorphism from X into X^{**}. We may regard X as a subset of X^{**}. Under the canonical mapping, if $X = X^{**}$, then it is called reflexive. If X is reflexive, then X^* is also reflexive.

Theorem A.7.2 (Reisz, Fréchet). Every bounded linear functional f on a Hilbert space can be represented in terms of the inner product $f(x) = \langle z, x\rangle$, where z depends on f, is uniquely determined by f and has norm $\|z\| = \|f\|$. Hence, a Hilbert space X and its dual X^* are isomorphic.

Definition A.7.6. Let $(X, \|\bullet\|)$ be a Banach space, and X^* be its dual. The norm topology is called the *strong topology*. Apart from this topology, the elements of X can also be used to generate another topology called the *weak topology*. A base for the weak topology consists of all neighborhoods of the form:

$$N(x_0, F^*, \varepsilon) = \{x \in X : |x^*(x - x_0)| < \varepsilon, x^* \in F^*\},$$

where $x_0 \in X$, F^* is any finite subset of X^*, and $\varepsilon > 0$.

Definition A.7.7. A sequence $\{x_n\}_{n=1}^{\infty}$ in a normed space X is said to *converge in the weak topology* to x (denoted by $x_n \xrightarrow{w} x$) if

$$\lim_{n \to \infty} |x^*(x_n - x)| = 0, \forall x^* \in X^*$$

Note that every weakly convergent sequence is bounded.

Definition A.7.8. A subset F of a normed space X is said to be *weakly closed* in X, if, for every sequence $\{x_n\}_{n=1}^{\infty} \subset F$ such that $x_n \xrightarrow{w} x$, then $x \in F$.

Since a strongly convergent sequence is weakly convergent, a weakly closed set is strongly closed. However, the converse is not necessarily true.

Theorem A.7.3 (Mazur). A convex subset of a normed space X is weakly closed if and only if it is strongly closed.

Theorem A.7.4 (Banach-Saks-Mazur). Let $(X, \|\bullet\|)$ be a normed space and $\{x_n\}_{n=1}^{\infty}$ be a sequence in X converging weakly to x. Then, there exists a sequence of finite convex combinations of $\{x_n\}_{n=1}^{\infty}$ that converges strongly to x.

For any Banach space X, two topologies have been introduced. Since under the canonical mapping, $X \subset X^{**}$, the topology induced by X on X^* is weaker than the weak topology. This topology is called the *weak* topology*. A base for the weak topology consists of all neighborhoods of the form:

$$N(x_0^*, F, \varepsilon) = \{x^* \in X^* : |x^*(x) - x_0^*(x)| < \varepsilon, x \in F\}$$

where $x_0^* \in X^*$, F is any finite subset of X, and $\varepsilon > 0$. If X is reflexive, then the weak and weak* topologies are equivalent.

Definition A.7.9. Let X^* be the dual space of a Banach space X. A sequence $\{x_n^*\}_{n=1}^{\infty}$ in X^* is said to *converge in the weak* topology* to x^* (denoted by $x_n^* \xrightarrow{w^*} x^*$) if

$$\lim_{n \to \infty} |x_n^*(x) - x^*(x)| = 0, \forall x \in X.$$

A.8 Measures and Integration

Definition A.8.1. A collection $\mathcal{A}$ of subsets of a set X is called an *algebra* if $X \in \mathcal{A}$ and $\mathcal{A}$ is closed under union and complementation, i.e. $A, B \in \mathcal{A} \Rightarrow A \cup B, A^c \in \mathcal{A}$.

Definition A.8.2. A collection $\mathcal{S}$ of subsets of a set X is called a *σ-algebra* if it is an algebra which is closed under countable union, i.e.

$$A_j \in \mathcal{S}, \forall j \in N \Rightarrow \bigcup_{j \in N} A_j \in \mathcal{S}.$$

Definition A.8.3. Let $(X, \mathcal{T})$ be a topological space. A *Borel σ-algebra* is the minimal σ-algebra containing $\mathcal{T}$, i.e. the intersection of all σ-algebras that contain the open sets (and also closed sets).

Definition A.8.4. A function $\mu : C \subset 2^X \to \boldsymbol{R}$ is *countably additive* if for any disjoint collection $\{A_j \in C, j \in N\}$ with $\bigcup_{j \in N} A_j \in C$, μ satisfies

$$\mu\left(\bigcup_{j \in N} A_j\right) = \sum_{j \in N} \mu(A_j).$$

Definition A.8.5.

(1) A *measure* μ is a countably additive, non-negative function whose domain is a σ-algebra and satisfies $\mu(\emptyset) = 0$.

(2) A *Borel measure* μ is a countably additive, non-negative function whose domain is a Borel σ-algebra and satisfies $\mu(\emptyset) = 0$.

Definition A.8.6. Let $\mathcal{S}$ be an σ-algebra subsets of a set X. The combination $(X, \mathcal{S})$ is called a *measurable space*. Let μ be a measure on $\mathcal{S}$. The combination $(X, \mathcal{S}, \mu)$ is called a *measure space*.

Definition A.8.7. Let $(X, \mathcal{S}, \mu)$ be a measure space with a topology $\mathcal{T}$. A set $A \in \mathcal{S}$ is said to be *regular* if

$$\mu(A) = \sup\{\mu(K): K \subset A, K \in \mathcal{S}, K \text{ compact}\}.$$

The measure μ is called a *regular measure* if every $A \in \mathcal{S}$ is regular.

Theorem A.8.1. If μ is a finite Borel measure on a metric space X, and X is regular, then μ is a regular measure.

Theorem A.8.2. If X is a complete, separable metric space, then any finite Borel measure μ on X is regular.

Definition A.8.8. Let $(X, \mathcal{S})$ be a measurable space, and let Y be a topological space with Borel σ-algebra $\mathcal{B}$. A function $f: X \to Y$ is call a *measurable function* if

$$\forall B \in \mathcal{B}, f^{-1}(B) \in \mathcal{S}.$$

Theorem A.8.3. Let $(X, \mathcal{S})$ be a measurable space, and let Y be a topological space with topology $\mathcal{T}$ and Borel σ-algebra $\mathcal{B}$. A function $f: X \to Y$ is measurable if and only if

$$\forall T \in \mathcal{T}, f^{-1}(T) \in \mathcal{S}.$$

Theorem A.8.4. Let $(X, \mathcal{S})$ be a measurable space, and let Y be a topological space with topology $\mathcal{T}$ and Borel σ-algebra $\mathcal{B}$. If $f_n: X \to Y$ is measurable and $f_n(x) \to f(x)$, for all $x \in X$, then f is measurable.

Definition A.8.9. A *simple function* on a measurable space $(X, \mathcal{S})$ is a function which can be written as

$$\varphi(x) = \sum_{j=1}^{n} a_j 1_{A_j}(x),$$

where the A_j's are disjoint sets and $1_{A_j}(x)$ is an indicator function.

Definition A.8.10. For a simple function $\varphi(x) = \sum_{j=1}^{n} a_j 1_{A_j}(x)$ on a measure space $(X, \mathcal{S}, \mu)$, the *integral* of φ with respect to the measure μ is defined as

$$\int \varphi d\mu = \sum_{j=1}^{n} a_j \mu(A_j).$$

It is easily seen that the integral of any simple function is well defined.

Definition A.8.11. Let $(X, \mathcal{S}, \mu)$ be a measure space and $f : X \to [0, \infty]$ be a measurable function. Then, the *integral* of f with respect to the measure μ is defined as

$$\int f d\mu = \sup\{\int \varphi d\mu : 0 \le \varphi \le f, \varphi \text{ simple}\},$$

and the *integral* of f on a set $E \in \mathcal{S}$ with respect to the measure μ is defined as

$$\int_E f d\mu = \int 1_E f d\mu.$$

Theorem A.8.5. Let $(X, \mathcal{S}, \mu)$ be a measure space and $f : X \to [0, \infty]$ be a measurable function. Then, there exists a non-decreasing sequence of non-negative simple functions $\{f_n\}_{n=1}^{\infty}$ converging pointwise to f and $\int f_n d\mu \uparrow \int f d\mu$.

Definition A.8.12. Let $(X, \mathcal{S}, \mu)$ be a measure space and $f : X \to [-\infty, \infty]$ be a measurable function. Then, the *integral* of f with respect to the measure μ is defined as

$$\int f d\mu = \int f^+ d\mu - \int f^- d\mu,$$

where $f^+ = \max\{f, 0\}$, $f^- = \max\{-f, 0\}$. The function f is said to be *integrable* if both $\int f^+ d\mu$ and $\int f^- d\mu$ are finite.

Definition A.8.13. Let $(X, \mathcal{S}, \mu)$ be a measure space. A statement $P(x)$ holds *almost everywhere* (a.e.) on X (with respect to the measure μ) if there exists a set $N \in \mathcal{S}$ such that $\mu(N) = 0$ and $P(x)$ holds for all $x \in X \backslash N$.

Definition A.8.14. Let $(X, \mathcal{S}, \mu)$ be a measure space and let f be a measureable function from X to $\boldsymbol{R}$. Then, f is said to be *p-integrable* if $\int |f|^p d\mu$ is finite. The class of p-integrable functions from X to $\boldsymbol{R}$ is denoted by $L_p(X, \mathcal{S})$

Lemma A.8.1 (Fatou's Lemma). Let $(X, \mathcal{S}, \mu)$ be a measure space. If $\{f_n\}_{n=1}^{\infty}$ is a sequence of non-negative integrable functions for which $\liminf_n \int f_n d\mu < \infty$, then the function f defined by $f(x) = \liminf_n f_n(x)$ is integrable and $\int f d\mu \le \liminf_n \int f_n d\mu$.

Theorem A.8.6 (Dominated Convergence Theorem). Let $(X, \mathcal{S}, \mu)$ be a measure space. If $\{f_n\}_{n=1}^{\infty}$ is a sequence of integrable functions which converges to f a.e., and if g is an integrable function such that

$$|f_n(x)| \le |g(x)| \text{ a.e. } \forall n \in N.$$

Then, f is integrable and the sequence $\{f_n\}_{n=1}^{\infty}$ converges to f in the mean i.e.

$$\int |f_n - f| d\mu \to 0, \text{ as } n \to \infty.$$

Definition A.8.15. A sequence $\{f_n\}_{n=1}^{\infty}$ is said to *converge in measure* to f if

$$\forall \varepsilon > 0, \lim_{n \to \infty} \mu(\{x : |f_n(x) - f(x)| > \varepsilon\}) = 0.$$

Definition A.8.16. A set F of measurable functions is said to be *uniformly integrable* on A if $\sup\{\int_A |f| d\mu : f \in F\} < \infty$ and

$$\forall \varepsilon > 0, \exists \delta(\varepsilon) \colon \mu(A) < \delta(\varepsilon) \Rightarrow \int_A |f| d\mu < \varepsilon, \forall f \in F.$$

Theorem A.8.7. Let $(X, \mathcal{S}, \mu)$ be a measure space with μ a finite measure, and let $\{f_n\}_{n=1}^{\infty}$ and f be integrable functions. Then,

$$\int |f_n - f| d\mu \to 0, \text{ as } n \to \infty$$

if and only if $\{f_n\}_{n=1}^{\infty}$ is uniformly integrable and converges to f in measure.

Definition A.8.17. Let $\mathcal{S}$ be a σ-algebra of a set X. A *signed measure* μ on $\mathcal{S}$ is a countably additive $[-\infty, \infty]$ valued function on $\mathcal{S}$. A signed measure ν on $\mathcal{S}$ is said to be *absolutely continuous* with respect the measure μ if $\nu(E) = 0$ whenever $\mu(E) = 0$.

Theorem A.8.8 (Radon-Nikodym). If $(X, \mathcal{S}, \mu)$ is a totally σ-finite measure space and if a σ-finite signed measure ν on $\mathcal{S}$ is absolutely continuous with respect to μ, then there exists a finite valued measureable function f on X such that

$$\nu(E) = \int_E f d\mu, \ \forall E \in \mathcal{S}$$

The function f is called the *Radon-Nikodym derivative* of ν with respect to μ. The Radon-Nikodym derivative f is unique in the sense that if also $\nu(E) = \int_E g d\mu$, $\forall E \in \mathcal{S}$, then $f = g$ a.e. with respect to μ.

Theorem A.8.9 (Hahn-Jordan Decomposition). Let μ be a signed measure on $(X, \mathcal{S})$. Then, there exists a $D \in \mathcal{S}$ such that

$$E \in \mathcal{S} \Rightarrow \mu(D \cap E) \geq 0, \mu(E \setminus D) \leq 0.$$

From Theorem A.8.9, we can decompose a signed measure μ into its *positive variation* μ^+ *and negative variation* μ^- as $\mu^+(E) = \mu(E \cap D)$, and $\mu^-(E) = -\mu(E \setminus D)$ so that $\mu^+ \perp \mu^-$ and

$$\mu(E) = \mu^+(E) - \mu^-(E).$$

The *total variation* of a signed measure is defined as

$$|\mu| = \mu^+ + \mu^-.$$

Definition A.8.18. Let $(X_1, \mathcal{S}_1, \mu_1)$ and $(X_2, \mathcal{S}_2, \mu_2)$ be measure spaces, and $\mathcal{S}_1 \times \mathcal{S}_2$ be the σ-algebra generated by $\{A \times B: A \in \mathcal{S}_1, B \in \mathcal{S}_2\}$. Then, there exists a unique measure on $\mathcal{S}_1 \times \mathcal{S}_2$, called the *product measure* $\mu_1 \times \mu_2$, such that

$$(\mu_1 \times \mu_2)(A \times B) = \mu_1(A)\mu_2(B)$$

Theorem A.8.10 (Fubini). If $f: X_1 \times X_2 \to [-\infty, \infty]$ is $\mu_1 \times \mu_2$ integrable, then

$$\int_{X_1 \times X_2} f d(\mu_1 \times \mu_2) = \int_{X_1} \left\{ \int_{X_2} f d\mu_2 \right\} d\mu_1 = \int_{X_2} \left\{ \int_{X_1} f d\mu_1 \right\} d\mu_2.$$

APPENDIX B OPTIMIZATION THEORY

The theory of optimization presented in this monograph is derived from a few simple, intuitive geometric relations, and their extensions to infinite-dimensional spaces.

B.1 Projection Theorem

Theorem B.1.1 (Minimizing vector theorem [58]*)*. Let X be an inner product space and $M \neq \varnothing$ be a convex complete subset of X. Then, for any $x \in X$, there exists a unique $\hat{y} \in M$ such that

$$\inf\{\|x-y\| : y \in M\} = \|x-\hat{y}\|.$$

Furthermore, suppose that M is a complete subspace of X and $x \in X$ is fixed. Then, $x - \hat{y} \perp M$, and $\hat{y}$ is called the orthogonal projection of x on M.

Theorem B.1.2 ([66]). Let x be a vector in a Hilbert space H and let K be a closed convex subset of H. Then, there is a unique vector $k_0 \in K$ such that

$$\|x - k_0\| \le \|x - k\|, \forall k \in K.$$

Furthermore, a necessary and sufficient condition that k_0 be the minimizing vector is

$$\langle x - k_0, x - k \rangle \le 0, \forall k \in K.$$

B.2 Hahn-Banach Theorem

The Hahn-Banach theorem can be stated in two forms, the extension form and the geometric form.

Definition B.2.1 Let $f : \mathcal{D}(f) \to \mathbf{R}$ be a linear functional defined on a subset of a vector space X. The *restriction* of f to a set $B \subset \mathcal{D}(f)$ is the linear functional $f|_B : B \to \mathbf{R}$ defined by $f|_B(x) = f(x)$ for all $x \in B$. An *extension* of f to a set $M \supset \mathcal{D}(f)$ is a linear functional $\tilde{f} : M \to \mathbf{R}$ such that $\tilde{f}|_{\mathcal{D}(f)} = f$, i.e. $\tilde{f}(x) = f(x)$ for all $x \in \mathcal{D}(f)$.

Definition B.2.2 A functional p on a vector space X is said to be *sub-linear* if

(1) $p(x_1 + x_2) \le p(x_1) + p(x_2), \forall x_1, x_2 \in X$, and

(2) $p(\alpha x) = \alpha p(x), \forall \alpha \ge 0, \forall x \in X$.

Note that any norm is a sub-linear functional.

Theorem B.2.1 (Hahn-Banach, extension form,[66]). Let X be a real normed space and p a continuous sub-linear functional on X. Let f be a linear functional defined on a subspace M of X satisfying $f(m) \le p(m)$ for all $m \in M$. Then, there is an extension $\tilde{f}$ of f from M to X such that $\tilde{f}(x) \le p(x)$ for all $x \in X$.

Corollary B.2.1. Let f be a bounded linear functional defined on a subspace M of a real normed space X. Then, there is a bounded linear functional $\tilde{f}$ defined on X which is an extension of f and which has norm equal to the norm of f on M.

Corollary B.2.2. Let x be an element of a normed space X. Then there is a non-zero bounded linear functional $x^* \in X^*$ such that $x^*(x) = \|x^*\|\|x\|$.

In general, for any $x \in X$ and $x^* \in X^*$, we have $x^*(x) \le \|x^*\|\|x\|$. In Hilbert space we have equality in this relation if and only if the functional is represented by a non-negative multiple of x. In this case, x and x^* are said to be *aligned.* When $x^*(x) = 0$, x and x^* are said to be orthogonal.

B.3 Positive Cones and Convex Mappings

The materials in this section are mainly taken from [96]

Definition B.3.1. A linear combination $\alpha_1 x_1 + \ldots + \alpha_m x_m$ is a conic combination if $\alpha_i \ge 0$, for each $i \in \{1, \ldots, m\}$, and a conic combination is a convex combination if $\alpha_1 + \ldots + \alpha_m = 1$.

Definition B.3.2. A set K in a vector space is said to be *convex* if, given any $x_1, x_2 \in K$, $\alpha x_1 + (1-\alpha)x_2 \in K$ for all $\alpha \in [0, 1]$.

Theorem B.3.1. Let K and L be convex sets in a vector space. Then,

(i) $\alpha K = \{\alpha k : k \in K\}$ is convex for any scalar α,

(ii) $K + L$ is convex,

(iii) The intersection of an arbitrary collection of convex sets is convex.

Theorem B.3.2. A set S in a linear vector space is convex if and only if it contains all convex combinations of its elements (i.e. closed under convex combination).

Definition B.3.3. Let S be an arbitrary set in a linear vector space. The *convex cover* or *convex hull,* denoted by $\text{co}(S)$, is the smallest convex set containing S. In other words, $\text{co}(S)$ is the intersection of all convex sets containing S.

Theorem B.3.3. $\text{co}(S)$ is the set of all convex combinations of elements of S.

Definition B.3.4. A set K in a linear vector space is said to be a *cone* with vertex at the origin if $x \in K$ implies that $\alpha x \in K$ for all $\alpha \geq 0$.

Theorem B.3.4. A set K in a linear vector space is a convex cone if and only if it contains all conic combinations of its elements (i.e. closed under conic combination).

Definition B.3.5. Let S be an arbitrary set in a linear vector space. The *convex conical hull*, denoted by $\mathrm{cone}(S)$, is the smallest convex cone containing S. In other words, $\mathrm{cone}(S)$ is the intersection of all convex cones containing S.

Theorem B.3.5. $\mathrm{cone}(S)$ is the set of all conic combinations of elements of S.

Lemma B.3.1 (Carathéodory [96]). Let S be an arbitrary set in $\boldsymbol{R}^n$. Then, any $x \in \mathrm{cone}(S)$ can be expressed as a conic combination of n elements of S.

Theorem B.3.6 (Carathéodory's Theorem [96]). Let S be an arbitrary set in $\boldsymbol{R}^n$. Then, $x \in \mathrm{co}(S)$ if and only if x can be expressed as a convex combination of $n+1$ elements of S.

Definition B.3.6. Let P be a convex cone in a vector space X. For $x, y \in X$, we write

$$x \preceq y \text{ (with respect to } P\text{) if } y - x \in P.$$

The relation $\preceq$ generates a *partial ordering* on X. (A partial ordering is defined to be a transitive and reflexive but not necessarily an antisymetric relation [19]. Caution: there are also definitions which require the relation to be antisymetric). The cone P defining this relation is called the *positive cone* in X. The cone $N = -P$ is called the *negative cone* in X and we write

$$x \preceq y \text{ for } x - y \in N.$$

The notation $x \prec y$ indicates that $y - x$ is an interior point of P.

Definition B.3.7. Given a positive cone P^+ in X, the *dual positive* cone $P^{\oplus}$ in the dual space X^* is defined by

$$P^{\oplus} = \{x^* \in X^* : x^* x \geq 0, \forall x \in P^+\}.$$

Definition B.3.8. Let X be a vector space and let Z be a vector space having a cone P specified as the positive cone. The function $G:X \to Z$ is said to be *convex* with respect to the partial ordering $\preceq$ if the domain Ω of G is a convex set and

$$G(\alpha u + (1-\alpha)v) \preceq \alpha G(u) + (1-\alpha)G(v),\ \forall u, v \in X, \alpha \in [0, 1].$$

If the ordering sign is reversed, then the function is said to be *concave*. Note that convexity is not an intrinsic property of a mapping but is dependent on the specified positive cone in the range space.

Let $G:X \to Z$ be a convex mapping. Then, for every $z \in Z$, $\{x \in \Omega : G(x) \preceq z\}$ is a convex set.

Theorem B.3.7([66]). Suppose that f is a convex functional defined on a convex subset $\mathcal{F}$ of a normed space. Let $\mu = \inf_{x \in \mathcal{F}} f(x)$. Then the subset $\mathcal{G}$ of $\mathcal{F}$ where $f(x) = \mu$ is convex. Moreover, if x^o is a local minimum of f, then $f(x^o) = \mu$, and hence x^o is a global minimum.

B.4 Gateaux and Fréchet Differentials

Let X be a vector space, Y a normed space and T a (possibly nonlinear) transformation defined on a domain $\mathcal{D}(T) \subset X$ and has range $\mathcal{R}(T) \subset Y$.

Definition B.4.1. Let $x_0 \in \mathcal{D}(T)$, and let h be an arbitrary point in X. If the limit

$$\delta T(x_0\,;h) = \lim_{\alpha \to 0} \frac{1}{\alpha}[T(x_0 + \alpha h) - T(x_0)]$$

exists, then it is called the *Gateaux differential* of T at x_0 with increment h (see [35], [49], [51], [77]). If the limit exists for each $h \in X$, the transformation T is said to be *Gateaux differentiable* at x_0 [97].

The Gateaux differential generalizes the concept of directional derivative (see [49], [66]) in finite-dimensional spaces. The existence of the Gateaux differential is a rather weak requirement, since the definition requires no norm on X. When X is normed a more satisfactory definition is given by the *Fréchet differential.*

Definition B.4.2 Let T be a transformation defined on an open domain $\mathcal{D}(T)$ in a normed space $(X, \|*\|_X)$ and has range $\mathcal{R}(T)$ in a normed space $(Y, \|*\|_Y)$. If for fixed $x_0 \in \mathcal{D}(T)$, there exists a bounded linear operator $(dT)_{x_0} \in \mathcal{L}(X, Y)$ such that

$$\forall \varepsilon > 0, \exists \delta > 0 \ni \|x - x_0\|_X \le \delta \Rightarrow \|T(x) - T(x_0) - (dT)_{x_0}(x - x_0)\|_Y \le \varepsilon \|x - x_0\|_X$$

that is

$$\lim_{\|h\|_X \to 0} \frac{\|T(x_0 + h) - T(x_0) - (dT)_{x_0} h\|_Y}{\|h\|_X} = 0$$

then T is said to be *Fréchet differentiable* at x_0, $(dT)_{x_0}$ is said to be the *Fréchet derivative* of T at x_0 and $(dT)_{x_0}h$ is called *Fréchet differential* of T at x_0 with increment h.

The notation $f'(x_0)$ is also used interchangeably with $(df)_{x_0}$. For the special case where f is a functional, $(df)_{x_0}$ can also be written as $\nabla f(x_0)$.

Note that if the transformation T has a Fréchet derivative, then it is unique. Furthermore, if the Fréchet derivative exists at x_0, then the Gateaux derivative exists at x_0 and they are equal.

If the transformation T defined on an open set $\mathcal{D}(T)$ in $(X, \|*\|_X)$ has a Fréchet differential at x_0, then T is continuous at x_0.

If the mapping $dT : X \to \mathcal{L}(X, Y)$ which maps x_0 to $(dT)_{x_0}$ is continuous at the point x_0, we say that the Fréchet derivative of T is continuous at x_0. This should not be confused with the statement that $(dT)_{x_0}$ is a continuous mapping from X to Y, a property that is basic in the definition of Fréchet derivative. If the Fréchet derivative of T is continuous on some open sphere S, then we say that T is continuously Fréchet differentiable on S.

The Fréchet derivative of linear combination of transformations is a linear combination of the Fréchet derivatives of the individual transformations, i.e.

$$\left(d\left(\sum_{i=1}^{m} T_i\right)\right)_x = \sum_{i=1}^{m} (dT_i)_x$$

Theorem B.4.1 (Chain rule). Let S be a transformation mapping an open set $\mathcal{D}(S) \subset X$ into an open set $\mathcal{R}(S) \subset Y$ and R be a transformation mapping $\mathcal{R}(S)$ into a normed space Z. Suppose that S is Fréchet differentiable at $x \in \mathcal{D}(S)$ and R is Fréchet differentiable at $S(x) \in \mathcal{R}(S)$. Then, $R \circ S$ is Fréchet differentiable at x and

$$(d(R \circ S))_x = (dR)_{S(x)} \circ (dS)_x .$$

It is relatively simple to apply the concepts of Gateaux and Fréchet differentials to the problem of minimizing a functional on a linear space. The technique leads quite naturally to the rudiments of calculus of variations. This section extends the familiar technique of minimizing or maximizing a function of a single variable by ordinary calculus to a similar technique based on more general differentials.

Definition B.4.3. Let f be a real-valued functional defined on a subset Ω of a normed space X. A point $x_0 \in \Omega$ is said to be a *relative minimum* of f on Ω if there exists an open ball N containing x_0 such that $f(x_0) \leq f(x)$ for all $x \in \Omega \cap N$. The point x_0 is said to be a *strict relative minimum* of f on Ω if $f(x_0) < f(x)$ for all $x \neq x_0$, $x \in \Omega \cap N$.

Relative maxima are defined similarly. The term *extrema* will be used to refer to either maxima or minima.

Theorem B.4.2 [97]. Let the real-valued functional f have a Gateaux differential on a vector space X. A necessary condition for f to have an extremum at $x_0 \in X$ is that $\delta f(x_0;h) = 0$ for all $h \in X$.

The point at which $\delta f(x_0;h) = 0$ for all $h \in X$ is called a *stationary point.*

Theorem B.4.3 [97]. Let the real-valued functional f have a Gateaux differential on a vector space X. Suppose that x_0 minimizes f on the convex set $\Omega \cap X$ and that f is Gateaux differentiable at x_0. Then, $\delta f(x_0;x - x_0) = 0$ for all $x \in \Omega$.

Subdifferentials and subgradients

We turn now to a satisfactory generalization of the notion of gradient in the case of convex functions which are not differentiable. The appropriate concept is a 'subgradient' of a given convex function. The theory of which has been extensively developed in [79], [80], [81], [97], [99].

Definition B.4.4. Let f be a convex functional on a vector space X. Any $\phi \in X^*$ satisfying

$$\phi(x - x_0) \leq f(x) - f(x_0) \text{ for any } x_0, x \in \mathcal{D}(f)$$

is called a *subgradient* of f at x_0. The set of all such ϕ is called the *sub-differential* of f at x_0 and is denoted by $\partial f(x_0)$. The mapping $x \to \partial f(x_0)$ of X into (possibly empty) subsets of X^* is the sub-derivative of f.

If $\nabla f(x_0)$ exists, then it obviously belongs to $\partial f(x_0)$. In this case, there can be no other subgradient at x_0.

If $x_0 \notin \mathcal{D}(f)$, then $\partial f(x_0)$ is void by definition. On the other hand, if $x_0 \in \mathcal{D}(f)$, then $\nabla f(x_0) \in X^*$ is a subgradient of f at x_0 if and only if $\nabla f(x_0)h \leq \delta f(x_0;h)$, $\forall h \in X$.

If f is subdifferentiable at x_0, which means that $\partial f(x_0)$ is not void, then f must be lower semi-continuous at x_0.

Theorem B.4.4 [78]. Let f be a convex functional on a Hausdorf space X. If f is continuous at x_0, then $\partial f(x_0)$ is a non-empty weak* compact convex subset of X^*.

Corollary B.4.1. Let f be a convex functional on a normed space $(X, \|\bullet\|_X)$. If f is continuous at x_0. Then, $\partial f(x_0)$ is a bounded subset of X^*.

Theorem B.4.5 [98]. Let X be a Hausdorf space and let f be a convex functional on X. Assume that f is continuous at x_0. Then, $\forall h \in X$,

$$\delta^+ f(x_0;h) \equiv \lim_{\alpha \to 0^+} \frac{f(x_0 + \alpha h) - f(x_0)}{\alpha} = \max_{\phi \in \partial f(x_0)} \phi(h)$$

Definition B.4.5. The normalized direction of steepest descent $\hat{d}$ of f at $x_0 \in X$ with respect to a norm $\|\bullet\|$ defined on X is given by

$$\hat{d} = \arg\min_{\|d\| \le 1} \delta^+ f(x_0; d)$$

Lemma B.4.1. Let f be a convex function on $\boldsymbol{R}^n$. Assume that f is continuous at $x_0 \in \boldsymbol{R}^n$. If $\hat{d}$ is the normalized direction of steepest descent of f at $x_0 \in X$, and $\hat{\phi}$ is a minimum norm element of $\partial f(x_0)$ then $\hat{d}$ is aligned with $\hat{\phi}$ i.e. $\hat{\phi}(\hat{d}) = \|\hat{\phi}\|$ and

$$\delta^+ f(x_0; \hat{d}) = -\|\hat{\phi}\|$$

Proof.

$$\begin{aligned} \min_{\|d\| \le 1} \delta^+ f(x_0; d) &= \min_{\|d\| \le 1} \max_{\phi \in \partial f(x_0)} \phi(d) \\ &= -\max_{\|d\| \le 1} \min_{\phi \in \partial f(x_0)} -\phi(d) \end{aligned}$$

Since the constraint sets on the RHS are compact and convex, we can interchange the min and max [104]. Hence,

$$\min_{\|d\| \le 1} \delta^+ f(x_0; d) = -\min_{\phi \in \partial f(x_0)} \max_{\|d\| \le 1} -\phi(d). \tag{B.4.1}$$

It now follows from the Hahn-Banach Theorem (Theorem B.2.1) that any vector d with $\|d\| \le 1$, which is aligned with $-\phi$, will maximize $-\phi(d)$ over $\|d\| \le 1$. Let $\hat{\phi}$ and $\hat{d}$ be the solution to the minimax problem (B.4.1). Then $\hat{d}$ is aligned with $-\hat{\phi}$. Moreover, from the Cauchy-Schwartz Inequality

$$\max_{\|d\| \le 1} -\phi(d) = \|\phi\|$$

Hence $\min_{\|d\| \le 1} \delta^+ f(x_0; d) = -\min_{\phi \in \partial f(x_0)} \|\phi\|$.

This Lemma states that the normalized direction of steepest descent at x_0 is any $\hat{d}$ which is aligned with $-\hat{\phi}$ where $\hat{\phi}$ is any minimum norm element of the subdifferential $\partial f(x_0)$. If f is differentiable at x_0 then the subdifferential $\partial f(x_0)$ consists of one element, the gradient of f at x_0, and the descent direction is just the negative direction as expected.

Theorem B.4.6 [97]. Let f be a convex functional on a Hausdorf space X, and K be a convex set in X. Then, $x_0 \in K$ is a solution of the convex program $\inf_{x \in K} f(x)$ if and only if there exists a $\phi \in \partial f(x_0)$ which attains its minimum over K at x_0.

B.5 Lagrange Multipliers

Classically, the method of Lagrange multipliers is a device for replacing a constrained optimization problem with an equivalent unconstrained problem.

Let $X, Y, Z_i, i = 1, \ldots, m$ be normed spaces. Suppose that each Z_i has a partial ordering, denoted by $\preceq$, generated by a positive cone P_i^+. Each dual space Z_i^* has a dual positive cone $P_i^{\oplus}$. Consider a general optimization problem (called the primal problem):

$$\min f(x)$$
$$\text{subject to } G_i(x) \preceq 0, i = 1, \ldots, m,$$

where $f : X \to \boldsymbol{R}$ is a functional and each $G_i : X \to Z_i$, $i = 1, \ldots, m$ is an operator.

Definition B.5.1. The *Lagrangian functional* $\mathcal{L} : (X \times Z_1^* \times \ldots \times Z_m^*) \to \boldsymbol{R}$ for the primal problem is defined as

$$\mathcal{L}(x; p_1, \ldots, p_m) \equiv f(x) + \sum_{i=1}^{m} p_i \bullet G_i(x).$$

Observe that if $p_i \in P_i^{\oplus}$ $i = 1, \ldots, m$ are given and the mappings $f, G_1, \ldots, G_m$ are convex, then the Lagrangian is a convex functional on X.

Theorem B.5.1. (Generalized Kuhn-Tucker Theorem [66]). Suppose that the mappings $f, G_1, \ldots, G_m$ are all convex. Suppose further that each positive cone P_i^+ has a non-empty interior and that there exists an $\bar{x} \in X$ for which $G_i(\bar{x}) \prec 0, i = 1, \ldots, m$. Then, x^o solves the primal problem if and only if

(a) x^o is feasible i.e. $G_i(x^o) \preceq 0, i = 1, \ldots, m$,

(b) there exists Lagrange multipliers $p_i^o \in P_i^{\oplus}, i = 1, \ldots, m$ such that x^o minimizes the Lagrangian functional

$$\mathcal{L}(x; p_1^o, \ldots, p_n^o) = f(x) + \sum_{i=1}^{m} p_i^o \bullet G_i(x) \tag{B.5.1}$$

over all $x \in X$, and

(c) $p_i^o \bullet G_i(x^o) = 0, i = 1, \ldots, m.$

If the mappings $f, G_1, \ldots, G_m$ are also differentiable, (B.5.1) can be replaced by

$$(d\mathcal{L}(\bullet, p_1^o, \ldots, p_m^o))_{x^o} = (df)_{x^o} + \sum_{i=1}^{m} p_i^o \bullet (dG_i)_{x^o} = 0, \tag{B.5.2}$$

In this form the theorem is often known as the Karush-Kuhn-Tucker theorem.

Definition B.5.2. The *dual functional* $\phi : Z_1^* \times \ldots \times Z_n^* \to \boldsymbol{R}$ for the primal problem is defined by

$$\phi(p_1, \ldots, p_m) \equiv \inf_x \mathcal{L}(x, p_1, \ldots, p_m)$$

and the *dual problem* is

$$\begin{gathered} \max \ \phi(p_1, \ldots, p_m) \\ \text{subject to } p_i \in P_i^{\oplus}, i = 1, \ldots, m \end{gathered}$$

Theorem B.5.2. (Generalized Duality Theorem [66]). Suppose that the mappings $f, G_1, \ldots, G_m$ are all convex and continuous. Suppose further that each positive cone P_i^+ has a non-empty interior and that there exists an $\bar{x} \in X$ for which $G_i(\bar{x}) \prec 0, i = 1, \ldots, m$. Then

$$\min_{G_i(x) \preceq 0, i = 1, \ldots, m} f(x) = \max_{p_i \in P_i^{\oplus}, i = 1, \ldots, m} \phi(p_1, \ldots, p_m) \tag{B.5.3}$$

Moreover, if x^o solves the primal problem and $p_i^o \in P_i^{\oplus}$, $i = 1, \ldots, m$ solves the dual problem, then

$$p_i^o \bullet G_i(x^o) = 0, i = 1, \ldots, m,$$

and x^o minimizes the Lagrangian $\mathcal{L}(x;p_1^o, \ldots, p_n^o)$ over all $x \in X$, i.e.

$$\phi(p_1^o, \ldots, p_m^o) = \mathcal{L}(x^o;p_1^o, \ldots, p_m^o).$$

Definition B.5.3. Let x be a feasible point. Then, d is called a *feasible direction* at x if there exists a $\sigma > 0$ such that $x + \tau d$ is feasible for any $\tau \in [0, \sigma]$ is

$$D(x) = \{d : \exists \sigma > 0 : \tau \in [0, \sigma] \Rightarrow x + \tau d \in \mathcal{F}\}$$

Definition B.5.4. The set $\mathcal{A}(x) = \{i : G_i(x) = 0\}$ is called the *active set.*

Let $D(x)$ be the set of all feasible directions at x and

$$\mathcal{D}(x) = \{d : (dG_i)_x d \geq 0, \forall i \in \mathcal{A}(x)\}.$$

Then, it can be shown that $\overline{D(x)} \subseteq \mathcal{D}(x)$. However, the converse does not necessarily hold, situations exists for which there are directions in $\mathcal{D}(x)$ which are not in $D(x)$. If x^* is optimal then *constraint qualification* is said to be satisfied if

$$\overline{D(x^*)} = \mathcal{D}(x^*).$$

References

[1] M. Aoki, *Introduction to Optimization Techniques*, MacMillan, 1971

[2] E. J. Anderson and A. B. Philpott (eds.), *Infinite Programming*, Springer Berlin, 1985,

[3] E. J. Anderson and P. Nash, *Linear Programming in Infinite Dimensional Spaces: Theory and Applications*, Wiley, Chichester, England 1987.

[4] R. B. Ash, Measure, *Integration and Functional Analysis*, Academic Press, 1972.

[5] Bell Communications, "DSX-3 Isolated Pulse Template and Equations", *Technical Reference TR-TSY-000499*, pp. 9-17, Issue 2, December 1988.

[6] J. Boot, *Quadratic Programming*, North-Holland Publishing Co. - Amsterdam, 1964.

[7] Z. L. Budrikis. "Visual fidelity criterion and modelling", *Proc. IEEE*, **Vol. 60**, No. 7, pp. 771-779, 1972.

[8] D. Burnside and T. Parks, "Optimal design of FIR filter with the complex Chebyshev error criteria", *IEEE Trans. Signal Processing*, **Vol. 43**, No. 3, pp. 605-616, 1995.

[9] M. Canon and C. Cullum, "A tight upper bound on the rate of convergence of the Frank-Wolfe algorithm", *SIAM, J. Control*, **Vol. 6**, No. 4, pp. 509-516, 1968.

[10] A. Cantoni and P. Butler, "Non-iterative automatic equalization", *IEEE Trans. Com.*, **Vol. 23**, pp. 621-633, June 1975.

[11] A. Cantoni, B. Vo, Z. Zang and K. L. Teo, "Analog envelope constrained filters for channel equalization", *Signal Processing in Telecommunications - Proc. 7th Tyrrhenian Int. Workshop on Com.*, E. Biglieri and M. Luise (eds.), Springer, London, pp. 397-408, 1995.

[12] CCITT, "Physical/Electrical characteristics of hierarchical digital interfaces", *G.703*, Fascicle III, 1984.

[13] C. K. Chui and G. Chen, *Signal Processing and Systems Theory*, Springer-Verlag, Berlin Heidelberg, 1992.

[14] R. E. Collin and F. J. Zucker, *Antenna Theory I, II*, McGraw-Hill, New York, 1969.

[15] C. E. Cook and M. Bernfeld, *Radar Signals*, Academic Press, New York, 1967.

[16] H. H. Dam, K. L. Teo, Y. Liu and S. Nordebo, "Optimum pole position for didgital Laguerre networks with least square error criterion", *Optimization Methods and Applications*, X. Q. Yang, K. L. Teo and L. Caccetta (eds.), Kluwer Academic publishers, 2001.

[17] J. E. Dennis Jr. and R. B. Schnabel, "A view of unconstrained optimization", *Handbooks in Operations Research and Management Science*, G. L. Nemhauser *et al.* (eds.), Elsevier Science Publishers B. V. (North-Holland) **Vol. 1**, pp. 1-72, 1989.

[18] J. E. Dennis Jr. and R. B. Schnabel, *Numerical Methods for Unconstrained Optimization and Nonlinear Equations*, Prentice-Hall, New Jersey, 1983.

[19] N. Dunford and J. T. Schwartz, *Linear Operators*, Part II, Chapter XI, Interscience Publishers, Wiley, New York, 1963.

[20] K. R. Erikson, F. J. Fry and J. P. Jones, "Ultra-sound in medicine-A review", *IEEE Trans. Sonics Ultrason.*, **Vol. SU-21**, pp. 144-170, July 1974.

[21] R. J. Evans, *Optimal Signal Processing with Constraints*, Ph.D Thesis, University of Newcastle, NSW, Australia, 1975.

[22] R. J. Evans, A. Cantoni, and T. E. Fortmann, "Envelope-constrained filter, Part II, adaptive structures", *IEEE Trans. Inform. Theory*, **Vol. 23**, pp. 435-444, July. 1977.

[23] R. J. Evans, T. E. Fortmann, and A. Cantoni, "Envelope-constrained filter, Part I, theory and applications", *IEEE Trans. Inform. Theory*, **Vol. 23**, pp. 421-434, July. 1977.

[24] R. J. Evans, A. Cantoni, and K. M. Ahmed, "Envelope-constrained filters with uncertain input", *Circuits System Signal processing.* **Vol. 2**, No. 2, pp. 131-154, 1983.

[25] Y. G. Evtushenko, *Numerical Optimization Techniques*, Optimization Software, New York, 1985.

[26] A. V. Fiacco and K. O. Kortanek (eds.), *Semi-infinite Programming and Applications*, Springer, Berlin, 1993

[27] R. Fletcher, *Practical Methods of Optimization*, 2nd ed., Wiley, New York, 1987.

[28] T. E. Fortmann and M. Athans, "Optimal filter design subject to output sidelobe constraints: Theoretical considerations", *J. Optimization Theory and Applications*, **Vol. 14**, No. 2, pp. 179-197, Aug. 1974.

[29] T. E. Fortmann, R. J. Evans, "Optimal filter design subject to output sidelobe constraints: Computational algorithm and numerical results", *J. Optimization. Theory and Applications*, **Vol. 14**, No. 3, pp. 271-290, Sept. 1974.

[30] B. A. Francis, *A Course in H-infinity Control Theory*, Lecture Notes in Control and Information Sciences, M. Thoma and A. Wyner (eds.), Springer Verlag, Berlin Heidelberg, 1987

[31] M. Frank and P. Wolfe, "An algorithm for quadratic programming", *Nav. Res Log. Quart*, **Vol. 1& 2**, pp. 95-110, 1956.

[32] L. Franks, *Signal Theory*, Prentice-Hall Inc. Englewood Cliffs, N. J. 1969.

[33] R. A. Gable and R. A. Roberts, *Signals and Linear Systems*, 3rd ed., New York: Wiley, 1987.

[34] W. Gardner, *Cyclostationarity in Communications and Signal Processing*, New York: IEEE Press, 1994.

[35] C. W. Gear, "Algorithm 407 DIFSUB for Solution of Ordinary Differential equations [D2]", *Communications of The Association for Computing Machinery*, **Vol. 14**, No. 3, 1971.

[36] A. Gersho, "Adaptive equalization of dispersive channels for data transmission", *B.S.T.J.*, **Vol. 48**, No. 1, January 1969, pp 55-70.

[37] P. E. Gill, W. Murray and M. H. Wright, *Practical optimization*, Academic Press, London, 1981.

[38] K. Glashoff and S. A. Gustafson, *Linear Optimization and Approximation*, Springer, New York, 1983.

[39] G. Gonzaga, E. Polak and R. Trahan, "An improved algorithm for optimization problems with functional inequality constraints", *IEEE Trans. Aut. Control*, **Vol. 25**, pp. 49-54, 1980.

[40] G. C. Goodwin, and K. S. Sin, *Adaptive Filtering, Prediction and Control*, Prentice-Hall, New Jersey, 1984.

[41] B. Gottfried and J. Weisman, *Introduction to Optimization Theory*, Prentice-Hall, Englewood Cliffs, N.J., 1973.

[42] A. Guida, "Optimum tapped delay line for digital signals", *IEEE Trans. Com.*, **Vol. 21**, No. 4, April 1973, pp. 277-283.

[43] M. Held, P. Wolfe, and P. Crowder, "Validation of subgradient optimization", *Mathematical Programming*, **Vol. 6**, pp. 62-88, 1974.

[44] C. W. Helstrom, *Statistical Theory of Signal Detection*, Pergamon Press, New York, 1960.

[45] D. Hertog, *Interior Point Approach to Linear, Quadratic and Convex Programming Algorithms and Complexity*, Kluwer Academic Publishers, Dordrecht, 1994.

[46] R. Hettich (ed.), *Semi-infinite Programming*, Springer, Berlin 1979.

[47] R. Hettich and K. O. Kortanek, "Semi-infinite programming: theory methods and applications", *SIAM review 35*, pp. 380-429, 1993.

[48] E. Hille, *Methods in Classical and Functional Analysis*, Addison-Wesley, 1972.

[49] R. B. Holmes, *A Course on Optimization and Best Approximations*, Lecture Notes in Mathematics 257, Springer-Verlag, 1972.

[50] L. S. Jennings and K. L. Teo, "A computational algorithm for functional inequality constrained optimization problems", *Automatica*, **Vol. 26**, pp. 371-375, 1990.

[51] D. H. Jacobson, and D. Q. Mayne, *Differential Dynamic Programming*, American Elsevier Publishing company, New York, 1970.

[52] F. John, "Extremum problems with inequalities as subsidiary conditions", in K. O. Friedrichs, O. E. Neugebauer, and J. J. Stoker (eds.), *Studies and Essays*, Courant Aniversary Volume, Wiley, New York, pp. 187-204, 1948.

[53] D. J. Jones, "Solutions of Troesch's two point boundary value problem by shooting methods", *Journal of Computational Physics*, **Vol. 12**, No. 3, 1973.

[54] S. A. Kassam and H. V. Poor, "Robust techniques for signal processing: a survey", *Proc. of the IEEE*, **Vol. 5**, pp. 433-481, 1985.

[55] W. H. Kautz, "Transient synthesis in the time domain", *IRE Trans. Circuit Theory*, **Vol. 1**, No. 3, pp. 29-38, 1954.

[56] E. J. Kelly, I. S. Reed and W. L. Root, "The detection of radar echoes in noise, Parts I and II", *SIAM Journal on Applied Mathematics*, **Vol. 8**, No. 2-3, 1960.

[57] R. E. King and P. N. Paraskevopoulous, "Digital Laguere filters", *Int. J. Circuit Theory Applic.*, **Vol. 5**, pp. 81-91, 1977.

[58] E. Kreyzig, *Introductory Functional Analysis with Applications*, Wiley, New York, 1978.

[59] J. W. Lechleider, ''Pulse envelope for digital systems'', *Proc IEEE. Int. Conf. Com.*, Atlanta, GA, **Vol. 2**, pp. 703-706, 1990.

[60] J. W. Lechleider, ''A new interpolation theorem with application to pulse transmission'', *IEEE. Trans. Com.*, **Vol. 39**, No. 10, pp. 1438-1444, 1991.

[61] Y. W. Lee, *Statistical Theory of Communications*, Wiley, NewYork, 1961.

[62] W. V. Lovitt, *Linear Integral Equations*, Dover, New York, 1950.

[63] R. Lucky, "Automatic equalization of digital communication", *B.S.T.J.*, **Vol. 44**, April 1965, pp 547-588.

[64] R. Lucky, J. Salz, E. J. Weldon, *Principles of Data Communication*, McGraw-Hill, New York, 1968.

[65] D. G. Luenberger, *Linear and Non-linear Programming*, 2nd ed., Adison-Wesley, Massachusset, 1989.

[66] D. G. Luenberger, *Optimization by Vector Space Methods*, Wiley, New York, 1969.

[67] P. M. Makila, "Approximation of of stable systems by Laguerre filters", *Automatica*, **Vol. 26**, No. 2, pp. 333-345, 1990.

[68] P. M. Makila, "Laguerre series approximation of infinite dimensional systems", *Automatica*, **Vol. 26**, No. 6, pp. 885-995, 1990.

[69] O. L. Mangasarian, *Nonlinear Programming*, McGraw-Hill, New York 1969.

[70] R. J. Marks, *Introduction to Shannon Sampling and Interpolation Theory*, Springer-Verlag, New York, 1991.

[71] M. A. Masnadi-Shirazi and N. Ahmed, ''Optimum Laguerre networks for a class of discrete-time systems'', *IEEE Trans. Signal Processing*, **Vol. 39**, No. 9, pp. 2104-2108, 1991.

[72] M. A. Masnadi-Shirazi and M. Ghasemi, ''Laguerre digital filter design'', *Proc. IEEE Int. Conf. Acoustics, Speech and Signal Processing*, Detroit Michigan, **Vol. 2**, pp. 1284-1287, May. 1995.

[73] R. J. McAulay, ''Numerical optimization techniques applied to PPM signal design'', *IEEE Trans. Inform. Theory*, **Vol. 14**, No. 5, pp 708-716, 1968.

[74] R. J. McAulay, ''Optimal control techniques applied to PPM signal design'', *Information & Control*, **Vol. 12**, No. 3, 1968.

[75] R. J. McAulay and J. R. Johnson, "Optimal mismatched filter design for radar ranging detection and resolution", *IEEE Trans. Inform. Theory*, **Vol. 17**, pp. 696-701, 1971.

[76] R. J. McAulay and L. P. Seidman, ''A usefull form of the Barankin lower bound and its application to PPM threshold analysis'', *IEEE Trans. Inform. Theory*, **Vol. 15**, No. 2, 1969.

[77] A. Miele, S. Naqvi, A. V. Levy, and R. R. Iyer, "Numerical solution of non-linear equations and non-linear two-point boundary-value problems", *Advances in Control Systems*, **Vol. 8**, C. T. Leondes (ed.), Academic Press, New York, 1972.

[78] G. Minty, "On the monotonicity of the gradient of a convex function", *Proc. J. Math.* 14, pp. 243-247, 1964.

[79] J. Moreau, Fonctionnelles sous-differentiables. C. R. Acad. Sci. pp. 4117-4119, Paris 257, 1963.

[80] J. Moreau, "Sous-differentiabilite", *Proc. Coll. Convexity*, pp. 185-201, Copenhagen 1965, 1967.

[81] J. Moreau, Fontionelles Convexes. Seminaire "Equations aux Derivee Partielles". College de France, 1966.

[82] R. A. Nobakht and M. R. Civanlar, "Optimal pulse shape design for digital communication systems by projections onto convex sets", *IEEE Trans. Com.*, **Vol. 43**, pp. 2874-2877, 1995.

[83] D. A. Pierre, *Optimization Theory with Applications*, Wiley, New York, 1969

[84] E. Polak, *Computational Methods in Optimization*, Academic Press, New York, 1971.

[85] E. Polak and D. Q. Mayne, "An algorithm for optimization problems with functional inequality constraints", *IEEE Trans. Aut. Control*, **Vol. 21**, pp. 184-193, 1976.

[86] E. Polak and Y. Wardi, "Nondifferential optimization algorithm for designing control systems having singular value inequalities", *Automatica*, **Vol. 18**(3), pp. 267-283, 1982.

[87] R. Price, "Non-linearly feedback equalized PAM versus capacity for noisy filter channels", *Proc. Int. Communication Conf., Philadelphia*, PA, pp. 19-21, June 1974.

[88] J. Proakis and D. Manolakis, Digital signal processing: principles, algorithms, and applications, Macmillan, New York, N.Y., 1992.

[89] J. Proaki and J. Miller, "An adaptive receiver for digital signalling through channels with intersymbol interference", *IEEE Trans. Inform. Theory*, **Vol. 15**, July 1969, pp. 484-497.

[90] B. N. Pshenichni and Y. M. Danilin, *Numerical Methods in Extremal Problems*, Mir Publishers, Moscow, 1978.

[91] L. R. Rabiner and B. Gold, *Theory and Application of Digital Signal Processing*, Prentice-Hall, New Jersey, 1975.

[92] R. Reemtsen and J.-J. Ruckmann (eds), *Semi-Infinite Programming*, Kluwer Academic Publishers, Dordrecht, 1998.

[93] D. Rhodes, "On an optimum line-source for maximum directivity", *IEEE Trans. Antennas and Prop.*, **Vol. AP-19**, No. 4 July 1971, pp 485-492.

[94] A. W. Rihaczek, *Principles of High-Resolution Radar*, McGraw-Hill, New York, 1969.

[95] S. M. Roberts and J. S. Shipman, "Solution of Troesch's two-point boundary value problem by a combination of techniques, *Journal of Computational Physics*, **Vol. 10**, No. 2, 1972.

[96] R. Rockafellar, *Convex Analysis*, Princeton University Press, New Jersey, 1970.

[97] R. Rockafellar, "Characterization of the subdifferentials of convex functions", *Pac. J. Math.*, **Vol. 17**, pp. 497-510, 1966.

[98] R. Rockafellar, "A dual approach to solving non-linear programming problems by unconstrained optimization", *Math. Prog.*, **Vol. 5**, pp. 354-373, 1973.

[99] R. Rockafellar, "An extension of Feuchel's duality theorem for convex functions", *Duke Math. J.*, **Vol. 33**, pp. 81-90, 1966.

[100] W. Rudin, *Real and Complex analysis*, McGraw-Hill Book Company, New York, 1966.

[101] T. L. Saaty and J. Bram, *Nonlinear Mathematics*, Dover, New York, 1981.

[102] R. W. H. Sargent and G. R. Sullivan, "The development of an efficient optimal control package", Int. Proc. *8th Conf. Optimization. Techniques*, Springer, Wurzberg, 1977.

[103] A. Seyler and J. Potter, "Waveform testing of television transmission facilities", *Proceedings of The IRE*, Australia, July 1960, pp. 470-478.

[104] S. Simons, "Maximinimax, minimax, and antiminimax theorems and a result of R. C. James", *Pac. J. Math.* 40, pp. 709-718, 1972.

[105] M. I. Skolnik, *Introduction to Radar Systems*, McGraw-Hill, New York, 1962.

[106] G. Szego, "Orthogonal polynomials", *American Mathematical Society Colloquium Publications*, **Vol. XXIII**, Rhode Island, 1939.

[107] C. Temes, V. Barcillon and F. Marshall, "Optimization of band limited systems", *Proc. IEEE*, **Vol. 61**, No. 2, Feb., pp. 196-234, 1973.

[108] K. L. Teo, A. Cantoni and X. G. Lin, ''A new approach to the optimization of envelope-constrained fillters with uncertain input'', *IEEE Trans. Signal processing*, **Vol. 2**, No. 2, pp. 426-429, 1994.

[109] K. L. Teo and C. J. Goh, "A simple computational procedure for optimization problems with functional inequality constraints", *IEEE Trans. Aut. Control*, **Vol. 32**, pp. 940-941, 1987.

[110] K. L. Teo C. J. Goh and K. H. Wong, *A Unified Approach to Optimal Control Problems*, Longman Scientific & Technical, London, 1991.

[111] K. L. Teo, V. Rehbock and L. S. Jennings, "A new computational algorithm for functional inequality constrained optimization problems", *Automatica*,**Vol. 29**, pp. 789-792, 1993.

[112] H. L. Van Trees, *Detection, Estimation and Modulation Theory, Part I*, Wiley, New York, 1968.

[113] C. H. Tseng, K. L. Teo, A. Cantoni, and Z. Zang "Envelope-Constrained Filters: Adaptive Algorithms", *IEEE Trans. Signal Processing*, **Vol. 48**, pp. 1597-1608, 2000.

[114] G. Ungerboeck, "Nonlinear equalization of binary bipolar signal in Gaussian noise", *IEEE Trans. Com. Tech.*, **Vol. 21**, December, 1971, pp. 979-990.

[115] B. Vo, *Optimum Envelope Constrained Filters*, Ph.D Dissertation, Curtin University of Technology, WA, Australia, 1997.

[116] B. Vo, A. Cantoni, and K. L. Teo, "Iterative algorithms for envelope constrained filter design", *Proc. IEEE Int. Conf. Acoustics, Speech and Signal Processing*, Detroit Michigan, **Vol. 2**, pp. 1288-1291, May. 1995.

[117] B. Vo, A. Cantoni, and K. L. Teo, "A penalty approach to iterative algorithms for envelope constrained filter design", *IEEE Trans. Signal Processing*, **Vol. 45**, pp. 1869-1873, 1997.

[118] B. Vo, A. Cantoni and K. L. Teo, "Envelope constrained filter with linear interpolator", *IEEE Trans. Signal Processing*, **Vol. 45**. No. 6, pp. 1405-1415, 1997.

[119] B. Vo, A. Cantoni and K. L. Teo, "Computational methods for a class of functional inequality constrained optimization problem", *World Scientific Series in Applicable Analysis - Recent Trends in Optimization Theory and Applications*, R. P. Agarwal (ed.), World Scientific Publishing, **Vol. 5**, pp. 447-465, Singapore, 1995.

[120] B. Vo, T. Ho, A. Cantoni and V. Sreeram, "FIR filters in continuous-time envelope constrained filter design", *Proc. IEEE Int. Conf. Acoustics, Speech and Signal Processing*, Munich Germany, **Vol. 3**, pp. 1905-1908, April. 1997.

[121] B. Vo and A. Cantoni, "Continuous-time envelope constrained filter design via DSP approach", *Submitted to IEE Proc. - Vision, Image and Signal Processing.*

[122] B. Vo and A. Cantoni, "Continuous-time envelope constrained filter design with input uncertainty", *IEEE Trans. Circuits and Systems I*, **Vol. 47**, No. 10, pp. 1445 -1454, Oct. 2000.

[123] B. Vo, Z. Zang, A. Cantoni and K. L. Teo, "Continuous-time envelope constrained filter design via orthonormal filters", *IEE Proc. - Vision, Image and Signal Processing*, **Vol. 142**, No. 6, pp. 389-394, 1995.

[124] B. Vo, W. X. Zheng, A. Cantoni and K. L. Teo, "Approximation of functional inequality constrained problem", *Proc. 17th IFIP TC7 Conf. System Modelling and Optimization,* pp. 347-350, Prague, 1995.

[125] B. Wahlberg, "System identification using Laguerre models", *IEEE Trans. Automatic Control,* **Vol. 36**, No. 5, pp. 551-562, 1991.

[126] N. Weiner, "The theory of prediction", *Modern Mathematics for the Engineer*, E. F. Beckenbach (ed.), McGraw-Hill, New York, 1956.

[127] B. Widrow, "Adaptive filters", *Aspects of Network and System Theory*, R. Kalman and N. DeClaris (eds.), Holt, Rinehart and Winston Inc., pp. 563-587, 1969.

[128] P. Wolfe, "A duality theorem for non linear programming", *Quarterly of Applied Mathematics*, **Vol. 10,** pp. 239-244, 1961.

[129] L. C. Wood and S. Treitel, "Seismic signal processing", *Proc. IEEE (Special issue on digital signal processing)*, **Vol. 63**, pp. 649-661, April 1975.

[130] J. M. Wozencraft and I. M. Jacobs, *Principles of Communication Engineering*, Wiley, New York, 1965.

[131] S. Wu, S. Boyd and L. Vandenberghe, "FIR filter design via spectral factorization and convex optimization", Biswa Datta (eds.), *Applied Computational Control, Signal and Communications*, Birkhauser, 1997.

[132] Z. Zang, B. Vo, A. Cantoni and K. L. Teo, "Applications of discrete-time Laguerre networks to envelope constrained filter design", *Proc. IEEE Int. Conf. Acoustics, Speech and Signal Processing*, Atlanta Georgia, **Vol. 3**, pp. 1363-1366, May. 1996.

[133] Z. Zang, B. Vo, A. Cantoni and K. L. Teo, "Iterative algorithms for Envelope Constrained recursive filter design via Laguerre functions", *IEEE Trans. Circuits and Systems I,* **Vol. 46**, No. 11, pp. 1342 -1348, 1999.

[134] Z. Zang, A. Cantoni and K. L. Teo. "Applications of H-infnity optimization method to envelope-constrained IIR filter design", *Nonlinear Analysis, Theory, Methods and Applications*, **Vol. 30**, pp. 4267-4276, 1997.

[135] Z. Zang, A. Cantoni and K.L. Teo, "Envelope-constrained IIR filter design via H-infinity optimization method", *IEEE Trans. Circuits & Systems I*, **Vol. 46**, No. 6, pp. 649-653, 1999.

[136] Z. Zang, A. Cantoni and K. L. Teo, "Continuous-time envelope constrained filter design via Laguerre filters and H-infinity optimization methods", *IEEE Trans. Signal Processing*, **Vol. 46**, No. 10, pp. 2601-2610, 1998

[137] W. Zangwill, *Non-linear Programming a Unified Approach*, Prentice-Hall, Englewood Cliffs, N.J., 1969.

[138] W. X. Zheng, A. Cantoni, B. Vo and K. L. Teo, "Recursive procedures for constrained optimization problems and its application in signal processing", *IEE Proc. - Vision, Image and Signal Processing*, **Vol. 142**, No. 3, pp. 161-168, Jun. 1995.

Index

Applied Optimization

1. D.-Z. Du and D.F. Hsu (eds.): *Combinatorial Network Theory*. 1996
ISBN 0-7923-3777-8
2. M.J. Panik: *Linear Programming: Mathematics, Theory and Algorithms*. 1996
ISBN 0-7923-3782-4
3. R.B. Kearfott and V. Kreinovich (eds.): *Applications of Interval Computations*. 1996 ISBN 0-7923-3847-2
4. N. Hritonenko and Y. Yatsenko: *Modeling and Optimimization of the Lifetime of Technology*. 1996 ISBN 0-7923-4014-0
5. T. Terlaky (ed.): *Interior Point Methods of Mathematical Programming*. 1996
ISBN 0-7923-4201-1
6. B. Jansen: *Interior Point Techniques in Optimization*. Complementarity, Sensitivity and Algorithms. 1997 ISBN 0-7923-4430-8
7. A. Migdalas, P.M. Pardalos and S. Storøy (eds.): *Parallel Computing in Optimization*. 1997 ISBN 0-7923-4583-5
8. F.A. Lootsma: *Fuzzy Logic for Planning and Decision Making*. 1997
ISBN 0-7923-4681-5
9. J.A. dos Santos Gromicho: *Quasiconvex Optimization and Location Theory*. 1998
ISBN 0-7923-4694-7
10. V. Kreinovich, A. Lakeyev, J. Rohn and P. Kahl: *Computational Complexity and Feasibility of Data Processing and Interval Computations*. 1998
ISBN 0-7923-4865-6
11. J. Gil-Aluja: *The Interactive Management of Human Resources in Uncertainty*. 1998
ISBN 0-7923-4886-9
12. C. Zopounidis and A.I. Dimitras: *Multicriteria Decision Aid Methods for the Prediction of Business Failure*. 1998 ISBN 0-7923-4900-8
13. F. Giannessi, S. Komlósi and T. Rapcsák (eds.): *New Trends in Mathematical Programming*. Homage to Steven Vajda. 1998 ISBN 0-7923-5036-7
14. Ya-xiang Yuan (ed.): *Advances in Nonlinear Programming*. Proceedings of the '96 International Conference on Nonlinear Programming. 1998 ISBN 0-7923-5053-7
15. W.W. Hager and P.M. Pardalos: *Optimal Control*. Theory, Algorithms, and Applications. 1998 ISBN 0-7923-5067-7
16. Gang Yu (ed.): *Industrial Applications of Combinatorial Optimization*. 1998
ISBN 0-7923-5073-1
17. D. Braha and O. Maimon (eds.): *A Mathematical Theory of Design: Foundations, Algorithms and Applications*. 1998 ISBN 0-7923-5079-0

Applied Optimization

18. O. Maimon, E. Khmelnitsky and K. Kogan: *Optimal Flow Control in Manufacturing.* Production Planning and Scheduling. 1998 ISBN 0-7923-5106-1
19. C. Zopounidis and P.M. Pardalos (eds.): *Managing in Uncertainty: Theory and Practice.* 1998 ISBN 0-7923-5110-X
20. A.S. Belenky: *Operations Research in Transportation Systems:* Ideas and Schemes of Optimization Methods for Strategic Planning and Operations Management. 1998 ISBN 0-7923-5157-6
21. J. Gil-Aluja: *Investment in Uncertainty.* 1999 ISBN 0-7923-5296-3
22. M. Fukushima and L. Qi (eds.): *Reformulation: Nonsmooth, Piecewise Smooth, Semismooth and Smooting Methods.* 1999 ISBN 0-7923-5320-X
23. M. Patriksson: *Nonlinear Programming and Variational Inequality Problems.* A Unified Approach. 1999 ISBN 0-7923-5455-9
24. R. De Leone, A. Murli, P.M. Pardalos and G. Toraldo (eds.): *High Performance Algorithms and Software in Nonlinear Optimization.* 1999 ISBN 0-7923-5483-4
25. A. Schöbel: *Locating Lines and Hyperplanes.* Theory and Algorithms. 1999 ISBN 0-7923-5559-8
26. R.B. Statnikov: *Multicriteria Design.* Optimization and Identification. 1999 ISBN 0-7923-5560-1
27. V. Tsurkov and A. Mironov: *Minimax under Transportation Constrains.* 1999 ISBN 0-7923-5609-8
28. V.I. Ivanov: *Model Development and Optimization.* 1999 ISBN 0-7923-5610-1
29. F.A. Lootsma: *Multi-Criteria Decision Analysis via Ratio and Difference Judgement.* 1999 ISBN 0-7923-5669-1
30. A. Eberhard, R. Hill, D. Ralph and B.M. Glover (eds.): *Progress in Optimization.* Contributions from Australasia. 1999 ISBN 0-7923-5733-7
31. T. Hürlimann: *Mathematical Modeling and Optimization.* An Essay for the Design of Computer-Based Modeling Tools. 1999 ISBN 0-7923-5927-5
32. J. Gil-Aluja: *Elements for a Theory of Decision in Uncertainty.* 1999 ISBN 0-7923-5987-9
33. H. Frenk, K. Roos, T. Terlaky and S. Zhang (eds.): *High Performance Optimization.* 1999 ISBN 0-7923-6013-3
34. N. Hritonenko and Y. Yatsenko: *Mathematical Modeling in Economics, Ecology and the Environment.* 1999 ISBN 0-7923-6015-X
35. J. Virant: *Design Considerations of Time in Fuzzy Systems.* 2000 ISBN 0-7923-6100-8

Applied Optimization

36. G. Di Pillo and F. Giannessi (eds.): *Nonlinear Optimization and Related Topics*. 2000 ISBN 0-7923-6109-1
37. V. Tsurkov: *Hierarchical Optimization and Mathematical Physics*. 2000 ISBN 0-7923-6175-X
38. C. Zopounidis and M. Doumpos: *Intelligent Decision Aiding Systems Based on Multiple Criteria for Financial Engineering*. 2000 ISBN 0-7923-6273-X
39. X. Yang, A.I. Mees, M. Fisher and L.Jennings (eds.): *Progress in Optimization*. Contributions from Australasia. 2000 ISBN 0-7923-6286-1
40. D. Butnariu and A.N. Iusem: *Totally Convex Functions for Fixed Points Computation and Infinite Dimensional Optimization*. 2000 ISBN 0-7923-6287-X
41. J. Mockus: *A Set of Examples of Global and Discrete Optimization*. Applications of Bayesian Heuristic Approach. 2000 ISBN 0-7923-6359-0
42. H. Neunzert and A.H. Siddiqi: *Topics in Industrial Mathematics*. Case Studies and Related Mathematical Methods. 2000 ISBN 0-7923-6417-1
43. K. Kogan and E. Khmelnitsky: *Scheduling: Control-Based Theory and Polynomial-Time Algorithms*. 2000 ISBN 0-7923-6486-4
44. E. Triantaphyllou: *Multi-Criteria Decision Making Methods*. A Comparative Study. 2000 ISBN 0-7923-6607-7
45. S.H. Zanakis, G. Doukidis and C. Zopounidis (eds.): *Decision Making: Recent Developments and Worldwide Applications*. 2000 ISBN 0-7923-6621-2
46. G.E. Stavroulakis: *Inverse and Crack Identification Problems in Engineering Mechanics*. 2000 ISBN 0-7923-6690-5
47. A. Rubinov and B. Glover (eds.): *Optimization and Related Topics*. 2001 ISBN 0-7923-6732-4
48. M. Pursula and J. Niittymäki (eds.): *Mathematical Methods on Optimization in Transportation Systems*. 2000 ISBN 0-7923-6774-X
49. E. Cascetta: *Transportation Systems Engineering: Theory and Methods*. 2001 ISBN 0-7923-6792-8
50. M.C. Ferris, O.L. Mangasarian and J.-S. Pang (eds.): *Complementarity: Applications, Algorithms and Extensions*. 2001 ISBN 0-7923-6816-9
51. V. Tsurkov: *Large-scale Optimization – Problems and Methods*. 2001 ISBN 0-7923-6817-7
52. X. Yang, K.L. Teo and L. Caccetta (eds.): *Optimization Methods and Applications*. 2001 ISBN 0-7923-6866-5
53. S.M. Stefanov: *Separable Programming Theory and Methods*. 2001 ISBN 0-7923-6882-7

Applied Optimization

54. S.P. Uryasev and P.M. Pardalos (eds.): *Stochastic Optimization: Algorithms and Applications.* 2001 ISBN 0-7923-6951-3
55. J. Gil-Aluja (ed.): *Handbook of Management under Uncertainty.* 2001 ISBN 0-7923-7025-2
56. B.-N. Vo, A. Cantoni and K.L. Teo: *Filter Design with Time Domain Mask Constraints: Theory and Applications.* 2001 ISBN 0-7923-7138-0
57. S. Zlobek: *Stable Parametric Programming.* 2001 ISBN 0-7923-7139-9
58. M.G. Nicholls, S. Clarke and B. Lehaney (eds.): *Mixed-Mode Modelling: Mixing Methodologies for Organisational Intervention.* 2001 ISBN 0-7923-7151-8

KLUWER ACADEMIC PUBLISHERS – DORDRECHT / BOSTON / LONDON

GPSR Compliance

The European Union's (EU) General Product Safety Regulation (GPSR) is a set of rules that requires consumer products to be safe and our obligations to ensure this.

If you have any concerns about our products, you can contact us on ProductSafety@springernature.com

In case Publisher is established outside the EU, the EU authorized representative is:

Springer Nature Customer Service Center GmbH
Europaplatz 3
69115 Heidelberg, Germany

Batch number: 10755284

Printed by Printforce, the Netherlands